Frontiers in Nucleic Acid Chemistry—in Memory of Professor Enrique Pedroso for His Outstanding Contributions to Nucleic Acid Chemistry

Frontiers in Nucleic Acid Chemistry—in Memory of Professor Enrique Pedroso for His Outstanding Contributions to Nucleic Acid Chemistry

Editors

Ramon Eritja
Daniela Montesarchio
Montserrat Terrazas

Basel • Beijing • Wuhan • Barcelona • Belgrade • Novi Sad • Cluj • Manchester

Editors

Ramon Eritja
Institute for Advanced
Chemistry of Catalonia (IQAC)
CSIC
Barcelona
Spain

Daniela Montesarchio
Department of Chemical
Sciences
University of Naples Federico II
Napoli
Italy

Montserrat Terrazas
Department of Inorganic and
Organic Chemistry
University of Barcelona
Barcelona
Spain

Editorial Office
MDPI
St. Alban-Anlage 66
4052 Basel, Switzerland

This is a reprint of articles from the Special Issue published online in the open access journal *Molecules* (ISSN 1420-3049) (available at: www.mdpi.com/journal/molecules/special_issues/nucleicacid_Chemistry).

For citation purposes, cite each article independently as indicated on the article page online and as indicated below:

Lastname, A.A.; Lastname, B.B. Article Title. *Journal Name* **Year**, *Volume Number*, Page Range.

ISBN 978-3-0365-9898-7 (Hbk)
ISBN 978-3-0365-9897-0 (PDF)
doi.org/10.3390/books978-3-0365-9897-0

© 2024 by the authors. Articles in this book are Open Access and distributed under the Creative Commons Attribution (CC BY) license. The book as a whole is distributed by MDPI under the terms and conditions of the Creative Commons Attribution-NonCommercial-NoDerivs (CC BY-NC-ND) license.

Contents

About the Editors vii

Montserrat Terrazas, Ramon Eritja and Daniela Montesarchio
Special Issue "Frontiers in Nucleic Acid Chemistry—In Memory of Professor Enrique Pedroso for His Outstanding Contributions to Nucleic Acid Chemistry"
Reprinted from: *Molecules* **2023**, *28*, 7278, doi:10.3390/molecules28217278 1

Xiaole Liu, Xinyue Zhao, Jinhan He, Sishi Wang, Xinfei Shen and Qingfeng Liu et al.
Advances in the Structure of GGGGCC Repeat RNA Sequence and Its Interaction with Small Molecules and Protein Partners
Reprinted from: *Molecules* **2023**, *28*, 5801, doi:10.3390/molecules28155801 6

Sibasish Paul and Marvin H. Caruthers
Synthesis of Backbone-Modified Morpholino Oligonucleotides Using Phosphoramidite Chemistry
Reprinted from: *Molecules* **2023**, *28*, 5380, doi:10.3390/molecules28145380 25

Nadezhda O. Kropacheva, Arseniy A. Golyshkin, Mariya A. Vorobyeva and Mariya I. Meschaninova
Convenient Solid-Phase Attachment of Small-Molecule Ligands to Oligonucleotides via a Biodegradable Acid-Labile P-N-Bond
Reprinted from: *Molecules* **2023**, *28*, 1904, doi:10.3390/molecules28041904 37

Mark L. Sowers, James W. Conrad, Bruce Chang-Gu, Ellie Cherryhomes, Linda C. Hackfeld and Lawrence C. Sowers
DNA Base Excision Repair Intermediates Influence Duplex–Quadruplex Equilibrium
Reprinted from: *Molecules* **2023**, *28*, 970, doi:10.3390/molecules28030970 50

Anna Clua, Santiago Grijalvo, Namrata Erande, Swati Gupta, Kristina Yucius and Raimundo Gargallo et al.
Correction: Clua et al. Properties of Parallel Tetramolecular G-Quadruplex Carrying N-Acetylgalactosamine as Potential Enhancer for Oligonucleotide Delivery to Hepatocytes. *Molecules* 2022, 27, 3944
Reprinted from: *Molecules* **2022**, *28*, 98, doi:10.3390/molecules28010098 68

Caroline P. Shepard, Raymond G. Emehiser, Saswata Karmakar and Patrick J. Hrdlicka
Factors Impacting Invader-Mediated Recognition of Double-Stranded DNA
Reprinted from: *Molecules* **2022**, *28*, 127, doi:10.3390/molecules28010127 71

Chiara Figazzolo, Frédéric Bonhomme, Saidbakhrom Saidjalolov, Mélanie Ethève-Quelquejeu and Marcel Hollenstein
Enzymatic Synthesis of Vancomycin-Modified DNA
Reprinted from: *Molecules* **2022**, *27*, 8927, doi:10.3390/molecules27248927 92

Chang Lu, Anand Lopez, Jinkai Zheng and Juewen Liu
Using the Intrinsic Fluorescence of DNA to Characterize Aptamer Binding
Reprinted from: *Molecules* **2022**, *27*, 7809, doi:10.3390/molecules27227809 106

Núria Escaja, Bartomeu Mir, Miguel Garavís and Carlos González
Non-G Base Tetrads
Reprinted from: *Molecules* **2022**, *27*, 5287, doi:10.3390/molecules27165287 115

Stefania Mazzini, Salvatore Princiotto, Roberto Artali, Loana Musso, Anna Aviñó and Ramon Eritja et al.
Exploring the Interaction of G-quadruplex Binders with a (3 + 1) Hybrid G-quadruplex Forming Sequence within the PARP1 Gene Promoter Region
Reprinted from: *Molecules* **2022**, *27*, 4792, doi:10.3390/molecules27154792 . . . . **129**

Virginia Martín-Nieves, Yogesh S. Sanghvi, Susana Fernández and Miguel Ferrero
Sustainable Protocol for the Synthesis of 2′,3′-Dideoxynucleoside and 2′,3′-Didehydro-2′,3′-dideoxynucleoside Derivatives
Reprinted from: *Molecules* **2022**, *27*, 3993, doi:10.3390/molecules27133993 . . . . **144**

Anna Clua, Santiago Grijalvo, Namrata Erande, Swati Gupta, Kristina Yucius and Raimundo Gargallo et al.
Properties of Parallel Tetramolecular G-Quadruplex Carrying *N*-Acetylgalactosamine as Potential Enhancer for Oligonucleotide Delivery to Hepatocytes
Reprinted from: *Molecules* **2022**, *27*, 3944, doi:10.3390/molecules27123944 . . . . **156**

Francesca Greco, Domenica Musumeci, Nicola Borbone, Andrea Patrizia Falanga, Stefano D'Errico and Monica Terracciano et al.
Exploring the Parallel G-Quadruplex Nucleic Acid World: A Spectroscopic and Computational Investigation on the Binding of the c-myc Oncogene NHE III1 Region by the Phytochemical Polydatin
Reprinted from: *Molecules* **2022**, *27*, 2997, doi:10.3390/molecules27092997 . . . . **176**

About the Editors

Ramon Eritja

Dr. Ramon Eritja is a Research Professor of the Nucleic Acids Chemistry group at the Institute for Advanced Chemistry of Catalonia (IQAC) of the Spanish National Research Council (CSIC). He obtained his PhD in chemistry from the University of Barcelona (UB) in 1983 under the supervision of Dr. Ernest Giralt. He was a postdoctoral fellow in Dr. Itakura's (Beckman Research Institute at City of Hope, USA, 1984–1988) and Dr. Caruthers's (University of Colorado at Boulder, USA, 1988–1989) laboratories. He moved to UB (1989–1990) and then CSIC, Barcelona (1990–1994). He was a Group Leader at EMBL, Germany (1994–1999). He returned to CSIC, Barcelona (1999), and was an invited member at the IRB Barcelona (2006–2012). He has been selected as a member of the Networking Center on Bioengineering, Biomaterials and Nanomedicine (CIBER-BBN) since 2006. He was Director of IQAC (2012–2017). His research interests are centred around the synthesis and properties of nucleic acid derivatives, including siRNA, antisense, aptamers, oligonucleotide conjugates, and DNA nanotechnology.

Daniela Montesarchio

Daniela Montesarchio is a full professor of organic chemistry at Federico II University of Naples, where she graduated in chemistry with honors in 1989. In 1993, she obtained her Ph.D. in chemical sciences with a thesis on the chemical synthesis of modified oligonucleotides and nucleoside analogs under the supervision of Prof. Ciro Santacroce. After her postdoctoral experience in the laboratories of Prof. George Just at McGill University, Montreal, she came back to Federico II University, where, in 1994, she became an assistant professor. Since 2005, she has been an associate professor, and since 2021, she has been a full professor. Her research interests are mainly devoted to the design and synthesis of hybrid systems at the interface between chemistry and biology, including oligonucleotide and nucleoside analogs, aptamers and peptido- and glycomimetics, as well as metal-based drugs for therapeutic/diagnostic applications. She is the co-author of >160 scientific articles published in international, peer-reviewed journals, as well as more than 100 oral and/or poster communications presented at international and national conferences. She is one of the co-founders of Women in Science in Naples, an association born in 2007 from the initiative of a group of researchers in Federico II University, SUN (Seconda Università di Napoli), INFN and CNR, aimed at promoting equal opportunities in scientific research and academia.

Montserrat Terrazas

Dr. Terrazas' research is focused on the design and synthesis of chemical nucleosides, nucleotides, and oligonucleotides of structural and biological interest. After graduating in Chemistry (2001), she completed her MSc and PhD studies at the Dpt. of Organic Chemistry of the Univ. of Barcelona under the supervision of Profs. Xavier Ariza and Jaume Vilarrasa. She worked on the development of synthetic methodologies for the labeling of nucleosides with ^{15}N at specific positions, as well as on the computational design of new anti-HIV-1 nucleoside analogues (completed in a 3-month stay at Dr. Botta's group (Univ. of Siena, Italy)). In 2006, she moved to the Univ. of Leuven (Belgium) to do a 1-year postdoc in the group of Prof. Herdewijn, working on the synthesis and biological evaluation of new nucleotide triphosphate analogues as potential antiviral agents.

In 2007, she moved to Stanford (Prof. Eric Kool's group) to focus on the field of therapeutic RNAs, constituting the basis for her present line of research: understanding the mechanism of RNA interference (induced by siRNAs) and improving the properties of siRNAs, as therapeutic tools, using chemical modifications. In 2009, she returned to Barcelona to join Prof. Eritja's group at the IQAC-CSIC, where she combined her background in RNAi, synthetic organic chemistry, and molecular modeling to design and develop new forms of chemically modified siRNAs with potential therapeutic applications. In 2014, she joined Prof. Orozco's group as a Miguel Servet Research Associate, where she started her own research line focused on the design and synthesis of new therapeutic siRNA tools to overcome drug resistance in cancer. She joined the Organic Chemistry Section of the Faculty of Chemistry (Univ. of Barcelona) as an Associate Professor in March 2021, where she is currently developing new multifunctional oligonucleotides for cancer treatment, as well as new delivery systems for selective drug delivery.

Editorial

Special Issue "Frontiers in Nucleic Acid Chemistry—In Memory of Professor Enrique Pedroso for His Outstanding Contributions to Nucleic Acid Chemistry"

Montserrat Terrazas [1,*], Ramon Eritja [2,3,*] and Daniela Montesarchio [4,*]

1. Department of Inorganic and Organic Chemistry, Organic Chemistry Section, Institute of Biomedicine of the University of Barcelona (IBUB), University of Barcelona, 08028 Barcelona, Spain
2. Institute for Advanced Chemistry of Catalonia (IQAC), CSIC, 08034 Barcelona, Spain
3. Centro de Investigación Biomédica en Red de Bioingeniería, Biomateriales y Nanomedicina (CIBER-BBN), Instituto de Salud Carlos III, 28029 Madrid, Spain
4. Department of Chemical Sciences, University of Naples Federico II, I-80126 Napoli, Italy
* Correspondence: montserrat.terrazas@ub.edu (M.T.); recgma@cid.csic.es (R.E.); daniela.montesarchio@unina.it (D.M.)

Citation: Terrazas, M.; Eritja, R.; Montesarchio, D. Special Issue "Frontiers in Nucleic Acid Chemistry—In Memory of Professor Enrique Pedroso for His Outstanding Contributions to Nucleic Acid Chemistry". *Molecules* 2023, 28, 7278. https://doi.org/10.3390/molecules28217278

Received: 13 October 2023
Accepted: 24 October 2023
Published: 26 October 2023

Copyright: © 2023 by the authors. Licensee MDPI, Basel, Switzerland. This article is an open access article distributed under the terms and conditions of the Creative Commons Attribution (CC BY) license (https://creativecommons.org/licenses/by/4.0/).

This Special issue is dedicated to the memory of Enrique Pedroso, Professor Emeritus of Organic Chemistry at University of Barcelona, who passed away at the age of 72 in September 2020. Professor Enrique Pedroso has been one of the pioneers of Nucleic Acids Chemistry in Spain, significantly contributing to the development of this highly interdisciplinary field which combines organic chemistry, biochemistry, pharmacology, materials chemistry, and biophysics. His major research achievements have been accomplished in the synthesis of modified oligonucleotides and especially conjugates and cyclic oligonucleotides, as well as their analogues, which opened new avenues in the search for novel applications of oligonucleotides.

His research activity started in 1981 at the Department of Organic Chemistry of University of Barcelona, where he spent all his intense academic and scientific career, covering almost four decades. After having deeply investigated the methods of solid phase peptide synthesis, providing significant progress in the field also in collaboration with Ernest Giralt, Fernando Albericio and Ramon Eritja of the same University, in 1990, he published his first article on the solid-phase synthesis of oligonucleotides [1]. This work was immediately followed by an important contribution on the synthesis and characterization of oligodeoxynucleotides containing the mutagenic base analogue 4-*O*-ethylthymine, carried out in collaboration with Ramon Eritja [2], with whom he always maintained strict research relationships.

After an enlightening research stay in University of Colorado in Boulder, hosted in the group of prof. Marvin H. Caruthers, he grew a solid and active research group starting from the early 1990s in collaboration with Anna Grandas—his closest collaborator in research and beloved partner in life. Together, they developed efficient approaches for the synthesis of hybrid compounds, such as nucleopeptides and peptide–oligonucleotide conjugates, among others (see: [3–6]).

In 1997, he published an effective method for the solid-phase synthesis of cyclic oligodeoxyribonucleotides, successfully engineering the functionalized solid support so to exploit phosphotriester chemistry for the crucial cyclization step of the target oligonucleotide, previously assembled via standard phosphoramidite chemistry protocols [7]. Later, this approach was extended to the synthesis of cyclic oligoribonucleotides [8]. Using this methodology, his group synthesized several small and medium-sized cyclic oligonucleotides, studying their peculiar properties and discovering unusual conformational motifs, also in collaboration with the research group of Carlos Gonzalez [9–11]. A more recent contribution revisited the thiol–maleimide condensation reaction to obtain cyclic oligonucleotide constructs [12].

If most of his scientific collaborations were with eminent Spanish researchers, relevant was also his international profile, being an invited speaker in many international conferences, Visiting Professor in many Universities outside Spain and having a number of fruitful international collaborations, e.g., with Eric T. Kool [13], Daniela Montesarchio [14] and Keith Fox [15], among others. In his last contribution, always collaborating with Anna Grandas, he explored the synthesis of oligonucleotide, peptide or PNA conjugates, successfully obtained via inverse electron-demand Diels−Alder cycloaddition [16].

In all his career, Enrique was deeply involved in the research and promotion of Nucleic Acid Chemistry. He was indeed an active member of the International Society for Nucleosides, Nucleotides and Nucleic Acids (IS3NA, www.is3na.org), which organizes International Roundtables on a biannual basis to present and discuss advances in chemistry and biology of nucleosides, nucleotides and nucleic acids, as well as of the Spanish Society for Nucleic Acids and Nucleosides, which gathers researchers every two years in the Spanish Nucleosides Nucleotides and Nucleic Acids meetings (RANN).

This Special issue comprises a collection of twelve research or review articles prepared by international experts in nucleic acids, including nucleoside and oligonucleotide synthesis, nucleic acids structural studies, DNA repair, and biophysical characterization of DNA-targeting ligands, especially G-quadruplex binding drugs.

A novel route for the synthesis of $2',3'$-dideoxy and $2',3'$-didehydro nucleosides, including several anti-HIV drugs such as stavudine, zalcitabine and didanosine, was presented by the group of Dr. Ferrero and Dr. Fernández (contribution 1) from the University of Oviedo. Starting from the protection at the $5'$-hydroxyl group of the corresponding ribonucleotides followed by the formation of the corresponding $2',3'$-bisxanthates, a key step in the synthesis of the target compounds was the radical deoxygenation of the bisxanthates, successfully realized using environmentally friendly and low-cost reagents.

Phosphorodiamidate morpholino oligomer (PMO) derivatives are extensively used in exon-skipping strategies for the treatment of Duchenne muscular dystrophy, but the preparation of these derivatives is not trivial at all. Prof. Caruthers provided a comprehensive review on the synthesis and properties of modified morpholino oligonucleotides developed by his group at the University of Colorado in Boulder (contribution 2). This work is a masterpiece in the field of nucleic acid chemistry showing the state-of-the art of phosphoramidite chemistry, nucleoside chemistry and development of novel protecting groups for oligonucleotide synthesis.

An exceptional development in siRNA therapeutics is the discovery and use of oligonucleotide conjugates carrying trivalent N-acetyl galactosamine (GalNAc) residues. These oligonucleotides are rapidly internalized via the clathrin-mediated pathway in hepatocytes due to the presence of an asialoglycoprotein receptor with high affinity for galactose glycoproteins and trivalent GalNAc oligonucleotides. Eritja et al. at the IQAC-CSIC in Barcelona demonstrated that the tetramerization of G-rich oligonucleotides may be a novel and simple route to obtain the beneficial effects of multivalent N-acetylgalactosamine functionalization (contribution 3,4).

Meschaninova et al. from Novosibirsk State University described novel methods for the $5'$-functionalization of oligonucleotides through acid labile phosphoramidate linkages (contribution 5). A wide variety of oligonucleotides $5'$-conjugated with ligands, such as cholesterol, oleylamine, and p-anisic acid, was described. The methodology was successfully applied to DNA, RNA, and $2'$-O-methyl-RNA oligonucleotides.

The possibility of using enzymes for the preparation of modified oligonucleotides was addressed by Hollenstein et al. from the University of Paris (contribution 6). In this work, the enzymatic synthesis of a modified nucleoside triphosphate equipped with a vancomycin moiety on the nucleobase was described, demonstrating that this nucleotide analogue is suitable for polymerase-mediated synthesis of modified DNA and compatible with the SELEX methodology for the production of aptamers suitable to fight bacterial resistance.

The impact of the duplex-G-quadruplex equilibrium in DNA Base Excision Repair was studied by Sowers et al. from the University of Texas Medical Branch (contribution 7), suggesting that DNA damage and repair intermediates can alter duplex-quadruplex equilibrium. To corroborate this hypothesis, the authors used G-quadruplex stabilizing compounds, such as pyridostatin, modified oligonucleotides containing uracil, 5-hydroxymethyluracil, 5-fluorouracil, as well as abasic sites as building blocks inserted into the loop region of a 22-base telomeric repeat sequence known to form stable and well-characterized G-quadruplex structure.

In addition to G-quadruplex structures formed by G-tetrads, other tetrads can be formed. These structures found in some constrained oligonucleotides are more frequent than expected. An excellent review by Gonzalez et al. summarized the present state-of-the-art on our knowledge on novel non-G-tetrads including homotetrads, as well as major and minor groove tetrads, emphasizing their peculiar structural features (contribution 8).

Strand-invading approaches using chemically modified oligonucleotides and nucleic acid mimics capable of unzipping Watson–Crick base pairs of dsDNA targets and forming new Watson–Crick base pairs between probe strands and the complementary DNA (cDNA) regions have been explored by Hrdlicka et al. from the University of Idaho (contribution 9). The use of densely modified oligonucleotides with 2′-O-(pyren-1-yl)methyl RNA pyrimidine building blocks is highly recommended to achieve satisfactory results.

Aptamers are nucleic acid molecules able to selectively recognize several substrates, including small molecules and proteins, acting in a similar way as antibodies. Liu et al. analyzed the intrinsic fluorescent properties of DNA for the characterization of the binding of several model aptamers to their targets, such as cortisol, Hg^{2+}, adenosine or caffeine (contribution 10). They found that some aptamers may induce changes in intrinsic fluorescence, but these changes cannot be used for the determination of binding constants.

G-quadruplexes are important non-canonical DNA structures that are present in G-rich regions such as telomeres and some promoter regions of oncogenes. The polymorphism of these structures and the possibility of stabilizing them using planar heterocyclic small molecules is one of the most intense areas of research in nucleic acid chemistry. Dallavalle et al. from the University of Milan explored the interaction of several G-quadruplex ligands such as Curaxin, CX 5461, BA41, TPHS 4, Pyridostatin or BMH 21 with a G-quadruplex sequence found in the PARP1 promoter region (contribution 11). Pyridostatin was found to be the best binder for this sequence, being able to adapt its planar but flexible conformation to the dynamic nature of the G-quadruplex, especially the hybrid 3 + 1 G-quadruplex structure found in the PARP 1 promoter region.

Oliviero et al. studied the interaction of the c-myc oncogene NHE III1 region with the 3-β-D-glucoside of trans-resveratrol (Polydatin) (contribution 12). The experimental and modelling data show that this compound may be involved in partial end-stacking to the terminal G-quartet. Moreover, H-bonding interactions between the sugar moiety of the ligand and deoxynucleotides not included in the G-tetrads are possible.

Expansion of short nucleotide repeats is one of the causes of genetic diseases and is dramatically crucial for neurological diseases such as amyotrophic lateral sclerosis and frontal temporal dementia. Wang et al. described the state-of-the art knowledge of the structural properties of GGGGCC repeats in RNA and their interaction with small molecules and proteins specifically binding to these repeats (contribution 13).

By collecting in these scientific works this Special Issue, prepared by international leaders in the fields of chemistry and biochemistry of nucleosides, nucleotides and nucleic acids, we hope to contribute to advance knowledge on these fascinating molecules and their almost infinite applications, following Enrique Pedroso's lifelong commitment and example.

Conflicts of Interest: The authors declare no conflict of interest.

List of Contributions

1. Martín-Nieves, V.; Sanghvi, Y.S.; Fernández, S.; Ferrero, M. Sustainable protocol for the synthesis of 2′,3′-dideoxynucleoside and 2′,3′-didehydro-2′,3′-dideoxynucleoside derivatives. *Molecules* **2022**, *27*, 3993.
2. Paul, S.; Caruthers, M.H. Synthesis of backbone-modified morpholino oligonucleotides using phosphoramidite chemistry. *Molecules* **2023**, *28*, 5380.
3. Clua, A.; Grijalvo, S.; Erande, N.; Gupta, S.; Yucius, K.; Gargallo, R.; Mazzini, S.; Manoharan, M.; Eritja, R. Properties of parallel tetramolecular G-quadruplex carrying N-acetylgalactosamine as potential enhancer for oligonucleotide delivery to hepatocytes. *Molecules* **2022**, *27*, 3944.
4. Clua, A.; Grijalvo, S.; Erande, N.; Gupta, S.; Yucius, K.; Gargallo, R.; Mazzini, S.; Manoharan, M.; Eritja, R. Correction. Clua et al. Properties of parallel tetramolecular G-quadruplex carrying N-acetylgalactosamine as potential enhancer for oligonucleotide delivery to hepatocytes. *Molecules* **2023**, *28*, 98.
5. Kropacheva, N.O.; Golyshkin, A.A.; Vorobyeva, M.A.; Meschaninova, M.I. Convenient solid-phase attachment of small-molecule ligands to oligonucleotides via a biodegradable acid-labile P-N-bond. *Molecules* **2023**, *28*, 1904.
6. Figazzolo, C.; Bonhomme, F.; Saidjalolov, S.; Ethève-Quelquejeu, M.; Hollenstein, M. Enzymatic synthesis of vancomycin-modified DNA. *Molecules* **2022**, *27*, 8927.
7. Sowers, M.L.; Conrad, J.W.; Chang-Gu, B.; Cherryhomes, E.; Hackfeld, L.C.; Sowers, L.C. DNA base excision repair intermediates influence duplex–quadruplex equilibrium. *Molecules* **2023**, *28*, 970.
8. Escaja, N.; Mir, B.; Garavís, M.; González, C. Non-G base tetrads. *Molecules* **2022**, *27*, 5287.
9. Shepard, C.P.; Emehiser, R.G.; Karmakar, S.; Hrdlicka, P.J. Factors impacting invader-mediated recognition of double-stranded DNA. *Molecules* **2023**, *28*, 127.
10. Lu, C.; Lopez, A.; Zheng, J.; Liu, J. Using the intrinsic fluorescence of DNA to characterize aptamer binding. *Molecules* **2022**, *27*, 7809.
11. Mazzini, S.; Princiotto, S.; Artali, R.; Musso, L.; Aviñó, A.; Eritja, R.; Gargallo, R.; Dallavalle, S. Exploring the interaction of G-quadruplex binders with a (3 + 1) hybrid G-quadruplex forming sequence within the PARP1 gene promoter region. *Molecules* **2022**, *27*, 4792.
12. Greco, F.; Musumeci, D.; Borbone, N.; Falanga, A.P.; D'Errico, S.; Terracciano, M.; Piccialli, I.; Roviello, G.N.; Oliviero, G. Exploring the parallel G-quadruplex nucleic acid world: A spectroscopic and computational investigation on the binding of the c-myc oncogene NHE III1 region by the phytochemical polydatin. *Molecules* **2022**, *27*, 2997.
13. Liu, X.; Zhao, X.; He, J.; Wang, S.; Shen, X.; Liu, Q.; Wang, S. Advances in the structure of GGGGCC repeat RNA sequence and its interaction with small molecules and protein partners. *Molecules* **2023**, *28*, 5801.

References

1. Bardella, F.; Giralt, E.; Pedroso, E. Polysytyrene-supported synthesis by the phosphite triester approach: An alternative for the large-scale synthesis of small oligodeoxyribonucleotides. *Tetrahedron Lett.* **1990**, *31*, 6231–6234. [CrossRef]
2. Fernandez-Forner, D.; Palom, Y.; Ikuta, S.; Pedroso, E.; Eritja, R. Synthesis and characterization of oligodeoxynucleotides containing the mutagenic base analogue 4-O-ethylthymine. *Nucleic Acids Res.* **1990**, *18*, 5729–5734. [CrossRef] [PubMed]
3. Robles, J.; Pedroso, E.; Grandas, A. Peptide–oligonucleotide hybrids with N-acylphosphoramidate linkages. *J. Org. Chem.* **1995**, *59*, 2482–2486. [CrossRef]
4. Robles, J.; Maseda, M.; Beltrán, M.; Concernau, M.; Pedroso, E.; Grandas, A. Synthesis and enzymatic stability of phosphodiester-linked peptide-oligonucleotide hybrids. *Bioconjug. Chem.* **1997**, *8*, 785–788. [CrossRef] [PubMed]
5. Marchan, V.; Ortega, S.; Pulido, D.; Pedroso, E.; Grandas, A. Diels-Alder cycloadditions in water for the straightforward preparation of peptide–oligonucleotide conjugates. *Nucleic Acids Res.* **2006**, *34*, e24. [CrossRef] [PubMed]
6. Grandas, A.; Marchán, V.; Debéthune, L.; Pedroso, E. Stepwise solid-phase synthesis of nucleopeptides. *Curr. Protoc. Nucleic Acid. Chem.* **2007**, *31*, 4–22. [CrossRef] [PubMed]

7. Alazzouzi, E.; Escaja, N.; Grandas, A.; Pedroso, E. A straightforward solid-phase synthesis of cyclic oligodeoxyribonucleotides. *Angew. Chem. Int. Ed. Engl.* **1997**, *36*, 1506–1508. [CrossRef]
8. Frieden, M.; Grandas, A.; Pedroso, E. Making cyclic RNAs easily available. *Chem. Commun.* **1999**, *16*, 1593–1594. [CrossRef]
9. Salisbury, S.A.; Wilson, S.E.; Powell, H.R.; Kennard, O.; Lubini, P.; Sheldrick, G.M.; Escaja, N.; Alazzouzi, E.; Grandas, A.; Pedroso, E. The bi-loop, a new general four-stranded DNA motif. *Proc. Natl. Acad. Sci. USA* **1997**, *94*, 5515–5518. [CrossRef] [PubMed]
10. González, C.; Escaja, N.; Rico, M.; Pedroso, E. NMR structure of two cyclic oligonucleotides. A monomer-dimer equilibrium between dumbbell and quadruplex structures. *J. Am. Chem. Soc.* **1998**, *120*, 2176–2177. [CrossRef]
11. Escaja, N.; Gelpí, J.L.; Orozco, M.; Rico, M.; Pedroso, E.; González, C. Four-stranded DNA structure stabilized by a novel G:C:A:T tetrad. *J. Am. Chem. Soc.* **2003**, *125*, 5654–5662. [CrossRef] [PubMed]
12. Sánchez, A.; Pedroso, E.; Grandas, A. Oligonucleotide cyclization: The thiol-maleimide reaction revisited. *Chem. Commun.* **2013**, *49*, 309–311. [CrossRef] [PubMed]
13. Frieden, M.; Pedroso, E.; Kool, E.T. Tightening the belt on polymerases: Evaluating the physical constraints on enzyme substrate size. *Angew. Chem. Int. Ed.* **1999**, *38*, 3654–3657. [CrossRef]
14. Di Fabio, G.; Randazzo, A.; D'Onofrio, J.; Ausín, C.; Pedroso, E.; Grandas, A.; De Napoli, L.; Montesarchio, D. Cyclic phosphate-linked oligosaccharides: Synthesis and conformational behavior of novel cyclic oligosaccharide analogues. *J. Org. Chem.* **2006**, *71*, 3395–3408. [CrossRef] [PubMed]
15. Casals, J.; Debéthune, L.; Alvarez, K.; Risitano, A.; Fox, K.R.; Grandas, A.; Pedroso, E. Directing quadruplex-stabilizing drugs to the telomere: Synthesis and properties of acridine-oligonucleotide conjugates. *Bioconjug. Chem.* **2006**, *17*, 1351–1359. [CrossRef] [PubMed]
16. Agramunt, J.; Ginesi, R.; Pedroso, E.; Grandas, A. Inverse electron-demand Diels-Alder bioconjugation reactions using 7-oxanorbornenes as dienophiles. *J. Org. Chem.* **2020**, *85*, 6593–6604. [CrossRef] [PubMed]

Disclaimer/Publisher's Note: The statements, opinions and data contained in all publications are solely those of the individual author(s) and contributor(s) and not of MDPI and/or the editor(s). MDPI and/or the editor(s) disclaim responsibility for any injury to people or property resulting from any ideas, methods, instructions or products referred to in the content.

Review

Advances in the Structure of GGGGCC Repeat RNA Sequence and Its Interaction with Small Molecules and Protein Partners

Xiaole Liu [1], Xinyue Zhao [1], Jinhan He [1], Sishi Wang [1], Xinfei Shen [1], Qingfeng Liu [1] and Shenlin Wang [1,2,*]

[1] State Key Laboratory of Bioreactor Engineering, East China University of Science and Technology, Shanghai 200237, China; y85210065@mail.ecust.edu.cn (X.L.); 20000624@mail.ecust.edu.cn (X.Z.); y85220142@mail.ecust.edu.cn (J.H.); 20000612@mail.ecust.edu.cn (S.W.); 20000603@mail.ecust.edu.cn (X.S.); 20000608@mail.ecust.edu.cn (Q.L.)
[2] Beijing NMR Center, Peking University, Beijing 100087, China
* Correspondence: wangshenlin@ecust.edu.cn

Citation: Liu, X.; Zhao, X.; He, J.; Wang, S.; Shen, X.; Liu, Q.; Wang, S. Advances in the Structure of GGGGCC Repeat RNA Sequence and Its Interaction with Small Molecules and Protein Partners. *Molecules* **2023**, *28*, 5801. https://doi.org/10.3390/molecules28155801

Academic Editors: Ramon Eritja, Daniela Montesarchio and Montserrat Terrazas

Received: 26 June 2023
Revised: 21 July 2023
Accepted: 21 July 2023
Published: 1 August 2023

Copyright: © 2023 by the authors. Licensee MDPI, Basel, Switzerland. This article is an open access article distributed under the terms and conditions of the Creative Commons Attribution (CC BY) license (https://creativecommons.org/licenses/by/4.0/).

Abstract: The aberrant expansion of GGGGCC hexanucleotide repeats within the first intron of the *C9orf72* gene represent the predominant genetic etiology underlying amyotrophic lateral sclerosis (ALS) and frontal temporal dementia (FTD). The transcribed r(GGGGCC)$_n$ RNA repeats form RNA foci, which recruit RNA binding proteins and impede their normal cellular functions, ultimately resulting in fatal neurodegenerative disorders. Furthermore, the non-canonical translation of the r(GGGGCC)$_n$ sequence can generate dipeptide repeats, which have been postulated as pathological causes. Comprehensive structural analyses of r(GGGGCC)$_n$ have unveiled its polymorphic nature, exhibiting the propensity to adopt dimeric, hairpin, or G-quadruplex conformations, all of which possess the capacity to interact with RNA binding proteins. Small molecules capable of binding to r(GGGGCC)$_n$ have been discovered and proposed as potential lead compounds for the treatment of ALS and FTD. Some of these molecules function in preventing RNA–protein interactions or impeding the phase transition of r(GGGGCC)$_n$. In this review, we present a comprehensive summary of the recent advancements in the structural characterization of r(GGGGCC)$_n$, its propensity to form RNA foci, and its interactions with small molecules and proteins. Specifically, we emphasize the structural diversity of r(GGGGCC)$_n$ and its influence on partner binding. Given the crucial role of r(GGGGCC)$_n$ in the pathogenesis of ALS and FTD, the primary objective of this review is to facilitate the development of therapeutic interventions targeting r(GGGGCC)$_n$ RNA.

Keywords: amyotrophic lateral sclerosis; frontotemporal dementia; *C9orf72*; GGGGCC; G4

1. Introduction

Amyotrophic lateral sclerosis (ALS) and frontal temporal dementia (FTD) are two neurodegenerative disorders characterized by progressive degeneration and dysfunction of neuronal architecture [1–5]. Both diseases have a fatality rate typically occurring within three to five years after the onset of symptoms [6,7]. ALS, affecting approximately two individuals per 100,000, is characterized by the degeneration of motor neurons, leading to muscle weakness and atrophy [8,9]. FTD, the second most prevalent form of dementia in individuals under the age of 65, is typically characterized by atrophy of the frontal and/or temporal lobes, manifesting as heterogeneous symptoms encompassing behavioral changes (behavioral variant FTD, bvFTD), language impairment (primary progressive aphasia, PPA), or deterioration in motor skills [10]. Despite considerable efforts, the development of efficacious therapeutic strategies for the treatment of ALS and FTD remains a challenge [11].

The etiologies of ALS and FTD are various. Sporadic ALS (sALS) accounts for 90% of the ALS patients. The remaining 10% of ALS patients are familial ALS (fALS) caused by mutations. The fALS can be caused by the dysfunction of mutated proteins, such as SOD1 mutations, FUS/TLS, and TDP-43 provoked by *TARDBP* mutations [12,13], which lead to neurotoxicity. The aberrant elongation of the hexanucleotide repeat GGGGCC in the non-coding region of *C9orf72* has also been demonstrated to be causally associated with a ALS and FTD [8,14–22]. Aberrant expansion of the GGGGCC repeats is observed in 8% of sALS patients, as well as in more than 40% of fALS cases [14]. Individuals affected by ALS exhibit an average repeat count ranging from 700 to 1600, whereas healthy individuals possess fewer than 25 repeats [23–26]. As for FTD, approximately one-third of FTDs are familial, with autosomal dominant mutations in three genes accounting for the majority of inheritance, including progranulin (GRN), *C9orf72*, and microtubule- associated protein tau (MAPT) [27]. The co-occurrence of these two disorders among families provides support for their genetic linkage [28].

Moreover, aberrant expansion of short nucleotide repeats has been observed in many neurodegenerative diseases. CTG triplet amplification in 3′-UTR may occur in dystrophia myotonica protein kinase (DMPK) gene, and alternative splicing of junctophilin (JPH) gene exon 2a and ataxin8 (ATXN8) gene, which can, respectively, result in Muscular dystrophy type 1 (DM1), Huntington disease-like 2 (HDL2), and Spinocerebellar Ataxia 8 (SCA8). The amplification of CGG triplet in 5′-UTR of the fragile X mental retardation 1 (FMR1) gene may lead to Fragile X disorders (FXTAS), while (CAG)n in the exon of ataxin3 (ATXN3) may cause Spinocerebellar Ataxia 3 (SCA3). In addition, (ATTCT)n within an intron of the ataxin 10 (ATXN10) gene and (CCTG)n in the first intron of the zinc finger protein 9 (ZNF9)gene may lead to Spinocerebellar Ataxia 10 (SAC10) and Muscular dystrophy type 2 (DM2), respectively [29]. Characterization of pathological mechanisms by which these short nucleotide repeats cause fatal diseases has been the research focus aiming in finding the treatments.

Three mechanisms have been proposed to elucidate the pathological underpinnings of aberrant GGGGCC expansion (Figure 1) [30]. Firstly, these abnormal expansions can lead to a gain or loss of function in the associated gene [7,31]. Secondly, the transcribed r(GGGGCC)$_n$ RNA forms RNA foci that recruit RNA binding proteins (RBPs), consequently impairing protein function and ultimately triggering intracellular cytotoxicity [7,32–36]. Lastly, non-ATG translation of the r(GGGGCC)$_n$ sequence produces dipeptide repeats (DPRs) that exert neurotoxic effects within the central nervous system [7,37–42]. Among the three proposed mechanisms, the formation of RNA foci and recruitment of RBPs have garnered the most attention. This process involves the spontaneous liquid–liquid separation of r(GGGGCC)$_n$, followed by a sol–gel phase transition by increased interactions [43]. The RNA foci can recruit various RBPs, including hnRNP H [44], Zfp106 [45], ADARB2 [46], Purα [47,48], and FUS [49,50], ultimately leading to disruptions in the intracellular environment [43,51]. The aberrant phase separation and spread of hnRNP H within r(GGGGCC)$_n$ in ALS patients are key features in the pathogenesis of the disease [52].

Consequently, the investigation of the r(GGGGCC)$_n$ RNA repeats structures and the interactions between r(GGGGCC)$_n$ and small molecules is currently a highly prominent area of research. The primary objective is to identify lead compounds for the treatment of FTD and ALS [53]. The r(GGGGCC)$_n$ sequence exhibits characteristic structural polymeric and has the ability to adopt either a hairpin or G-quadruplex (G4) structure [54]. Several small molecules have been discovered to possess strong binding affinity for r(GGGGCC)$_n$. A majority of these compounds incorporate polyaromatic ring conjugated systems that effectively stabilize the G4 structures of r(GGGGCC)$_n$, thereby inhibiting phase separation, disrupting protein–RNA interactions [55] and/or preventing non-ATG translation of DPRs [53]. Additionally, two drugs, riluzole and edaravone, have received approval for the treatment of ALS [56]. While these drugs can delay disease progression, they do not specifically target the r(GGGGCC)$_n$ RNA [56]. Moreover, recent research by Meijboom and colleagues used the adeno-associated virus vector system to deliver CRISPR/Cas9 gene editing system into neuron cells, and successfully removed the hexanuclear repeat expansion from the *C9orf72* gene in the mouse model (500–600 repeats), as well as the patient-derived Induced Pluripotent Stem Cell (iPSC) motor neuron and brain organoid (450 repeats). This led to a reduction in RNA foci, DPRs and haploinsufficiency, major hallmarks of C9-ALS/FTD, making this a promising therapeutic approach to ALS/FTD diseases [57].

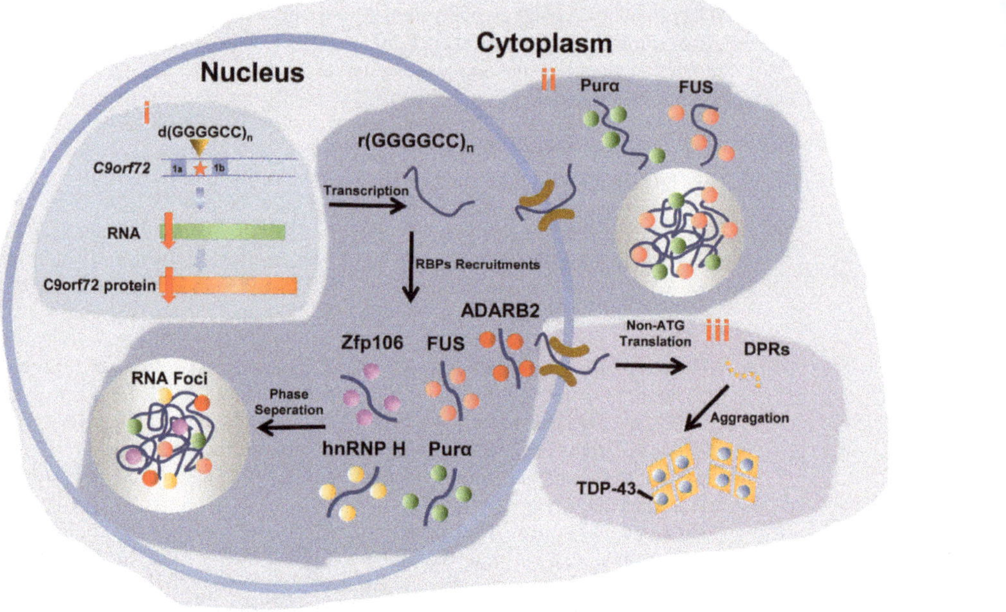

Figure 1. Three pathogenic mechanisms associated with *C9orf72*-related FTD and ALS. (i) Aberrant expansion of the d(GGGGCC)n of *C9orf72* within contains two non-coding exons (1a and 1b), and under pathological conditions, repress transcription, resulting in the reduction in *C9orf72* protein. (ii) The transcribed r(GGGGCC)n aggregates in the nucleus to form RNA foci that recruit RBPs, affecting the intra cellular functions of RBPs, i.e., splicing. (iii) The r(GGGGCC)n RNA is transported into the cytoplasm and undergoes repeat-associated non-ATG translation, resulting in the synthesis of DPRs. The DPRs forms aggregation and associate TDP-43, which induce cytotoxic effects in cells. The labels (i–iii) correspond to the three pathological mechanisms. The red star highlights the hexanucleotide repeat GGGGCC in the non-coding region of *C9orf72*.

This review provides a comprehensive overview of the recent progress made in understanding the structures of r(GGGGCC)$_n$, and the interactions between r(GGGGCC)$_n$ and small molecules and between r(GGGGCC)$_n$ and protein partners. We focus on elucidating the structural diversity of r(GGGGCC)$_n$ and its implications for partner binding. Given the crucial role of r(GGGGCC)$_n$ in the pathogenesis of ALS and FTD, the primary objective of this review is to support the development of drugs targeting r(GGGGCC)$_n$ RNA.

2. The Structure of r(GGGGCC)$_n$ RNA Repeats and the RNA within the RNA Foci of r(GGGGCC)$_n$

2.1. The Solution Structures of r(GGGGCC)$_n$ RNA

The r(GGGGCC)$_n$ RNA is a guanine-rich sequence, which promotes the formation of G4 structures. While the tertiary structure of r(GGGGCC)$_n$ remains to be fully elucidated, the secondary structures have been extensively studied. Circular Dichroism (CD) spectra are commonly used to demonstrate the G4 structures and the topology of G4, which for instance, the spectral patterns would provide the evidence of parallel, antiparallel, or other types of topologies (Figure 2). Nuclear Magnetic Resonance (NMR) spectra are able to provide more structural details, some of which yield full structural determination or provide evidence of co-existence of different conformations (Figure 2a). Depending on the sequence length and solution conditions, r(GGGGCC)$_n$ can adopt different secondary structures, including G4 [58,59] and hairpin conformations [60]. In 2012, Adrian M. Isaacs and colleagues demonstrated that r(GGGGCC)$_3$GGGGC can fold into G4 or double-stranded structures, with the topology being influenced by the presence of cation ions in the solution [61]. In a K$^+$ buffer, it forms a stable parallel intramolecular G4 structure, while it becomes less stable in Na$^+$ and Li$^+$ solutions.

Further investigations by Pearson and colleagues employed circular dichroism (CD) spectroscopy (Figure 2a) and gel-shift assays, revealing that r(GGGGCC)$_n$ (n = 2, −5, −6, and −8) predominantly adopt highly stable uni- and multi-molecular parallel G4 structures [62]. The abundance of G4 structures is influenced by the repeat number and RNA concentrations, with the proportion of multi-molecular G4 structures increasing as the number of repetitions rises.

The equilibrium between G4 and hairpin structures has also been observed in r(GGGGCC)$_n$. In the absence of K$^+$ ions, r(GGGGCC)$_4$ RNA forms a hairpin conformation [62], featuring single-stranded bulges within the RNA chain. However, in a K$^+$ buffer (Figure 2c), it adopts a parallel G4 structures (Figure 2g) [59]. This equilibrium between hairpin and G4 structures is suggested to be linked to the presence of an abortive transcript containing hexanucleotide repeats [55]. The G4 structure may hinder the transcription of full-length RNA and recruit RBPs in cells, contributing to disease pathogenesis. The equilibrium is biased towards the hairpin conformation with a higher repeat number of r(GGGGCC)$_n$. Specifically, r(GGGGCC)$_4$ predominantly adopts a G4 topology, while r(GGGGCC)$_8$ RNA exhibits both G4 and hairpin structures, even in a K$^+$ buffer, as confirmed by various biophysical methods. However, in a Na$^+$ buffer, r(GGGGCC)$_8$ RNA solely adopts a hairpin structure [55]. Furthermore, r(GGGGCC)$_4$ undergoes a monomer-dimer equilibrium in a pH-dependent manner. At pH 6.0 and 25 °C, it exists as both a homodimer and a hairpin structure. Decreasing the temperature increases the population of dimeric RNA, which exhibits distinct structural differences compared to G4 structures in the presence of K$^+$ [63]. Conversely, at neutral pH, r(GGGGCC)$_4$ primarily adopts a hairpin conformation.

2.2. Structure of d(GGGGCC)$_n$ DNA

High-resolution structures of d(GGGGCC)$_n$ have been successfully determined [64,65]. Janez Plavec and colleagues utilized NMR spectroscopy to elucidate the structure of d[(GGGGCC)$_3$GG^{Br}GG] (represented by PDB codes 2N2D) [66]. The incorporation of a bromine-substituted guanine residue (G^{Br}) contributed to the stabilization of the conformation, leading to a more rigid structure amenable to structural analysis. The d[(GGGCCC)$_3$GG^{Br}GG] sequence adopted an antiparallel G4 topology (Figure 2b) [67].

In 2015, Guang Zhu and colleagues employed CD, NMR, and native polyacrylamide gel electrophoresis (PAGE) to investigate the structures of d(GGGGCC)$_n$ repeats. Their studies revealed distinct G4 folding patterns in the presence of K$^+$ ions. Notably, d(GGGGCC)GGGG, d(GGGGCC)$_2$, and d(GGGGCC)$_3$ did not exhibit stable G4 structures. Instead, d(GGGGCC)$_2$ and d(GGGGCC)$_3$ displayed mixed forms of parallel and antiparallel G4 folding. On the other hand, d(GGGGCC)$_4$ and d(GGGGCC)$_5$ formed stable G4 structures. Specifically, d(GGGGCC)$_5$ exhibited a combination of parallel and antiparallel G4 folds, while d(GGGGCC)$_4$ adopted a homogeneous monomeric form characterized by a chair-type G4 structure [10]. In 2021, this group determined the crystal structure of d(GGGGCC)$_2$ in both Ba^{2+} and K$^+$ solutions, revealing an eight-layer parallel G4 structure for d(GGGGCC)$_2$ (represented by PDB codes 7ECF and 7ECG) (Figure 2e) [68]. Jiou Wang and colleagues (Figure 2d) also confirmed that d(GGGGCC)$_4$ adopts an antiparallel G4 (Figure 2f) [59].

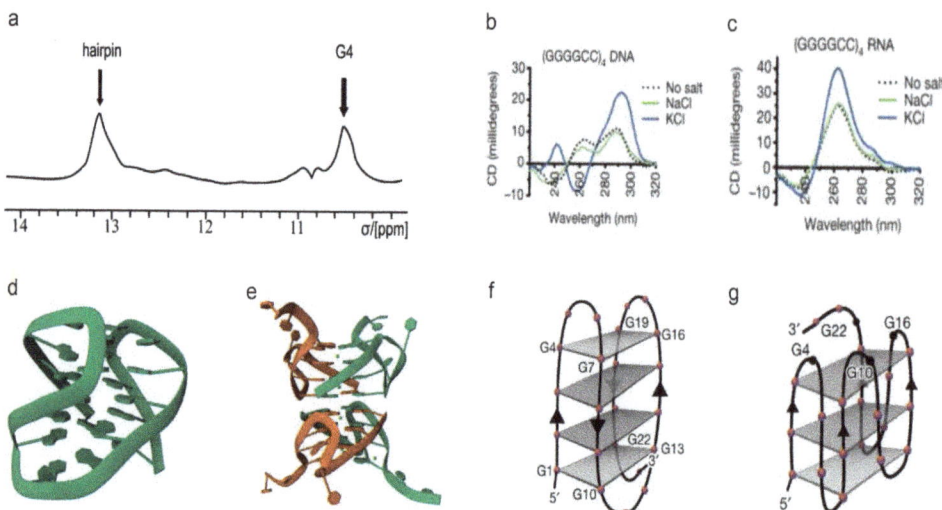

Figure 2. Biophysical methods in studying the r(GGGGCC)$_n$ and the d(GGGGCC)$_n$ structures. (**a**) ^{1}H NMR spectra of r(GGGGCC), with highlighting the cross-peaks of the hairpin and G4 structures. (**b,c**) The CD spectra with the characteristic antiparallel G4 topology of d(GGGGCC)$_4$ (**b**) and the parallel G4 for the r(GGGGCC)$_4$ in the presence of 100 mM KCl. The ions dependent of G4 structures were shown. (**d,e**) The high-resolution crystal structures of d[(GGGGCC)$_3$GG^{Br}GG] (**d**) and d(GGGGCC)$_2$ (**e**). (**f,g**) The proposed topology for the antiparallel DNA G4 formed by (GGGGCC)$_4$ (**f**) and the parallel G4 topology formed by the r(GGGGCC)$_4$ RNA (**g**). The (**b,c,f,g**) were reprinted from the reference [59]. (**d,e**) were reprinted from the reference [67,68], respectively.

2.3. Biological Phase Separation and Transition of r(GGGGCC)$_n$

Biological liquid–liquid phase separation is a widely observed phenomenon in cells and plays a critical role in the formation of membraneless organelles, signal transduction, and DNA packaging [69–72]. As the strength of interactions in phase separation systems increases, a transition from a liquid to a solid state often occurs, resulting in the formation of insoluble gel-like states, many of which are associated with diseases [73]. Jain and colleagues demonstrated that r(GGGGCC)$_n$ can undergo phase separation both in vivo and in vitro [43]. They found that phase separation of r(GGGGCC)$_n$ occurs once a specific threshold of repeat value is reached, leading to a solution-gel phase transition as the strength of multi-base interactions increases (Figure 3). The formation of RNA foci is dependent on solution conditions and is reinforced by Mg^{2+} but impaired by monovalent cations such as K$^+$ or Na$^+$. The authors proposed that inter-chain hydrogen bonds stabilize intermolecular G4s, which serve as the building blocks of RNA foci. However, direct evidence of the secondary structure of r(GGGGCC)$_n$ within RNA foci is still lacking.

Christopher E. Shaw and colleagues discovered that r(GGGGCC)$_n$ RNA foci were detected in neuronal cell lines and zebrafish embryos expressing 38 or 72 repeats but not in those expressing 8 repeats [6]. This finding indicates that longer r(GGGGCC)$_n$ sequences lead to nuclear retention of transcripts and the formation of RNA foci, which are resistant to the enzyme ribonuclease (RNase) [6,52]. Extended r(GGGGCC)$_n$ sequences exhibit significant neurotoxicity and bind to hnRNP H and other RBPs. RNA toxicity and sequestration of RBPs may impair RNA processing and contribute to neurodegenerative diseases. In a study conducted by Simon Alberti and colleagues, it was demonstrated that RNA plays a crucial role in regulating the phase behavior of prion-like RBPs [74]. Lower RNA to protein ratios promote the separation of RBPs into liquid droplets, whereas higher ratios prevent droplet formation in vitro. When nuclear RNA levels are reduced or RNA binding is genetically ablated, excessive phase separation occurs, leading to the formation of cytotoxic solid-like assemblies in cells. The researchers proposed that the nucleus functions as a buffered system, with high RNA concentrations maintaining RBPs in a soluble state. Disruptions in RNA levels or the RNA binding abilities of RBPs result in abnormal phase transitions [75].

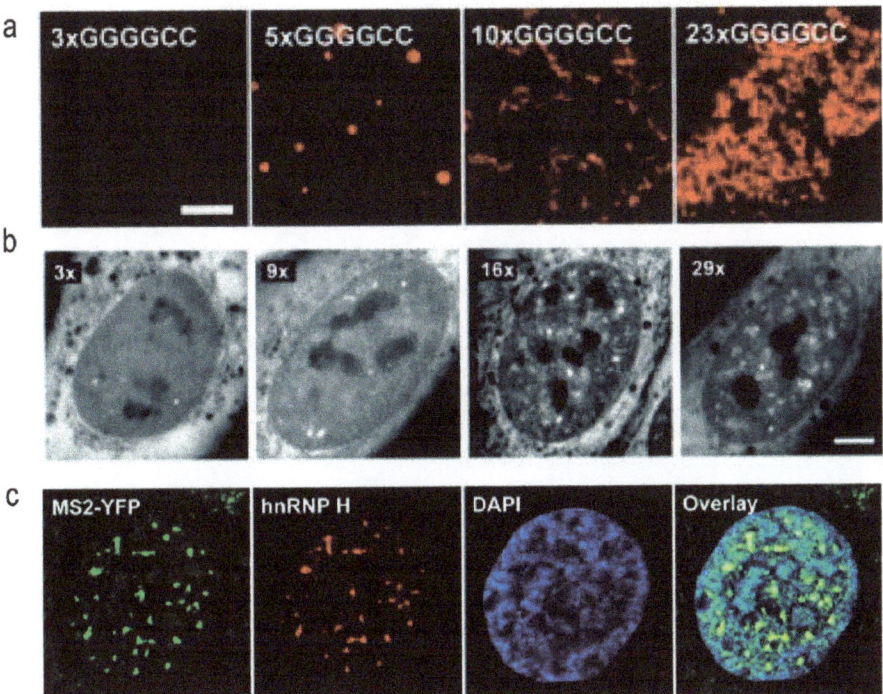

Figure 3. Representative results of biological phase separation of r(GGGGCC)$_n$. (a) The in vitro fluorescence imaging of r(GGGCC)$_n$ RNA clusters at indicated number of r(GGGCC)$_n$. (b) Representative fluorescence micrographs and corresponding quantification of the total volume of foci per cell in U-2OS cells transduced with r(GGGCC)$_n$ RNA with the indicated number of r(GGGCC)$_n$. (c) Representative immuno fluorescence images illustrating that the r(GGGCC)$_{29}$ recruited endogenous hnRNP H. Figure 3 was reprinted from the reference [43].

3. Disease Related RBPs That Bind to r(GGGGCC)$_n$

3.1. hnRNP H and TDP-43

Heterogeneous nuclear ribonucleoprotein H (hnRNP H) is a member of the hnRNP family and functions as a multifunctional RBP involved in mRNA maturation at various stages [76]. It contains a modular domain consisting of tandem quasi-RNA recognition motifs (HqRRM1,2) at the N-terminus and a third qRRM3 at the C-terminus, situated between two glycine-rich segments [44,77,78]. The hnRNP H has the ability to bind G-rich RNA sequences containing at least three consecutive guanines [44]. In the brain cells of ALS patients, hnRNP H has been found associated with insoluble aggregation of r(GGGGCC)$_n$, leading to aberrant alternative splicing [52]. This phenotype has been utilized as a biomarker for disease diagnosis. Furthermore, ALS/FTD patients exhibit splicing alterations in several key targets and insoluble hnRNP H, indicating that modifications along this axis are critical aspects of disease etiology [52].

James L. Manley and colleagues demonstrated that hnRNP H binds to r(GGGGCC)$_n$ in vitro, and this interaction is dependent on the formation of G4s. The hnRNP H colocalizes with G4 aggregates in C9 patient-derived fibroblasts and astrocytes, but not in control cells, as proven by imaging on BG4, a G4 structure-specific antibody (Figure 4) [79]. Another study by Donald C. Rio and colleagues revealed that in sporadic ALS/FTD patients, insolubility of hnRNP H was associated with altered splicing of a wide range of targets [52]. Numerous ALS/FTD brains show high levels of insoluble hnRNP H sequestered in r(GGGGCC)$_4$ RNA foci, resulting from RNA splicing defects involving intron retention [52]. These findings highlight previously unreported splicing abnormalities in extremely insoluble hnRNP H-related ALS brains, suggesting a potential feedback relationship between effective RBP concentrations and protein quality control in all ALS/FTD cases.

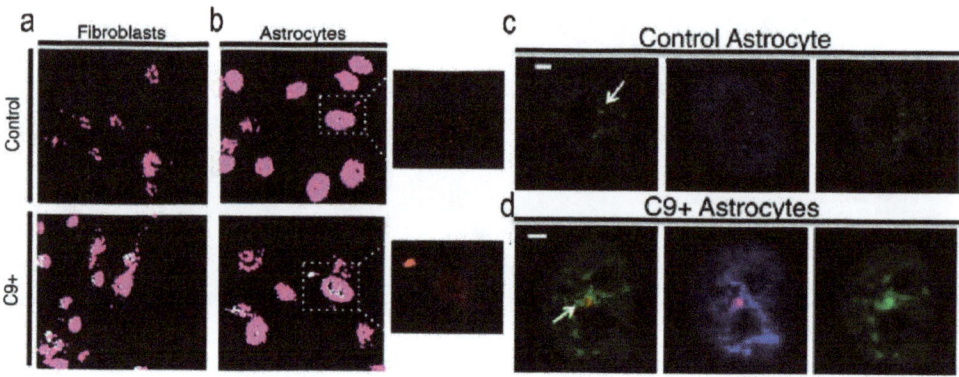

Figure 4. Quantification of stained BG4-foci and area was performed in fibroblasts and astrocytes derived from patients with ALS/FTD and healthy controls. Representative images of non-ALS and ALS fibroblasts (**a**) and astrocytes (**b**) are shown, with the 'BG4 Count' projection representing all stained areas above the determined threshold (showed in red), and areas of particularly dense staining shown in white. The inset displays the source image, highlighting only the red channel corresponding to BG4-FLAG staining. (**c**) Control astrocytes exhibit single, small nuclear hnRNP H/BG4 foci. (**d**) Patient astrocytes demonstrate nuclear hnRNP H/BG4 foci. Figure 4 was reprinted from the reference [79].

TAR DNA binding protein 43 (TDP-43), another member of the hnRNP family, possesses two RNA recognition motifs (RRMs), a nuclear localization signal (NLS), and a prion-like domain at the C-terminus [80]. Numerous mutations in TDP-43 have been associated with ALS and FTD [81,82]. The accumulation of TDP-43 is a major pathological feature of ALS and FTD [83–85], and inclusion bodies are observed in patients with abnormal expansions of r(GGGGCC)$_n$, serving as a histopathological marker in 97% of ALS cases and 45% of FTD cases.

In contrast to hnRNP H, which directly associates with r(GGGGCC)$_n$, the pathogenic mechanism of TDP-43 in ALS/FTD is believed to involve its interaction with DPRs, which are non-ATG translation products of r(GGGGCC)$_n$ [15,86]. Edward B. Lee and colleagues discovered that DPRs induce TDP-43 protein lesions in an ALS/FTD model and trigger the onset and progression of FTD [81]. The amount and characteristics of produced DPRs, rather than the length of r(GGGGCC)$_n$ repeats, determine the duration and severity of TDP-43 dysfunction.

3.2. FUS

Sarcoma fusion protein (FUS) is a 526-amino acid residue protein. [87] It is predominantly expressed in neurons and is involved in DNA and RNA metabolism through its interactions with motor proteins kinesin [88] and myosin-Va [89]. Missense mutations in the FUS gene have been associated with ALS [90,91], although the prevalence of FUS gene variants in the familial ALS population is low.

Sua Myong and colleagues conducted investigations on the binding of wild-type FUS to single-stranded RNAs, including r(GGGGCC)$_4$, in a length-dependent manner. They observed the formation of a highly dynamic protein–RNA complex. The FUS–RNA interaction involves two mechanisms: (i) stable binding of FUS monomers to single-stranded RNA (ssRNA), and (ii) weak interaction of two FUS units with RNA, resulting in a highly dynamic interaction.

Higuro and workers observed the formation and phase transition of FUS condensates in vitro using purified full-length wild-type and mutant FUS proteins and r(GGGGCC)$_4$. They found that FUS specifically forms complexes with r(GGGGCC)$_4$ in a G4 structure-dependent manner, leading to a transition from liquid–liquid separation to liquid–solid transitions. Importantly, amino acid mutations associated with ALS significantly impact G4-dependent FUS condensation. These findings provide insights into the relationship between protein aggregation and dysfunction of FUS in ALS [49].

3.3. Zfp106

Zfp106 is a C2H2 zinc finger protein characterized by the presence of seven WD40 domains and four putative zinc fingers [92]. It plays a crucial role in maintaining neuromuscular signaling. Knockout mice exhibit gene expression patterns indicative of neuromuscular degeneration in their muscles and spinal cords. Interestingly, this phenotype can be reversed through motor neuron-specific repair of the Zfp106 transgene, highlighting its essential role in biological processes [93]. The functional acquisition model of *C9orf72* neurodegeneration has been investigated in a *Drosophila* model [94], where Zfp106 effectively mitigates the neurotoxicity associated with the expression of GGGGCC repeat in *C9orf72* ALS *Drosophila*. This suggests that Zfp106 acts as a repressor of neurodegeneration in *C9orf72* ALS models and demonstrates a functional interaction between Zfp106 and the r(GGGGCC)$_n$ sequence. Furthermore, Brian L. Black and colleagues conducted pull-down assays and mobility shift assays, providing evidence that Zfp106 specifically binds to r(GGGGCC)$_8$ but not to the sequence of r(AAAACC)$_8$. The ability of Zfp106 to regulate normal cellular functions and inhibit ALS by binding to r(GGGGCC)$_n$ makes it a potential drug target for treating ALS [45]. However, the mechanisms through which Zfp106 regulates normal cellular processes via RNA binding and how it inhibits ALS progression by interacting with r(GGGGCC)$_n$ are still being investigated to guide drug design efforts [45].

3.4. ADARB2

ADARB2 is a member of the CNS-rich adenosine deaminase family, known for its role in mediating A-to-I (adenosine to inosine) editing of RNA [95]. It consists of two double-stranded-specific adenosine deaminase repeats, three double-stranded RNA-binding domains, and one editase domain spanning from the N- to C-terminus. The A-to-I editing activity primarily occurs within the 16–130 nucleotide interval. This enzyme selectively deaminates adenosine (A) residues in the double-stranded region of mRNA, converting them to inosine (I), which is recognized as guanine by the cellular translation machinery, resulting in codon alterations within the synthesized protein [46] (Figure 5).

Jeffrey D. Rothstein and colleagues conducted RNA fluorescence in situ hybridization (RNA FISH) and immunofluorescence labeling of RBP simultaneously in the induced pluripotent stem neuron (IPSN) cell line derived from *C9orf72*-related cases. Their study revealed the co-localization of ADARB2 protein with nuclear r(GGGGCC)$_n$ RNA foci, while mRNA levels remained unchanged. Co-precipitation of ADARB2 with r(GGGGCC)$_n$ repeats was also observed in vivo.

In vitro investigations utilizing recombinant ADARB2 through gel shift assays clearly demonstrated its binding to r(GGGGCC)$_n$, implying the possible formation of ADARB2-RNA complexes. These collective findings indicate a strong binding between ADARB2 and r(GGGGCC)$_n$. Furthermore, this team verified in vivo that the formation of r(GGGGCC)$_n$ RNA foci requires the involvement of ADARB2 protein. Treatment of the IPSN line with specific siRNA targeting ADARB2 significantly reduced the number of RNA foci. However, further experimental evidence is still needed to fully elucidate ADARB2's in vivo function [96]. Another unresolved aspect of ADARB2 function is the speculation that ADARB2 may lose its editing activity upon interaction with r(GGGGCC)$_n$, although experimental validation of its downstream editing effects is currently lacking.

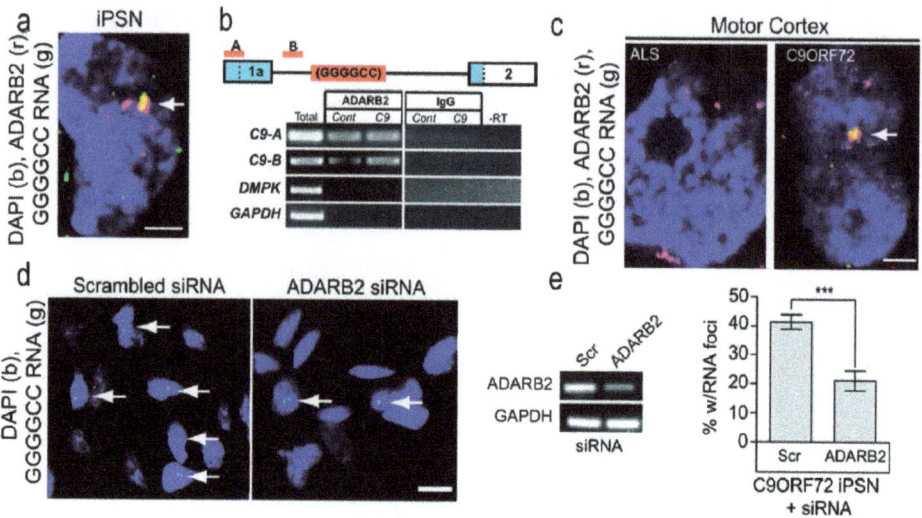

Figure 5. ADARB2 Protein Binds to the r(GGGGCC)$_n$. (**a**) Colocalization of r(GGGGCC)$_n$ RNA foci with ADARB2 signal in IPSN cells. (**b**) Co-immunoprecipitation (co-IP) of ADARB2-bound RNA isolated from control and *C9orf72*-induced cell lines. RT-PCR of the co-IP RNA using two primer sets (A and B, red), located upstream of the r(GGGGCC)$_n$ repeat, demonstrated ADARB2 binding to *C9orf72* RNA in both control and *C9orf72* cell lines. (**c**) Colocalization of r(GGGGCC)$_n$ RNA foci and ADARB2 was observed in postmortem motor cortex tissue from *C9orf72* patients. (**d,e**) Knockdown of ADARB2 using siRNA significantly reduced the percentage of nuclear RNA foci (indicated by arrows). siRNA knockdown of ADARB2 results in a significant reduction in the percent of iPSNs with nuclear RNA foci (arrows). Data in (E) indicate mean ±SEM (*** $p < 0.001$). Figure 5 was reprinted from the reference [46].

3.5. Purα

Pur-alpha (Purα) is a highly conserved DNA and RNA binding protein in eukaryotic cells [97]. It performs diverse physiological functions, including transcription activation or inhibition, cell growth, and translation [98,99]. While predominantly localized in the nucleus, Purα is also widely distributed in the cytoplasm of neurons, particularly in synaptic branches [88]. In the nucleus, Purα stimulates gene transcription by binding to mRNA transcripts and accompanying them to the cytoplasm. It remains associated with the mRNA during transport over considerable distances and functions at specific sites of mRNA translation [100]. The absence of Purα can lead to various neurological disorders [101,102].

The r(GGGGCC)$_n$ repeat can sequester Purα, thereby impairing its normal functions such as gene transcription and mRNA translation, ultimately resulting in cell death [103]. In an ALS/FTD zebrafish model, Swinnen and colleagues demonstrated that the Pur2 domain of Purα binds to r(GGGGCC)$_{90}$ repeat RNA [37]. Peng Jin and colleagues conducted studies on the pathogenesis of ALS/FTD, revealing that r(GGGGCC)$_{10}$ can sequester Purα, a major component of RBPs, from the whole-cell lysate of mouse spinal cord [47]. Rossi and colleagues found that Purα can aggregate into cytosolic and nuclear granules in HeLa cells transiently transfected with a plasmid expressing r(GGGGCG)$_{31}$. Nonetheless, due to the specific interaction between Purα and r(GGGGCC)$_n$, it is conceivable that Purα may influence the outcome of RAN translation. Consequently, in ALS, reduced protein levels amplify certain cellular characteristics. Over-expression of Purα in mammalian and *Drosophila* model systems can rescue r(GGGGCC)$_n$ repeat-induced neurodegeneration [47].

Furthermore, Purα also interacts with the C-terminal region of FUS, another protein recruited by r(GGGGCC)$_n$ [104]. In vivo expression of Purα in various *Drosophila* tissues significantly exacerbates neurodegeneration caused by mutated FUS. Conversely, reducing Purα expression in neurons expressing mutated FUS significantly improves the climbing ability of *Drosophila* flies. This suggests that downregulation of Purα ameliorates locomotion defects, a classical symptom of ALS resulting from mutant FUS expression. These findings indicate that Purα may contribute to the pathogenesis of ALS mediated by FUS. However, it remains unclear which functional domains or subdomains of Purα are involved in mediating its interaction with FUS [105].

Binding of Purα to other cellular proteins can directly impact the expression of the *PURA* gene. Purα itself can bind to GC/GA-rich sequences in its own promoter and inhibit gene expression [106]. Similarly, binding of Purα to expanded polynucleotide repeat RNA may also affect the expression of the *PURA* gene. In both scenarios, the mechanism of action may involve the combination of Purα with cellular components, resulting in a reduction in effective intracellular Purα levels. The reduction in Purα could trigger a feedback mechanism of the *PURA* gene, although it is unknown whether this compensates for Purα sequestration [100].

4. Lead Small Molecules Binds to r(GGGGCC)$_n$

Given the pharmacological advantages of r(GGGGCC)$_n$ formation of RNA foci and their recruitment of RBPs, small molecules present an attractive option for targeting r(GGGGCC)$_n$. Therefore, it is interesting to investigate the binding of r(GGGGCC)$_n$ to small molecules (Figure 6). Currently, a number of the small molecules contain aromatic rings have been found to bind to r(GGGGCC)$_n$.

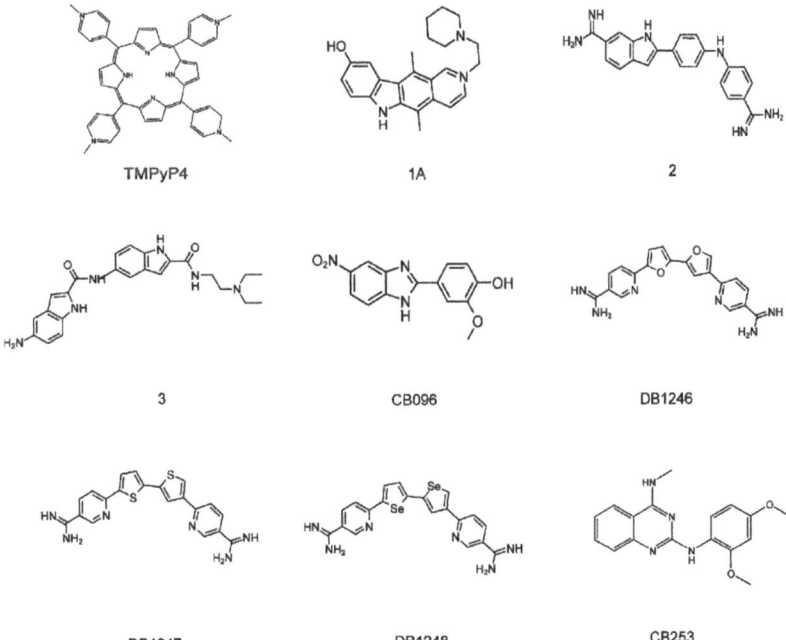

Figure 6. r(GGGGCC)$_n$ small molecular structure bound to small molecules.

4.1. Binding of r(GGGGCC)$_8$ with the TMPyP4

The G4 structure has been shown to bind to 5,10,15,20-tetra(N-methyl-4-pyridyl) porphyrin (TMPyP4), as demonstrated before [107,108]. TMPyP4 binds a variety of G4 structures of DNA or RNA [109,110]. In 2014, Christopher E. Pearson and colleagues found that TMPyP4 could bind and distort the G4 formed by r(GGGGCC)$_8$, inhibiting the interaction of some proteins with the repeat [23]. Several studies have shown that TMPyP4 disrupts the binding of hnRNPA1 to the r(GGGGCC)$_8$ repeat, that are supposed to link to ALS/FTD pathogenesis [23]. Therefore, it may be possible to develop therapeutic treatments using TMPyP4 to disrupt the interaction of RBPs. However, TMPyP4 may either stabilize or destabilize RNA G4. Kelly and colleagues used molecule dynamics simulations to analyze RNA G4 structure and speculated that TMPyP4 might interact with RNA G4 in three different ways: top-stacking, bottom-stacking, and side-binding, maintaining stability under certain conditions [111]. However, the specific structure and binding mode of the complex have not been reported. Therefore, further study on the interaction between TMPyP4 and r(GGGGCC)$_n$ RNA, as well as the destruction of RBPs binding which may cause toxicity, will be one of the directions for the development of related small molecule drugs.

4.2. Binding of r(GGGGCC)$_8$ with Other Liands

Matthew D. Disney and colleagues has discovered three lead compounds, **1a**, **2**, and **3**, that bind with r(GGGGCC)$_8$ in vitro, with Kds of 9.7, 10, and 16 µM, respectively [55]. These three small molecules were obtained by Hoechst or *bis*-benzimidazole query, and were derived from the small molecule library established by chemical similarity search. This library is enriched in compounds that have the potential to recognize RNA 1×1 nucleotide internal loops, among which 1a has been proven to bind 1×1 GG internal loops present in r(CGG)^{exp}, and improve fragile X-associated tremor/ataxia syndrome (FXTAS)-associated defects [112].

As r(GGGGCC)$_8$ RNA experiences dynamical equilibrium between hairpin and parallel G4 structure in solution, the binding constants of these lead compounds with RNA were evaluated in either K$^+$ containing buffer (favorable for G4 structure) or Na$^+$ buffer (favorable for hairpin). The 3–10 times higher Kds of **1a** and **3** were obtained in the presence of K$^+$ than the Na$^+$ buffer, demonstrating their favor binding to G4 structures of r(GGGGCC)$_8$. In contrast, a Na$^+$-dependent affinity of **2** was not affected by r(GGGGCC)$_8$, but it significantly decreased with K$^+$, showing the specific binding with hairpin structures. The optical melting data further demonstrated that compound **3** has no influence on the stability of r(GGGGCC)$_8$, while compounds **1a** and **2** improve it.

The effects of three ligands on non-ATG translation of r(GGGGCC)$_n$ were tested in HEK293 cells expressing r(GGGGCC)$_{66}$ [55]. It was found that poly(GP) and poly(GA) proteins, but not poly(GR) proteins, were produced in the system. Compound **3** (100 µM, 24 h) was shown to moderately limit poly(GP) synthesis while having no effect on poly(GA). Compounds **1a** and **2**, on the other hand, drastically reduced the amounts of GP and GA proteins, which dramatically lowered the percentage of positive cells in the lesions. This suggests that ligand binding to r(GGGGCC)$_n$ could be a potentially effective cure for FTD/ALS.

4.3. Binding of r(GGGGCC)$_8$ with CB096

Disney and colleagues discovered a benzimidazole derivative CB096 that binds to r(GGGGCC)$_n$. NMR, structure–activity relationship (SAR) studies, and molecular dynamics (MD) simulations with r(GGGGCC)$_n$ hairpin structure have been used to determine the molecular interaction between CB096 and r(GGGGCC)$_n$ (Figure 7) [113]. When r(GGGGCC)$_n$ is folded, CB096 can specifically bind to the repeating 1×1 GG inner ring structure of 5'CGG\3'GGC. The TO-PRO-1 (TO-1) fluorescent dye replacement assay and microscale thermoelectrophoresis (MST) were used to screen the ligands bound to the r(GGGGCC)$_8$ hairpin. CB096 binds to 5'CGG/3'GGC of the r(GGGGCC)$_n$ hairpin and breaks the base pair as shown by NMR. To bind to the r(GGGGCC)$_n$ hairpin structure, the chemical 5′s-NO2 group and 2-methoxyphenyl are crucial. In ALS/HEK293T FTD's cells, CB096 slowed RAN translation and reduced poly(GP) DPR formation, but did not affect r(GGGGCC)$_{66}$ mRNA levels. In conclusion, the researchers showed that CB096 binds particularly to the 1 × 1 GG inner ring 5′CGG\3′GGC generated during the expansion of r(GGGGCC)$_n$.

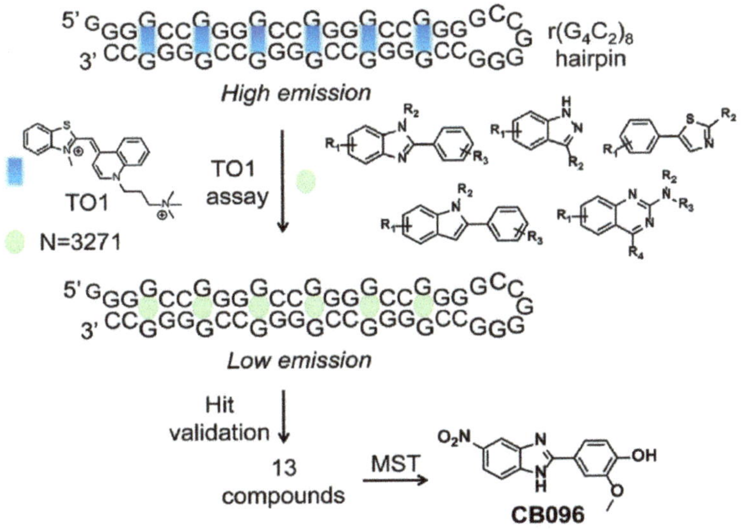

Figure 7. CB096 specifically bound to the repeated 1 × 1 GG inner loop structure 5′CGG/3′GGC in the r(GGGGCC)$_8$ hairpin structures. Figure 7 was reprinted from the reference [113].

4.4. Binding of r(GGGGCC)$_n$ with DB1246, DB1247, and DB1273

Isaacs and colleagues screened a chemical library of small molecules to find the r(GGGGCC)$_4$ binding ligands [53]. They identified 44 hits out of 138 small molecules by a FRET-based G4 melting assay. Among those hitting compounds, three molecules are structurally similar (DB1246, DB1247, and DB1273) and have the ability to bind and stabilize G4s structure, as shown by temperature dependent CD spectroscopy [53]. Treatment with these compounds led to a significant reduction in both RNA foci formation and dipeptide repeat protein levels in *Drosophila* carrying r(GGGGCC)$_{36}$ and improved survival in vivo [53]. These findings suggest that targeting the r(GGGGCC)$_n$ G4 using small molecules may be a promising therapeutic approach to alleviate two key pathologies associated with FTD/ALS.

4.5. Binding of r(GGGGCC)$_n$ with CB253

Andrei and colleagues incorporated ^{19}F modified nucleotides to replace the C6 residue in r(GGGGCC)$_2$ duplex model (5′CCGGGG/3′GGGGCC) to investigate the binding mechanism of CB253 to r(GGGGCC)$_n$ (Figure 8) [114]. The replacement of ^{19}F nucleotide enables the use of ^{19}F NMR spectroscopy to investigate the structure and interactions. Two types of inner ring, 1 × 1 GG and 2 × 2 GG, were detected and verified in the r(GGGGCC)$_2$ hairpin structure. Among them, the 1 × 1 GG was the main conformation, and the two conformations could slowly transform into each other to achieve an equilibrium. Addition of CB253 stabilizes the 2 × 2 GG inner ring structure of r(GGGGCC)$_2$ duplex, which becomes a stable dominant conformation. CB253 can form key interactions with N1-H of G3 and combine with r(GGGGCC)$_2$ at a 2:1 ratio. The precise 2,4-diamino substitution pattern within CB253's quinazoline scaffold is crucial for binding the r(GGGGCC)$_n$ hairpin RNA. In HEK293T and lymphoblastoid cells from *C9orf72* patients, CB253 reduced the formation of stress granules induced by r(GGGGCC)$_{66}$ and inhibited RAN translation in a dose-dependent manner, leading to a significant reduction in poly(GP) DPR levels. These findings indicate that CB253 is a promising chemical probe that can specifically bind to and stabilize the 2 × 2 GG inner ring of r(GGGGCC)$_n$ hairpin structure, and inhibit various *C9orf72*-specific pathological mechanisms by directly engaging r(GGGGCC)$_n$.

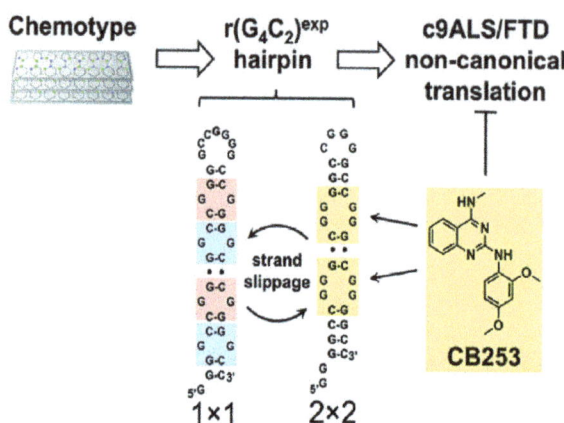

Figure 8. CB253 that selectively binds the hairpin form of r(GGGGCC)$_n$. Figure 8 was reprinted from the reference [114].

5. Summary and Perspective

In this review, we provide a comprehensive overview of the advancements in understanding the structure of r(GGGGCC)$_n$ and d(GGGGCC)$_n$, the phase separation and transition of r(GGGGCC)$_n$, the interactions of r(GGGGCC)$_n$ with RBPs, and the discovered ligands capable of inhibiting the non-ATG translations of r(GGGGCC)$_n$ and/or the interactions between r(GGGGCC)$_n$ and RBPs.

The relationship between the fatal neurodegenerative diseases ALS/FTD, the structure of r(GGGGCC)$_n$ RNA, and their interactions have garnered significant research attention. When the repeat number exceeds the threshold, r(GGGGCC)$_n$ RNA undergoes phase separation and transition, leading to the formation of nuclear RNA foci. These RNA foci recruit RBPs, disrupting the physiological functions of RNA splicing and maturation. Another pathogenic mechanism by which r(GGGGCC)$_n$ contributes to ALS or FTD is the cytotoxicity of repetitive dipeptide proteins generated through non-ATG translation. Aggregates of these repetitive dipeptide proteins, can recruit numerous 26S proteasome complexes and stabilize a transient substrate-processing conformation of the 26S proteasome, suggesting impaired degradation processes [115].

Characterizing the repeat structure of r(GGGGCC)$_n$ RNA and elucidating the structure-function relationship are key areas of research in understanding the pathogenic causes. r(GGGGCC)$_n$ can adopt diverse structures, including hairpin and parallel G4 topologies, with equilibrium between them depending on solution conditions. However, the three-dimensional structures of r(GGGGCC)$_n$ RNA are still unknown. Achieving a dominant conformation for structural studies may require sequence and solution condition optimization. Another challenging aspect is determining the secondary structures of r(GGGGCC)$_n$ within RNA foci or gel-like states. Due to the non-crystalline solid state and heterogeneous nature of RNA foci, commonly used high-resolution structure determination methods such as X-ray crystallography or solution NMR are not applicable [116,117]. To date, the RNA structures within RNA foci remain unidentified. Advancements in RNA structure determination methodologies, such as solid-state NMR [118–120], are needed to overcome this limitation.

Several small molecules that bind to r(GGGGCC)$_n$ have been discovered to block RBP interactions, inhibit phase separation, and/or hinder non-ATG translation, as evidenced both in vivo and in vitro. Understanding the structural details of the interactions between r(GGGGCC)$_n$ RNA and ligands is crucial for facilitating the design of lead compounds to treat ALS/FTD. Similar to the challenges faced in studying r(GGGGCC)$_n$ RNA, the complex structure determination of r(GGGGCC)$_n$ RNA repeats and small molecules is lacking, necessitating further developments to gain insights into drug design.

Another known treatment approach for ALS/FTD involves the use of antisense RNA. Single-dose injections of antisense oligonucleotides (ASOs) targeting repeat-containing RNAs, while preserving mRNA levels encoding *C9orf72*, have resulted in sustained reductions in RNA foci and dipeptide-repeat proteins, leading to the amelioration of behavioral deficits. These efforts have identified the gain of toxicity as a central disease mechanism caused by repeat-expanded *C9orf72* and established the feasibility of ASO-mediated therapy [16]. ALS brains treated with ASO therapeutics targeting the *C9orf72* transcript or repeat expansion showed mitigation despite the presence of repeat-associated non-ATG translation products [46]. Moreover, the introduction of mRNA that encodes r(GGGGCC)$_n$ binding proteins into ALS/FTD cells has the potential to restore RBP functions by augmenting the intracellular pool of RBPs recruited by RNA foci. This approach represents an alternative strategy for treating ALS by targeting r(GGGGCC)$_n$ RNA. Lastly, gene editing system by CRISPR/Cas9 has successfully removed the GGGGCC repeat expansion in *C9orf72*, leading to reduction in RNA foci and DPR formations, proving a promising approach in ALS treatments.

Author Contributions: Conceptualization, S.W. (Shenlin Wang); writing-original draft preparation: X.L., X.Z., J.H., X.S., Q.L., S.W. (Sishi Wang) and S.W. (Shenlin Wang); visualizaion, X.L., J.H. and X.Z.; writing-review and editing, X.L., J.H. and S.W. (Shenlin Wang); supervision, S.W. (Shenlin Wang); funing acquisiton, S.W. (Shenlin Wang). All authors have read and agreed to the published version of the manuscript.

Funding: The work was supported by the National Natural Science Foundation of China (22274050) and the Fundamental Research Funds for the Central Universities.

Institutional Review Board Statement: Not applicable.

Informed Consent Statement: Not applicable.

Data Availability Statement: Not applicable.

Conflicts of Interest: The authors declare no conflict of interest.

Sample Availability: Not applicable.

References

1. Ling, S.-C.; Polymenidou, M.; Cleveland, D.W. Converging Mechanisms in ALS and FTD: Disrupted RNA and Protein Homeostasis. *Neuron* **2013**, *79*, 416–438. [CrossRef]
2. Sagui, C. Structure and Dynamics of DNA and RNA Double Helices Obtained from the GGGGCC and CCCCGG Hexanucleotide Repeats That Are the Hallmark of C9FTD/ALS Diseases. *ACS Chem. Neurosci.* **2017**, *8*, 578–591.
3. Ash, P.E.A.; Bieniek, K.F.; Gendron, T.F.; Caulfield, T.; Lin, W.-L.; DeJesus-Hernandez, M.; van Blitterswijk, M.M.; Jansen-West, K.; Paul, J.W., III; Rademakers, R.; et al. Unconventional Translation of *C9ORF72* GGGGCC Expansion Generates Insoluble Polypeptides Specific to c9FTD/ALS. *Neuron* **2013**, *77*, 639–646. [CrossRef] [PubMed]
4. van der Ende, E.L.; Jackson, J.L.; White, A.; Seelaar, H.; van Blitterswijk, M.; Van Swieten, J.C. Unravelling the clinical spectrum and the role of repeat length in *C9ORF72* repeat expansions. *J. Neurol. Neurosurg. Psychiatry* **2021**, *92*, 502–509. [CrossRef]
5. Wang, C.; Chen, Z.; Yang, F.; Jiao, B.; Peng, H.; Shi, Y.; Wang, Y.; Huang, F.; Wang, J.; Shen, L.; et al. Analysis of the GGGGCC Repeat Expansions of the *C9orf72* Gene in SCA3/MJD Patients from China. *PLoS ONE* **2015**, *10*, e0130336. [CrossRef]
6. Lee, Y.-B.; Chen, H.-J.; Peres, J.N.; Gomez-Deza, J.; Attig, J.; Štalekar, M.; Troakes, C.; Nishimura, A.L.; Scotter, E.L.; Vance, C.; et al. Hexanucleotide repeats in ALS/FTD form length-dependent RNA foci, sequester RNA binding proteins, and are neurotoxic. *Cell Rep.* **2013**, *5*, 1178–1186. [CrossRef]
7. Balendra, R.; Isaacs, A.M. *C9orf72*-mediated ALS and FTD: Multiple pathways to disease. *Nat. Rev. Neurol.* **2018**, *14*, 544–558. [CrossRef] [PubMed]

8. DeJesus-Hernandez, M.; Mackenzie, I.R.; Boeve, B.F.; Boxer, A.L.; Baker, M.; Rutherford, N.J.; Nicholson, A.M.; Finch, N.A.; Flynn, H.; Adamson, J.; et al. Expanded GGGGCC hexanucleotide repeat in noncoding region of *C9ORF72* causes chromosome 9p-linked FTD and ALS. *Neuron* **2011**, *72*, 245–256. [CrossRef] [PubMed]
9. Dodd, K.C.; Power, R.; Ealing, J.; Hamdalla, H. FUS-ALS presenting with myoclonic jerks in a 17-year-old man. *Amyotroph. Lateral Scler. Front. Degener.* **2019**, *20*, 278–280. [CrossRef] [PubMed]
10. Zhou, B.; Liu, C.; Geng, Y.; Zhu, G. Topology of a G-quadruplex DNA formed by *C9orf72* hexanucleotide repeats associated with ALS and FTD. *Sci. Rep.* **2015**, *5*, 16673. [CrossRef]
11. Calcoen, D.; Elias, L.; Yu, X. What does it take to produce a breakthrough drug? *Nat. Rev. Drug Discov.* **2015**, *14*, 161–162.
12. Tamaki, Y.; Urushitani, M. Molecular Dissection of TDP-43 as a Leading Cause of ALS/FTLD. *Int. J. Mol. Sci.* **2022**, *23*, 12508. [CrossRef]
13. van Zundert, B.; Brown, R.H., Jr. Silencing strategies for therapy of SOD1-mediated ALS. *Neurosci. Lett.* **2017**, *636*, 32–39. [CrossRef]
14. Renton, A.E.; Majounie, E.; Waite, A.; Simon-Saánchez, J.; Rollinson, S.; Gibbs, J.R.; Schymick, J.C.; Laaksovirta, H.; van Swieten, J.C.; Myllykangas, L.; et al. A Hexanucleotide Repeat Expansion in *C9ORF72* Is the Cause of Chromosome 9p21-Linked ALS-FTD. *Neuron* **2011**, *72*, 257–268. [CrossRef] [PubMed]
15. O'rourke, J.G.; Bogdanik, L.; Muhammad, A.; Gendron, T.F.; Kim, K.J.; Austin, A.; Cady, J.; Liu, E.Y.; Zarrow, J.; Grant, S.; et al. *C9orf72* BAC Transgenic Mice Display Typical Pathologic Features of ALS/FTD. *Neuron* **2015**, *88*, 892–901.
16. Jiang, J.; Zhu, Q.; Gendron, T.F.; Saberi, S.; McAlonis-Downes, M.; Seelman, A.; Stauffer, J.E.; Jafar-Nejad, P.; Drenner, K.; Schulte, D.; et al. Gain of Toxicity from ALS/FTD-Linked Repeat Expansions in *C9ORF72* Is Alleviated by Antisense Oligonucleotides Targeting GGGGCC-Containing RNAs. *Neuron* **2016**, *90*, 535–550. [CrossRef] [PubMed]
17. Zhu, Q.; Jiang, J.; Gendron, T.F.; McAlonis-Downes, M.; Jiang, L.; Taylor, A.; Garcia, S.D.; Dastidar, S.G.; Rodriguez, M.J.; King, P.; et al. Reduced C9ORF72 function exacerbates gain of toxicity from ALS/FTD-causing repeat expansion in *C9orf2*. *Nat. Neurosci.* **2020**, *23*, 615–624. [CrossRef]
18. Rodriguez, C.; Todd, P. New pathologic mechanisms in nucleotide repeat expansion disorders. *Neurobiol. Dis.* **2019**, *130*, 104515. [CrossRef] [PubMed]
19. Zhang, K.; Daigle, J.G.; Cunningham, K.M.; Coyne, A.N.; Ruan, K.; Grima, J.C.; Bowen, K.E.; Wadhwa, H.; Yang, P.; Rigo, F.; et al. Stress Granule Assembly Disrupts Nucleocytoplasmic Transport. *Cell* **2018**, *173*, 958–971. [PubMed]
20. Zaepfel, B.L.; Zhang, Z.; Maulding, K.; Coyne, A.N.; Cheng, W.; Hayes, L.R.; Lloyd, T.E.; Sun, S.; Rothstein, J.D. UPF1 reduces *C9orf72* HRE-induced neurotoxicity in the absence of nonsense-mediated decay dysfunction. *Cell Rep.* **2021**, *34*, 108925. [CrossRef]
21. Wen, X.; Westergard, T.; Pasinelli, P.; Trotti, D. Pathogenic determinants and mechanisms of ALS/FTD linked to hexanucleotide repeat expansions in the *C9orf72* gene. *Neurosci. Lett.* **2017**, *636*, 16–26. [CrossRef]
22. Mizielinska, S.; Grönke, S.; Niccoli, T.; Ridler, C.E.; Clayton, E.L.; Devoy, A.; Moens, T.; Norona, F.E.; Woollacott, I.O.C.; Pietrzyk, J.; et al. *C9orf72* repeat expansions cause neurodegeneration in Drosophila through arginine-rich proteins. *Science* **2014**, *345*, 1192–1194. [CrossRef]
23. Zamiri, B.; Reddy, K.; Macgregor, R.B., Jr.; Pearson, C.E. TMPyP4 porphyrin distorts RNA G-quadruplex structures of the disease-associated r(GGGGCC)n repeat of the *C9orf72* gene and blocks interaction of RNA-binding proteins. *J. Biol. Chem.* **2014**, *289*, 4653–4659. [CrossRef]
24. West, R.J.; Sharpe, J.L.; Voelzmann, A.; Munro, A.L.; Hahn, I.; Baines, R.A.; Pickering-Brown, S. Co-expression of *C9orf72* related dipeptide-repeats over 1000 repeat units reveals age- and combination-specific phenotypic profiles in Drosophila. *Acta Neuropathol. Commun.* **2020**, *8*, 158. [CrossRef]
25. Nordin, A.; Akimoto, C.; Wuolikainen, A.; Alstermark, H.; Jonsson, P.; Birve, A.; Marklund, S.L.; Graffmo, K.S.; Forsberg, K.; Brännström, T.; et al. Extensive size variability of the GGGGCC expansion in *C9orf72* in both neuronal and non-neuronal tissues in 18 patients with ALS or FTD. *Hum. Mol. Genet.* **2015**, *24*, 3133–3142. [CrossRef]
26. Goodman, L.D.; Bonini, N.M. New Roles for Canonical Transcription Factors in Repeat Expansion Diseases. *Trends Genet.* **2020**, *36*, 81–92. [CrossRef]
27. Wang, J.; Wang, B.; Zhou, T. The Advance on Frontotemporal Dementia (FTD)'s Neuropathology and Molecular Genetics. *Mediat. Inflamm.* **2022**, *2022*, 5003902. [CrossRef]
28. Abramzon, Y.A.; Fratta, P.; Traynor, B.J.; Chia, R. The Overlapping Genetics of Amyotrophic Lateral Sclerosis and Frontotemporal Dementia. *Front. Neurosci.* **2020**, *14*, 42. [CrossRef]
29. Echeverria, G.V.; Cooper, T.A. RNA-binding proteins in microsatellite expansion disorders: Mediators of RNA toxicity. *Brain Res.* **2012**, *1462*, 100–111. [CrossRef]
30. Gitler, A.D.; Tsuiji, H. There has been an awakening: Emerging mechanisms of *C9orf72* mutations in FTD/ALS. *Brain Res.* **2016**, *1647*, 19–29. [CrossRef]
31. Gijselinck, I.; Van Langenhove, T.; van der Zee, J.; Sleegers, K.; Philtjens, S.; Kleinberger, G.; Janssens, J.; Bettens, K.; Van Cauwenberghe, C.; Pereson, S.; et al. A *C9orf72* promoter repeat expansion in a Flanders-Belgian cohort with disorders of the frontotemporal lobar degeneration-amyotrophic lateral sclerosis spectrum: A gene identification study. *Lancet Neurol.* **2012**, *11*, 54–65. [CrossRef] [PubMed]
32. Fay, M.M.; Anderson, P.J.; Ivanov, P. ALS/FTD-Associated C9ORF72 Repeat RNA Promotes Phase Transitions In Vitro and in Cells. *Cell Rep.* **2017**, *21*, 3573–3584. [CrossRef]

33. Mehta, A.R.; Selvaraj, B.T.; Barton, S.K.; McDade, K.; Abrahams, S.; Chandran, S.; Smith, C.; Gregory, J.M. Improved detection of RNA foci in *C9orf72* amyotrophic lateral sclerosis post-mortem tissue using BaseScope™ shows a lack of association with cognitive dysfunction. *Brain Commun.* **2020**, *2*, fcaa009. [CrossRef]
34. Malnar, M.; Rogelj, B. SFPQ regulates the accumulation of RNA foci and dipeptide repeat proteins from the expanded repeat mutation in *C9orf72*. *J. Cell Sci.* **2021**, *134*, jcs256602. [CrossRef]
35. Nedelsky, N.B.; Taylor, J.P. Bridging biophysics and neurology: Aberrant phase transitions in neurodegenerative disease. *Nat. Rev. Neurol.* **2019**, *15*, 272–286. [CrossRef]
36. Wang, X.; Goodrich, K.J.; Conlon, E.G.; Gao, J.; Erbse, A.; Manley, J.L.; Cech, T.R. *C9orf72* and triplet repeat disorder RNAs: G-quadruplex formation, binding to PRC2 and implications for disease mechanisms. *RNA* **2019**, *25*, 935–947. [CrossRef]
37. Swinnen, B.; Bento-Abreu, A.; Gendron, T.F.; Boeynaems, S.; Bogaert, E.; Nuyts, R.; Timmers, M.; Scheveneels, W.; Hersmus, N.; Wang, J.; et al. A zebrafish model for *C9orf72* ALS reveals RNA toxicity as a pathogenic mechanism. *Acta Neuropathol.* **2018**, *135*, 427–443. [CrossRef]
38. Tao, Z.; Wang, H.; Xia, Q.; Li, K.; Li, K.; Jiang, X.; Xu, G.; Wang, G.; Ying, Z. Nucleolar stress and impaired stress granule formation contribute to *C9orf72* RAN translation-induced cytotoxicity. *Hum. Mol. Genet.* **2015**, *24*, 2426–2441. [CrossRef]
39. Mori, K.; Weng, S.-M.; Arzberger, T.; May, S.; Rentzsch, K.; Kremmer, E.; Schmid, B.; Kretzschmar, H.A.; Cruts, M.; Van Broeckhoven, C.; et al. The *C9orf72* GGGGCC Repeat Is Translated into Aggregating Dipeptide-Repeat Proteins in FTLD/ALS. *Science* **2013**, *339*, 1335–1338. [CrossRef]
40. Liu, H.; Lu, Y.-N.; Paul, T.; Periz, G.; Banco, M.T.; Ferré-D'amaré, A.R.; Rothstein, J.D.; Hayes, L.R.; Myong, S.; Wang, J. A Helicase Unwinds Hexanucleotide Repeat RNA G-Quadruplexes and Facilitates Repeat-Associated Non-AUG Translation. *J. Am. Chem. Soc.* **2021**, *143*, 7368–7379. [CrossRef]
41. Flores, B.N.; Dulchavsky, M.E.; Krans, A.; Sawaya, M.R.; Paulson, H.L.; Todd, P.K.; Barmada, S.J.; Ivanova, M.I. Distinct *C9orf72*-Associated Dipeptide Repeat Structures Correlate with Neuronal Toxicity. *PLoS ONE* **2016**, *11*, e0165084. [CrossRef]
42. Cheng, W.; Wang, S.; Mestre, A.A.; Fu, C.; Makarem, A.; Xian, F.; Hayes, L.R.; Lopez-Gonzalez, R.; Drenner, K.; Jiang, J.; et al. C9ORF72 GGGGCC repeat-associated non-AUG translation is upregulated by stress through eIF2α phosphorylation. *Nat. Commun.* **2018**, *9*, 51. [CrossRef]
43. Jain, A.; Vale, R.D. RNA phase transitions in repeat expansion disorders. *Nature* **2017**, *546*, 243–247. [CrossRef]
44. Penumutchu, S.R.; Chiu, L.-Y.; Meagher, J.L.; Hansen, A.L.; Stuckey, J.A.; Tolbert, B.S. Differential Conformational Dynamics Encoded by the Linker between Quasi RNA Recognition Motifs of Heterogeneous Nuclear Ribonucleoprotein H. *J. Am. Chem. Soc.* **2018**, *140*, 11661–11673. [CrossRef]
45. Celona, B.; Dollen, J.V.; Vatsavayai, S.C.; Kashima, R.; Johnson, J.R.; Tang, A.A.; Hata, A.; Miller, B.L.; Huang, E.J.; Krogan, N.J.; et al. Suppression of *C9orf72* RNA repeat-induced neurotoxicity by the ALS-associated RNA-binding protein Zfp106. *eLife* **2017**, *6*, e19032. [CrossRef]
46. Donnelly, C.J.; Zhang, P.W.; Pham, J.T.; Haeusler, A.R.; Mistry, N.A.; Vidensky, S.; Daley, E.L.; Poth, E.M.; Hoover, B.; Fines, D.M.; et al. RNA Toxicity from the ALS/FTD C9ORF72 Expansion Is Mitigated by Antisense Intervention. *Neuron* **2013**, *80*, 415–428. [CrossRef]
47. Xu, Z.; Poidevin, M.; Li, X.; Li, Y.; Shu, L.; Nelson, D.L.; Li, H.; Hales, C.M.; Gearing, M.; Wingo, T.S.; et al. Expanded GGGGCC repeat RNA associated with amyotrophic lateral sclerosis and frontotemporal dementia causes neurodegeneration. *Proc. Natl. Acad. Sci. USA* **2013**, *110*, 7778–7783. [CrossRef]
48. Shen, J.; Zhang, Y.; Zhao, S.; Mao, H.; Wang, Z.; Li, H.; Xu, Z. Purα Repaired Expanded Hexanucleotide GGGGCC Repeat Noncoding RNA-Caused Neuronal Toxicity in Neuro-2a Cells. *Neurotox. Res.* **2018**, *33*, 693–701. [CrossRef]
49. Ishiguro, A.; Lu, J.; Ozawa, D.; Nagai, Y.; Ishihama, A. ALS-linked FUS mutations dysregulate G-quadruplex-dependent liquid-liquid phase separation and liquid-to-solid transition. *J. Biol. Chem.* **2021**, *297*, 101284. [CrossRef]
50. Murakami, T.; Qamar, S.; Lin, J.Q.; Schierle, G.S.K.; Rees, E.; Miyashita, A.; Costa, A.R.; Dodd, R.B.; Chan, F.T.S.; Michel, C.H.; et al. ALS/FTD Mutation-Induced Phase Transition of FUS Liquid Droplets and Reversible Hydrogels into Irreversible Hydrogels Impairs RNP Granule Function. *Neuron* **2015**, *88*, 678–690. [CrossRef]
51. Cooper-Knock, J.; Walsh, M.J.; Higginbottom, A.; Robin Highley, J.; Dickman, M.J.; Edbauer, D.; Ince, P.G.; Wharton, S.B.; Wilson, S.A.; Kirby, J.; et al. Sequestration of multiple RNA recognition motif-containing proteins by *C9orf72* repeat expansions. *Brain* **2014**, *137*, 2040–2051. [CrossRef]
52. Wang, Q.; Conlon, E.G.; Manley, J.L.; Rio, D.C. Widespread intron retention impairs protein homeostasis in *C9orf72* ALS brains. *Genome Res.* **2020**, *30*, 1705–1715. [CrossRef] [PubMed]
53. Simone, R.; Balendra, R.; Moens, T.G.; Preza, E.; Wilson, K.M.; Heslegrave, A.; Woodling, N.S.; Niccoli, T.; Gilbert-Jaramillo, J.; Abdelkarim, S.; et al. G-quadruplex-binding small molecules ameliorate *C9orf72* FTD/ALS pathology in vitro and in vivo. *EMBO Mol. Med.* **2018**, *10*, 22–31. [CrossRef] [PubMed]
54. Millevoi, S.; Moine, H.; Vagner, S. G-quadruplexes in RNA biology. *Wiley Interdiscip. Rev. RNA* **2012**, *3*, 495–507. [CrossRef]
55. Su, Z.; Zhang, Y.; Gendron, T.F.; Bauer, P.O.; Chew, J.; Yang, W.-Y.; Fostvedt, E.; Jansen-West, K.; Belzil, V.V.; Desaro, P.; et al. Discovery of a biomarker and lead small molecules to target r(GGGGCC)-associated defects in c9FTD/ALS. *Neuron* **2014**, *83*, 1043–1050. [CrossRef]
56. Jaiswal, M.K. Riluzole and edaravone: A tale of two amyotrophic lateral sclerosis drugs. *Med. Res. Rev.* **2019**, *39*, 733–748. [CrossRef]

57. Meijboom, K.E.; Abdallah, A.; Fordham, N.P.; Nagase, H.; Rodriguez, T.; Kraus, C.; Gendron, T.F.; Krishnan, G.; Esanov, R.; Andrade, N.S.; et al. CRISPR/Cas9-mediated excision of ALS/FTD-causing hexanucleotide repeat expansion in *C9ORF72* rescues major disease mechanisms in vivo and in vitro. *Nat. Commun.* **2022**, *13*, 6286. [CrossRef]
58. Maity, A.; Winnerdy, F.R.; Chen, G.; Phan, A.T. Duplexes Formed by G_4C_2 Repeats Contain Alternate Slow- and Fast-Flipping G.G Base Pairs. *Biochemistry* **2021**, *60*, 1097–1107. [CrossRef]
59. Haeusler, A.R.; Donnelly, C.J.; Periz, G.; Simko, E.A.; Shaw, P.G.; Kim, M.-S.; Maragakis, N.J.; Troncoso, J.C.; Pandey, A.; Sattler, R.; et al. C9orf72 nucleotide repeat structures initiate molecular cascades of disease. *Nature* **2014**, *507*, 195–200. [CrossRef]
60. Brčić, J.; Plavec, J. G-quadruplex formation of oligonucleotides containing ALS and FTD related GGGGCC repeat. *Front. Chem. Sci. Eng.* **2016**, *10*, 222–237. [CrossRef]
61. Fratta, P.; Mizielinska, S.; Nicoll, A.J.; Zloh, M.; Fisher, E.M.C.; Parkinson, G.; Isaacs, A.M. C9orf72 hexanucleotide repeat associated with amyotrophic lateral sclerosis and frontotemporal dementia forms RNA G-quadruplexes. *Sci. Rep.* **2012**, *2*, srep01016. [CrossRef]
62. Reddy, K.; Zamiri, B.; Stanley, S.Y.; Macgregor, R.B.; Pearson, C.E. The disease-associated r(GGGGCC)n repeat from the *C9orf72* gene forms tract length-dependent uni- and multimolecular RNA G-quadruplex structures. *J. Biol. Chem.* **2013**, *288*, 9860–9866. [CrossRef]
63. Božič, T.; Zalar, M.; Rogelj, B.; Plavec, J.; Šket, P. Structural Diversity of Sense and Antisense RNA Hexanucleotide Repeats Associated with ALS and FTLD. *Molecules* **2020**, *25*, 525. [CrossRef]
64. Brčić, J.; Plavec, J. ALS and FTD linked GGGGCC-repeat containing DNA oligonucleotide folds into two distinct G-quadruplexes. *Biochim. Biophys. Acta (BBA)-Gen. Subj.* **2017**, *1861*, 1237–1245. [CrossRef]
65. Thys, R.G.; Wang, Y.-H. DNA Replication Dynamics of the GGGGCC Repeat of the *C9orf72* Gene. *J. Biol. Chem.* **2015**, *290*, 28953–28962. [CrossRef]
66. Brčić, J.; Plavec, J. Solution structure of a DNA quadruplex containing ALS and FTD related GGGGCC repeat stabilized by 8-bromodeoxyguanosine substitution. *Nucleic Acids Res.* **2015**, *43*, 8590–8600. [CrossRef]
67. Brčić, J.; Plavec, J. NMR structure of a G-quadruplex formed by four $d(G_4C_2)$ repeats: Insights into structural polymorphism. *Nucleic Acids Res.* **2018**, *46*, 11605–11617. [CrossRef]
68. Geng, Y.; Liu, C.; Cai, Q.; Luo, Z.; Miao, H.; Shi, X.; Xu, N.; Fung, C.P.; Choy, T.T.; Yan, B.; et al. Crystal structure of parallel G-quadruplex formed by the two-repeat ALS- and FTD-related GGGGCC sequence. *Nucleic Acids Res.* **2021**, *49*, 5881–5890. [CrossRef]
69. Gao, Z.; Zhang, W.; Chang, R.; Zhang, S.; Yang, G.; Zhao, G. Liquid-Liquid Phase Separation: Unraveling the Enigma of Biomolecular Condensates in Microbial Cells. *Front. Microbiol.* **2021**, *12*, 751880. [CrossRef]
70. Boeynaems, S.; Alberti, S.; Fawzi, N.L.; Mittag, T.; Polymenidou, M.; Rousseau, F.; Schymkowitz, J.; Shorter, J.; Wolozin, B.; van den Bosch, L.; et al. Protein Phase Separation: A New Phase in Cell Biology. *Trends Cell Biol.* **2018**, *28*, 420–435. [CrossRef]
71. Bertrand, E.; Demongin, C.; Dobra, I.; Rengifo-Gonzalez, J.C.; Singatulina, A.S.; Sukhanova, M.V.; Lavrik, O.I.; Pastré, D.; Hamon, L. FUS fibrillation occurs through a nucleation-based process below the critical concentration required for liquid–liquid phase separation. *Sci. Rep.* **2023**, *13*, 7772. [CrossRef] [PubMed]
72. Watanabe, S.; Inami, H.; Oiwa, K.; Murata, Y.; Sakai, S.; Komine, O.; Sobue, A.; Iguchi, Y.; Katsuno, M.; Yamanaka, K. Aggresome formation and liquid-liquid phase separation independently induce cytoplasmic aggregation of TAR DNA-binding protein 43. *Cell Death Dis.* **2020**, *11*, 909. [CrossRef] [PubMed]
73. Shorter, J. Phase separation of RNA-binding proteins in physiology and disease: An introduction to the JBC Reviews thematic series. *J. Biol. Chem.* **2019**, *294*, 7113–7114. [CrossRef]
74. Portz, B.; Lee, B.L.; Shorter, J. FUS and TDP-43 Phases in Health and Disease. *Trends Biochem. Sci.* **2021**, *46*, 550–563. [CrossRef]
75. Maharana, S.; Wang, J.; Papadopoulos, D.K.; Richter, D.; Pozniakovsky, A.; Poser, I.; Bickle, M.; Rizk, S.; Guillén-Boixet, J.; Franzmann, T.M.; et al. RNA buffers the phase separation behavior of prion-like RNA binding proteins. *Science* **2018**, *360*, 918–921. [CrossRef] [PubMed]
76. Markovtsov, V.; Nikolic, J.M.; Goldman, J.A.; Turck, C.W.; Chou, M.Y.; Black, D.L. Cooperative Assembly of an hnRNP Complex Induced by a Tissue-Specific Homolog of Polypyrimidine Tract Binding Protein. *Mol. Cell. Biol.* **2000**, *20*, 7463–7479. [CrossRef]
77. Wang, E.; Cambi, F. Heterogeneous Nuclear Ribonucleoproteins H and F Regulate the Proteolipid Protein/DM20 Ratio by Recruiting U1 Small Nuclear Ribonucleoprotein through a Complex Array of G Runs. *J. Biol. Chem.* **2009**, *284*, 11194–11204. [CrossRef]
78. Van Dusen, C.M.; Yee, L.; McNally, L.M.; McNally, M.T. A glycine-rich domain of hnRNP H/F promotes nucleocytoplasmic shuttling and nuclear import through an interaction with transportin 1. *Mol. Cell. Biol.* **2010**, *30*, 2552–2562. [CrossRef]
79. Conlon, E.G.; Lu, L.; Sharma, A.; Yamazaki, T.; Tang, T.; Shneider, N.A.; Manley, J.L. The *C9ORF72* GGGGCC expansion forms RNA G-quadruplex inclusions and sequesters hnRNP H to disrupt splicing in ALS brains. *eLife* **2016**, *5*, e17820. [CrossRef]
80. Liao, Y.-Z.; Ma, J.; Dou, J.-Z. The Role of TDP-43 in Neurodegenerative Disease. *Mol. Neurobiol.* **2022**, *59*, 4223–4241. [CrossRef]
81. Lee, E.B.; Lee, V.M.-Y.; Trojanowski, J.Q. Gains or losses: Molecular mechanisms of TDP43-mediated neurodegeneration. *Nat. Rev. Neurosci.* **2011**, *13*, 38–50. [CrossRef] [PubMed]
82. Conicella, A.E.; Zerze, G.H.; Mittal, J.; Fawzi, N.L. ALS Mutations Disrupt Phase Separation Mediated by α-Helical Structure in the TDP-43 Low-Complexity C-Terminal Domain. *Structure* **2016**, *24*, 1537–1549. [CrossRef] [PubMed]

83. Chew, J.; Cook, C.; Gendron, T.F.; Jansen-West, K.; del Rosso, G.; Daughrity, L.M.; Castanedes-Casey, M.; Kurti, A.; Stankowski, J.N.; Disney, M.D.; et al. Aberrant deposition of stress granule-resident proteins linked to C9orf72-associated TDP-43 proteinopathy. *Mol. Neurodegener.* **2019**, *14*, 9. [PubMed]
84. Solomon, D.A.; Stepto, A.; Au, W.H.; Adachi, Y.; Diaper, D.C.; Hall, R.; Rekhi, A.; Boudi, A.; Tziortzouda, P.; Lee, Y.B.; et al. A feedback loop between dipeptide-repeat protein, TDP-43 and karyopherin-α mediates C9orf72-related neurodegeneration. *Brain* **2018**, *141*, 2908–2924. [CrossRef]
85. Mori, K.; Lammich, S.; Mackenzie, I.R.A.; Forné, I.; Zilow, S.; Kretzschmar, H.; Edbauer, D.; Janssens, J.; Kleinberger, G.; Cruts, M.; et al. hnRNP A3 binds to GGGGCC repeats and is a constituent of p62-positive/TDP43-negative inclusions in the hippocampus of patients with C9orf72 mutations. *Acta Neuropathol.* **2013**, *125*, 413–423.
86. Peters, O.M.; Cabrera, G.T.; Tran, H.; Gendron, T.F.; McKeon, J.E.; Metterville, J.; Weiss, A.; Wightman, N.; Salameh, J.; Kim, J.; et al. Human C9ORF72 Hexanucleotide Expansion Reproduces RNA Foci and Dipeptide Repeat Proteins but Not Neurodegeneration in BAC Transgenic Mice. *Neuron* **2015**, *88*, 902–909.
87. Lagier-Tourenne, C.; Polymenidou, M.; Cleveland, D.W. TDP-43 and FUS/TLS: Emerging roles in RNA processing and neurodegeneration. *Hum. Mol. Genet.* **2010**, *19*, R46–R64. [CrossRef]
88. Kanai, Y.; Dohmae, N.; Hirokawa, N. Kinesin transports RNA: Isolation and characterization of an RNA-transporting granule. *Neuron* **2004**, *43*, 513–525. [CrossRef]
89. Yoshimura, A.; Fujii, R.; Watanabe, Y.; Okabe, S.; Fukui, K.; Takumi, T. Myosin-Va facilitates the accumulation of mRNA/protein complex in dendritic spines. *Curr. Biol.* **2006**, *16*, 2345–2351.
90. Vance, C.; Rogelj, B.; Hortobágyi, T.; De Vos, K.J.; Nishimura, A.L.; Sreedharan, J.; Hu, X.; Smith, B.; Ruddy, D.; Wright, P.; et al. Mutations in FUS, an RNA Processing Protein, Cause Familial Amyotrophic Lateral Sclerosis Type 6. *Science* **2009**, *323*, 1208–1211.
91. Niaki, A.G.; Sarkar, J.; Cai, X.; Rhine, K.; Vidaurre, V.; Guy, B.; Hurst, M.; Lee, J.C.; Koh, H.R.; Guo, L.; et al. Loss of Dynamic RNA Interaction and Aberrant Phase Separation Induced by Two Distinct Types of ALS/FTD-Linked FUS Mutations. *Mol. Cell* **2020**, *77*, 82–94. [CrossRef] [PubMed]
92. Grasberger, H.; Bell, G.I. Subcellular recruitment by TSG118 and TSPYL implicates a role for zinc finger protein 106 in a novel developmental pathway. *Int. J. Biochem. Cell Biol.* **2005**, *37*, 1421–1437. [CrossRef] [PubMed]
93. Anderson, D.M.; Cannavino, J.; Li, H.; Anderson, K.M.; Nelson, B.R.; McAnally, J.; Bezprozvannaya, S.; Liu, Y.; Lin, W.; Liu, N.; et al. Severe muscle wasting and denervation in mice lacking the RNA-binding protein ZFP106. *Proc. Natl. Acad. Sci. USA* **2016**, *113*, E4494–E4503. [CrossRef] [PubMed]
94. reibaum, B.D.; Lu, Y.; Lopez-Gonzalez, R.; Kim, N.C.; Almeida, S.; Lee, K.-H.; Badders, N.; Valentine, M.; Miller, B.L.; Wong, P.C.; et al. GGGGCC repeat expansion in C9orf72 compromises nucleocytoplasmic transport. *Nature* **2015**, *525*, 129–133.
95. Hideyama, T.; Yamashita, T.; Aizawa, H.; Tsuji, S.; Kakita, A.; Takahashi, H.; Kwak, S. Profound downregulation of the RNA editing enzyme ADAR2 in ALS spinal motor neurons. *Neurobiol. Dis.* **2012**, *45*, 1121–1128. [CrossRef]
96. Chen, C.-X.; Cho, D.-S.C.; Wang, Q.; Lai, F.; Carter, K.C.; Nishikura, K. A third member of the RNA-specific adenosine deaminase gene family, ADAR3, contains both single- and double-stranded RNA binding domains. *RNA* **2000**, *6*, 755–767.
97. White, M.K.; Johnson, E.M.; Khalili, K. Multiple roles for Pur-α in cellular and viral regulation. *Cell Cycle* **2009**, *8*, 414–420. [CrossRef]
98. Johnson, E.M.; Kinoshita, Y.; Weinreb, D.B.; Wortman, M.J.; Simon, R.; Khalili, K.; Winckler, B.; Gordon, J. Role of Purα in targeting mRNA to sites of translation in hippocampal neuronal dendrites. *J. Neurosci. Res.* **2006**, *83*, 929–943. [CrossRef]
99. Daniel, D.C.; Wortman, M.J.; Schiller, R.J.; Liu, H.; Gan, L.; Mellen, J.S.; Chang, C.F.; Gallia, G.L.; Rappaport, J.; Khalili, K.; et al. Coordinate effects of human immunodeficiency virus type 1 protein Tat and cellular protein Purα on DNA replication initiated at the JC virus origin. *J. Gen. Virol.* **2001**, *82*, 1543–1553. [CrossRef]
100. Daniel, D.C.; Johnson, E.M. PURA, the gene encoding Pur-alpha, member of an ancient nucleic acid-binding protein family with mammalian neurological functions. *Gene* **2018**, *643*, 133–143. [CrossRef]
101. Khalili, K.; Del Valle, L.; Muralidharan, V.; Gault, W.J.; Darbinian, N.; Otte, J.; Meier, E.; Johnson, E.M.; Daniel, D.C.; Kinoshita, Y.; et al. Purα is essential for postnatal brain development and developmentally coupled cellular proliferation as revealed by genetic inactivation in the mouse. *Mol. Cell. Biol.* **2003**, *23*, 6857–6875. [CrossRef]
102. Barbe, M.F.; Krueger, J.J.; Loomis, R.; Otte, J.; Gordon, J. Memory Deficits, Gait Ataxia and Neuronal Loss in the Hippocampus and Cerebellum in Mice That Are Heterozygous for Pur-Alpha. *Neuroscience* **2016**, *337*, 177–190. [CrossRef]
103. Rossi, S.; Serrano, A.; Gerbino, V.; Giorgi, A.; Di Francesco, L.; Nencini, M.; Bozzo, F.; Schininà, M.E.; Bagni, C.; Cestra, G.; et al. Nuclear accumulation of mRNAs underlies G_4C_2 repeat-induced translational repression in a cellular model of C9orf72 ALS. *J. Cell Sci.* **2015**, *128*, 1787–1799. [CrossRef]
104. Daigle, J.G.; Krishnamurthy, K.; Ramesh, N.; Casci, I.; Monaghan, J.; McAvoy, K.; Godfrey, E.W.; Daniel, D.C.; Johnson, E.M.; Monahan, Z.; et al. Pur-alpha regulates cytoplasmic stress granule dynamics and ameliorates FUS toxicity. *Acta Neuropathol.* **2016**, *131*, 605–620. [CrossRef]
105. Di Salvio, M.; Piccinni, V.; Gerbino, V.; Mantoni, F.; Camerini, S.; Lenzi, J.; Rosa, A.; Chellini, L.; Loreni, F.; Carrì, M.T.; et al. Pur-alpha functionally interacts with FUS carrying ALS-associated mutations. *Cell Death Dis.* **2015**, *6*, e1943. [CrossRef]
106. Muralidharan, V.; Sweet, T.; Nadraga, Y.; Amini, S.; Khalili, K. Regulation of Purα gene transcription: Evidence for autoregulation of Purα promoter. *J. Cell. Physiol.* **2001**, *186*, 406–413. [CrossRef]

107. Martino, L.; Pagano, B.; Fotticchia, I.; Neidle, S.; Giancola, C. Shedding Light on the Interaction between TMPyP4 and Human Telomeric Quadruplexes. *J. Phys. Chem. B* **2009**, *113*, 14779–14786. [CrossRef] [PubMed]
108. Morris, M.J.; Wingate, K.L.; Silwal, J.; Leeper, T.C.; Basu, S. The porphyrin TmPyP4 unfolds the extremely stable G-quadruplex in MT3-MMP mRNA and alleviates its repressive effect to enhance translation in eukaryotic cells. *Nucleic Acids Res.* **2012**, *40*, 4137–4145. [CrossRef] [PubMed]
109. Zahler, A.M.; Williamson, J.R.; Cech, T.R.; Prescott, D.M. Inhibition of telomerase by G-quartet DNA structures. *Nature* **1991**, *350*, 718–720. [CrossRef] [PubMed]
110. Izbicka, E.; Wheelhouse, R.; Raymond, E.; Davidson, K.K.; Lawrence, R.A.; Sun, D.; Windle, B.E.; Hurley, L.H.; Von Hoff, D.D. Effects of cationic porphyrins as G-quadruplex interactive agents in human tumor cells. *Cancer Res.* **1999**, *59*, 639–644.
111. Mulholland, K.; Sullivan, H.J.; Garner, J.; Cai, J.; Chen, B.; Wu, C. Three-Dimensional Structure of RNA Monomeric G-Quadruplex Containing ALS and FTD Related G_4C_2 Repeat and Its Binding with TMPyP4 Probed by Homology Modeling based on Experimental Constraints and Molecular Dynamics Simulations. *ACS Chem. Neurosci.* **2020**, *11*, 57–75. [CrossRef]
112. Disney, M.D.; Liu, B.; Yang, W.-Y.; Sellier, C.; Tran, T.; Charlet-Berguerand, N.; Childs-Disney, J.L. A small molecule that targets r(CGG)^{exp} and improves defects in fragile X-associated tremor ataxia syndrome. *ACS Chem. Biol.* **2012**, *7*, 1711–1718. [CrossRef]
113. Ursu, A.; Wang, K.W.; Bush, J.A.; Choudhary, S.; Chen, J.L.; Baisden, J.T.; Zhang, Y.J.; Gendron, T.F.; Petrucelli, L.; Yildirim, I.; et al. Structural Features of Small Molecules Targeting the RNA Repeat Expansion That Causes Genetically Defined ALS/FTD. *ACS Chem. Biol.* **2020**, *15*, 3112–3123. [CrossRef] [PubMed]
114. Ursu, A.; Baisden, J.T.; Bush, J.A.; Taghavi, A.; Choudhary, S.; Zhang, Y.J.; Gendron, T.F.; Petrucelli, L.; Yildirim, I.; Disney, M.D. A Small Molecule Exploits Hidden Structural Features within the RNA Repeat Expansion That Causes c9ALS/FTD and Rescues Pathological Hallmarks. *ACS Chem. Neurosci.* **2021**, *12*, 4076–4089. [CrossRef]
115. Guo, Q.; Lehmer, C.; Martínez-Sánchez, A.; Rudack, T.; Beck, F.; Hartmann, H.; Pérez-Berlanga, M.; Frottin, F.; Hipp, M.S.; Hartl, F.U.; et al. In Situ Structure of Neuronal *C9orf72* Poly-GA Aggregates Reveals Proteasome Recruitment. *Cell* **2018**, *172*, 696–705. [CrossRef] [PubMed]
116. Zhao, S.; Yang, Y.; Zhao, Y.; Li, X.; Xue, Y.; Wang, S. High-resolution solid-state NMR spectroscopy of hydrated non-crystallized RNA. *Chem. Commun.* **2019**, *55*, 13991–13994. [CrossRef] [PubMed]
117. Marchanka, A.; Simon, B.; Althoff-Ospelt, G.; Carlomagno, T. RNA structure determination by solid-state NMR spectroscopy. *Nat. Commun.* **2015**, *6*, 7024. [CrossRef] [PubMed]
118. Zhao, S.; Li, X.; Wen, Z.; Zou, M.; Yu, G.; Liu, X.; Mao, J.; Zhang, L.; Xue, Y.; Fu, R.; et al. Dynamics of base pairs with low stability in RNA by solid-state nuclear magnetic resonance exchange spectroscopy. *iScience* **2022**, *25*, 105322. [CrossRef]
119. Yang, Y.; Wang, S. RNA Characterization by Solid-State NMR Spectroscopy. *Chemistry* **2018**, *24*, 8698–8707. [CrossRef]
120. Yang, Y.; Xiang, S.; Liu, X.; Pei, X.; Wu, P.; Gong, Q.; Li, N.; Baldus, M.; Wang, S. Proton-detected solid-state NMR detects the inter-nucleotide correlations and architecture of dimeric RNA in microcrystals. *Chem. Commun.* **2017**, *53*, 12886–12889. [CrossRef]

Disclaimer/Publisher's Note: The statements, opinions and data contained in all publications are solely those of the individual author(s) and contributor(s) and not of MDPI and/or the editor(s). MDPI and/or the editor(s) disclaim responsibility for any injury to people or property resulting from any ideas, methods, instructions or products referred to in the content.

Review

Synthesis of Backbone-Modified Morpholino Oligonucleotides Using Phosphoramidite Chemistry

Sibasish Paul [1] and Marvin H. Caruthers [2,*]

[1] Nucleic Acid Solutions Division, Agilent Technologies, Boulder, CO 80301, USA
[2] Department of Biochemistry, University of Colorado, Boulder, CO 80303, USA
* Correspondence: marvin.caruthers@colorado.edu

Abstract: Phosphorodiamidate morpholinos (PMOs) are known as premier gene knockdown tools in developmental biology. PMOs are usually 25 nucleo-base-long morpholino subunits with a neutral phosphorodiamidate linkage. PMOs work via a steric blocking mechanism and are stable towards nucleases' inside cells. PMOs are usually synthesized using phosphoramidate P(V) chemistry. In this review, we will discuss the synthesis of PMOs, phosphoroamidate morpholinos (MO), and thiophosphoramidate morpholinos (TMO).

Keywords: phosphorodiamidate morpholinos (PMO); phosphoroamidate morpholinos (MO); thiophosphoramidate morpholinos (TMO); phosphoramidites

1. Introduction

Among the various oligonucleotide analogues that have been used as therapeutic drugs, the phosphorodiamidate morpholino oligonucleotide (PMO; 3) remains one of the most successful (Figure 1). Of the 16 oligonucleotide therapeutics approved so far by the US Food and Drug Administration (FDA), 4 are PMO-based drugs. These are Eteplirsen (2016), Golodirsen (2019), Viltolarsen (2020), and Casimersen (2021) for the treatment of Duchenne Muscular Dystrophy (DMD) [1–4]. Eteplirsen and Golodirsen are used for the treatment of DMD, targeting exons 51 and 53, respectively, of the dystrophin mRNA. The other dystrophin drugs are Viltolarsen and Casimersen, which target exons 53 and 45, respectively. DMD is a rare genetic disease of the dystrophin protein and predominantly affects males. In healthy muscle, dystrophin interacts with other proteins at the cell membrane to stabilize and protect the cell during regular activity involving muscle contraction and relaxation. Individuals with DMD produce little or no dystrophin in their muscle. Without dystrophin, normal activity causes excessive damage to muscle cells and over time is replaced with fat and fibrotic tissue. PMOs bind to a target out-of-frame dystrophin pre-mRNA and alter splicing, which results in an in-frame mRNA. Consequently, a truncated but functional dystrophin protein is produced.

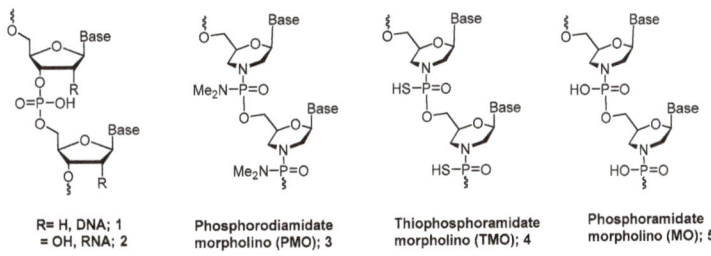

Figure 1. Chemical structure of morpholino oligonucleotides.

The basic PMO structure consists of the six-membered morpholine ring, instead of ribose, attached to the same nucleobases as RNA [5]. PMOs possess a neutral backbone that is completely resistant to nucleases. They are water soluble and display good cell permeability without assistance from cellular factors. They also hybridize to RNA but do not activate RNase H [6,7].

PMOs are currently synthesized with phosphoramidate P(V) chemistry on a solid support using chlorophosphoramidate building blocks, as described in Scheme 1. The first step is condensing a 6'-hydroxyl-N-tritylmorpholino nucleoside 6 with N,N-dimethylaminodichlorophosphoramidate in order to generate an N,N-dimethylamino chlorophosphoramidate synthon 7. Coupling the base of this synthon to a morpholino nucleoside linked to a support through the 6'-hydroxyl 8 generates the dimer 9 attached to the resin. Further detritylation with an acid salt yields a product 10 that can be elongated by repeating the cycle [7]. Several limitations of this synthesis approach have been reported [8]. PMOs are synthesized in a 5' to 3' direction, whereas almost all other oligonucleotide synthesis chemistries take place in a 3' to 5' direction [9]. Thus, the synthesis strategy for PMOs is not compatible with most other chemistries, which severely limits the ability to synthesize chimeras having PMOs and other analogues. Additionally, the active chlorophosphoramidate synthon is a P(V) compound which is less reactive and therefore leads to significantly lower yields than the P(III) oligonucleotide synthesis strategy.

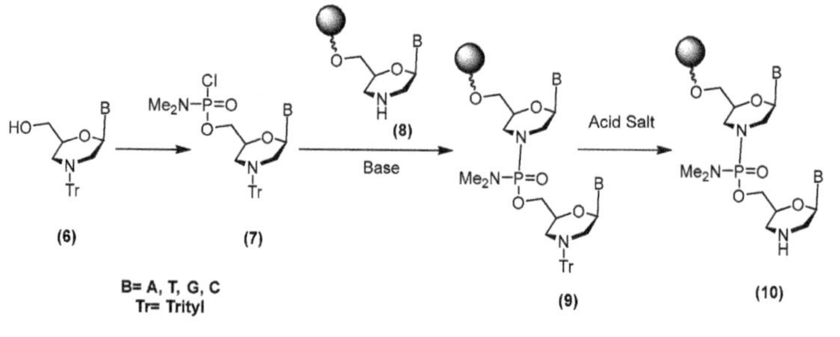

Scheme 1. Method for synthesis of PMOs using chlorophosphoramidate chemistry.

Because of the versatility of P(III) chemistry, several DNA/RNA derivatives in addition to natural DNA and RNA can be synthesized efficiently [10]. Thus, to develop an orthogonal synthetic approach for PMOs, we have explored the use of P(III) chemistry for preparing the PMO analogue and its derivatives on a solid support. If successful, a P(III) strategy would lead to procedures for preparing many oligonucleotide derivatives. In this short review, we will discuss the synthesis of PMOs and additional PMO-type analogues (MOs and TMOs, Figure 1) using phosphoramidite P(III) chemistry. Also, the biological properties of the newly developed TMOs will be discussed.

2. Initial Work

In 2015, we published our research regarding the conversion of boranephosphonate DNA to phosphoramidate DNA [11], as outlined in Scheme 2. The approach first involved the conversion of an internucleotide phosphite linkage to boranephosphonate using the BH_3-THF complex in THF to yield 11. This condensation was followed with 30 equivalents of iodine and a 0.4 M solution of methylamine in tetrahydrofuran, which led to the formation of methylphosphoramidate 14. This breakthrough chemistry enabled us to synthesize phorphoramidate oligodeoxyribonucleotides starting from boranephosphonate DNA. Of particular interest was the observation that a diastereomerically pure borane phosphonate isomer gave rise to the formation of one diastereomer via a stereospecific reaction [11]. This work encouraged us to attempt the synthesis of PMOs using the same iodine oxidation chemistry.

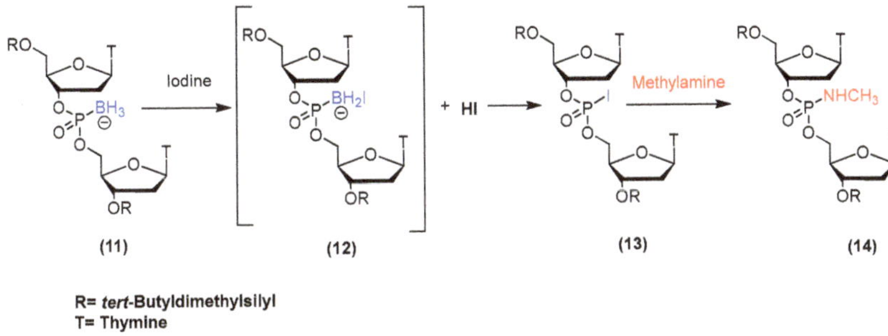

Scheme 2. Synthesis scheme for phosphoramidate DNA from boranephosphonate DNA.

3. Synthesis of PMOs and PMO/DNA Chimeras

Zhang and collaborators reported the synthesis of chimeric oligonucleotides having 2′-deoxyribonucleoside-3′-phosphate and morpholinonucleoside 3′-phosphoramidate internucleotide linkages [12,13]. They used morpholino uridine and thymidine phosphorodiamidites (Figure 2) to synthesize these oligonucleotides at a 0.1 µmol scale on a DNA synthesizer using phosphoramidite chemistry and 5-ethylthio-1*H*-tetrazole (ETT) as the activator. Additionally, oligonucleotides with morpholino phosphoramidate modifications in the 3′-terminal ends were synthesized using the universal CPG support. Aqueous ammonia at 55 °C for 12 h was then used to separate the oligonucleotides from the support and to remove base protecting groups. Crude oligonucleotides were purified using HPLC with yields of approximately 30% and analyzed via HPLC and MALDI-TOF mass spectroscopy. Using this approach, siRNAs containing up to four morpholino uridine phosphoroamidate linkages were synthesized. Although several interesting biological properties were reported (resistance to nuclease, silencing activity in HeLa cells) no further research has been reported. These results encouraged us to explore the synthesis of PMOs on solid supports using the P(III) chemistry we had developed previously for the synthesis of the corresponding 2′-deoxyoligonucleotide analogues [11].

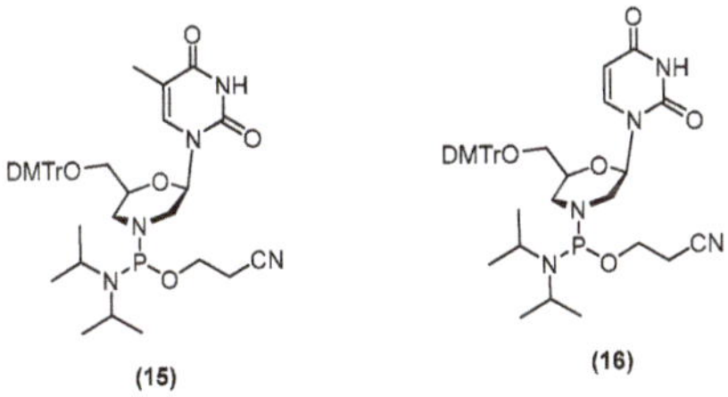

Figure 2. Phosphorodiamidite morpholino building blocks.

Our first step toward the synthesis of PMOs using a P(III) phosphoramidite approach was to synthesize all four morpholino phosphorodiamidite building blocks (Scheme 3). Our overall strategy was to use these building blocks for the preparation of morpholino borane phosphonate derivates and then convert these linkages to PMOs using iodine oxidation [14,15]. However, the standard amide protecting groups routinely used for oligonucleotide synthesis cannot be used with cytosine, adenine, and guanine because boronation reduces these amides to amines, which consequently cannot be removed following synthesis. Therefore, we turned to the previously developed bis(tert-butyl)isobutylsilyl (BIBS) group [11] for protection of the amino groups on these three bases because the BIBS protecting group was stable to P(III) chemistry synthesis procedures, including boronation, and could be readily removed with fluoride.

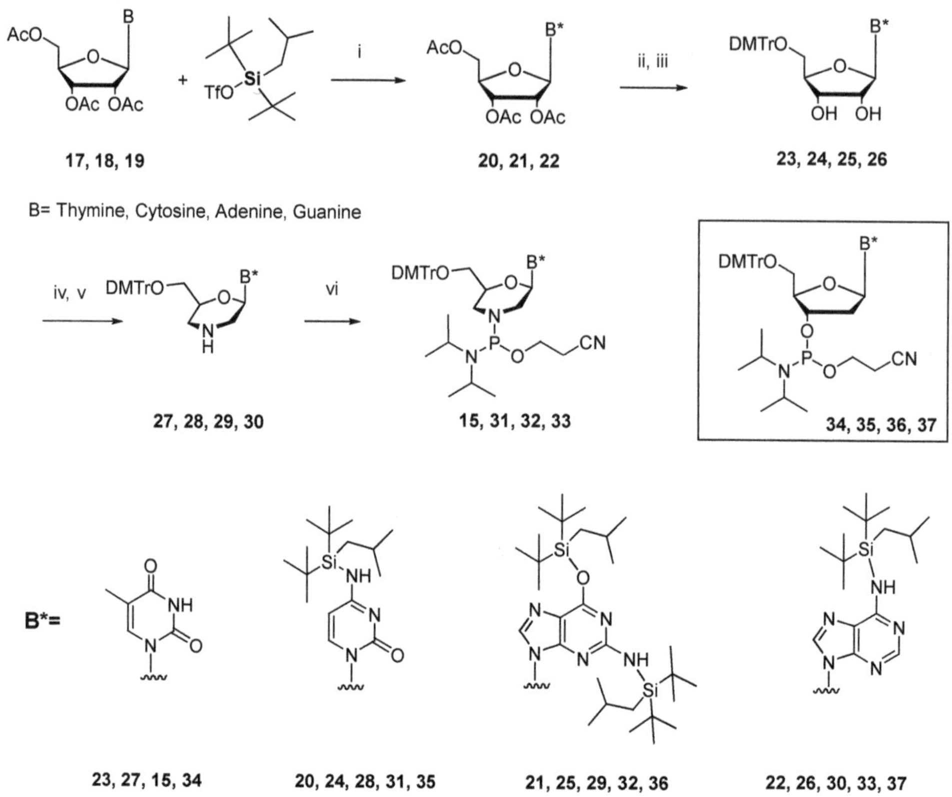

Reagents: (i) Lutidine (4.0 equiv)/1,4-Dioxane; (ii) 1.0 M NH$_3$/MeOH; (iii) DMTr-Cl (1.2 equiv)/Pyridine; (iv) NaIO$_4$ (1.1 equiv)/(NH$_4$)$_2$B$_4$O$_7$ (1.1equiv)/CH$_3$OH; (v) NaCNBH$_3$ (2.0 equiv)/AcOH (2.0 equiv)/CH$_3$OH; (vi) P(OCH$_2$CH$_2$CN)[N(^{i}Pr)$_2$]$_2$ (1.2 equiv)/4,5-Dicyanoimidazole (0.5 equiv)/CH$_2$Cl$_2$. DMTr: 4, 4'-Dimethoxytrityl.

Scheme 3. Synthesis of phosphorodiamidite building blocks for the preparation of PMOs.

The preparations of these four building blocks, as outlined in Scheme 3, were based upon modifications of the originally published procedures [16,17]. 5'-O-Dimethoxytrityl ribothymidine **23** was treated with sodium periodate and then ammonium biborate to afford the dihydroxythymine morpholino monomer. Reduction using sodium cyanoborohydride under mild acidic conditions yields the thymine morpholino monomer **27** [17]. The morpholino–thymine synthon **15** was then prepared in an 86% yield via the phosphitylation of **27** with 2-cyanoethyl-N,N,N',N'-tetraisopropylphosphordiamidite and 4,5-dicyanoimidazole (DCI) in dichloromethane under an argon atmosphere [12]. Synthesis of the cytosine, adenine, and guanine monomers began by first protecting the nucleoside amino groups with BIBS. The reaction of 5',3',2'-tri-O-acetylcytidine **17** with BIBS-OTf (Tf = triflate) in the presence of 2,6-lutidine gave a 76% yield of **20** after 2 h of stirring at 60 °C under argon. The synthesis of **21** and **22** from **18** and **19**, respectively, required much longer reaction times (3 days), with yields of 25–31% and 74%. These silylated ribonucleosides were then treated with ammonium hydroxide to remove the acetyl protecting groups, converted them to the 5'-dimethoxytrityl compounds and then to the morpholino derivatives (**28, 29,** and **30**) using the same chemistry as outlined for the preparation of

27. These compounds were used to generate the phosphorodiamidite building blocks **31**, **32**, and **33**. For the synthesis of PMO−DNA chimeras, 5′-O-(4,4′-dimethoxytrityl)-2′-deoxyribonucleoside-3′-phosphoramidite building blocks **35**, **36**, and **37** were prepared following a literature protocol [14,15], whereas compound **34** was obtained from commercially available sources.

The synthesis cycle for the preparation of PMOs using P(III) chemistry is outlined in Scheme 4 [15]. The synthesis begins with commercially available 5′-DMT-2′-deoxyribothymidine linked to a polystyrene support. The standard glass synthesis support (CPG) could not be used because this support was not compatible with fluoride reagents that were required to remove N-silyl ether protecting groups from the bases. Compound (**A**) can be any of the 2′-deoxyribonucleosides, but the cytosine, adenine, and guanine bases must be protected with the BIBS protecting group. The DMT group was removed with 0.5% trifluoroacetic acid in chloroform containing 10% trimethyl phosphite–borane (TMPB). To completely remove this detritylation solution, it was important to use a methanol wash [14]. The 5′-unprotected-2′-deoxyribonucleoside (**A**) was then reacted with **15**, **31**, **32**, or **33** in anhydrous acetonitrile containing 4,5-dicyanoimidazole (DCI) to generate a dimer having a phosphoramidite diester internucleotide linkage (**B**). Boronation was next carried out using borane–THF to generate a P(IV) morpholino compound (**C**) prior to the capping step. Post-boronation, the support was washed with acetonitrile, failure sequences were capped using acetic anhydride, and detritylation was carried out using a solution of 10% TMBP and 0.5% TFA in chloroform in order to generate (**D**). Following a methanol wash, multiple repetitions of this cycle yielded a product ready for further conversion to the PMOs (Table 1). Post-synthesis, supports were washed with acetonitrile, treated with a 1:1 mixture of triethylamine/acetonitrile to remove the cyanoethyl protecting group, washed with acetonitrile and dichloromethane to remove residual triethylamine, and dried. For conversion to the N,N-dimethylamino PMOs, the morpholino boranephosphoroamidates were treated for 2 h with a solution of 0.05 M iodine and 2.0 M dimethylamine in tetrahydrofuran. Using the procedure outlined in Scheme 4, a trimer having only thymine nucleoside bases was synthesized, converted to the N,N-dimethylamino PMO derivative with iodine/N,N-dimethylamine, removed from the support with ammonia, and analyzed by LC−MS. A 94% yield of the trimer PMO was calculated. PMOs having oligothymidine and all four bases have been synthesized using this approach and characterized by LC−MS.

For the synthesis of PMO−DNA chimeras, 5-ethylthio-1H-tetrazole (0.25 M and 180 s coupling time) was used as an activator for the 5′-O-(4,4′-dimethoxytrityl)-2′-deoxyribonucleoside-3′-phosphoramidites (**34**, **35**, **36**, or **37**). Following the addition of the 2′-deoxyribonucleotide, iodine oxidation converted the phosphite triester to the corresponding phosphate linkage. Since PMO−DNA chimeras are new to the scientific community and could prove to be useful for various research projects, several were synthesized. Initially, these chimeras were a series of 21mer oligothymidines containing four N,N-dimethylamino PMO linkages. These encouraging results were followed by the synthesis of PMO−DNA chimeras containing all four nucleobases with variable locations and a number of PMO linkages [15].

In addition to testing this new synthetic route by synthesizing PMO analogues having the N,N-dimethylamino–phosphorodiamidate linkage, we investigated other amines, which could be used in order to generate several new PMO−DNA derivatives. An oligothymidine 21mer having four morpholino boranephosphoroamidate linkages near the center of this oligomer was synthesized. The support containing this oligonucleotide was divided into three samples that were treated with N-methylamine, ammonia, and morpholine under iodine oxidation conditions and then purified using reverse-phase column chromatography. Additionally, mixed-sequence PMO−DNA chimers having all four bases and amino-phosphorodiamidate internucleotide linkages were synthesized, where the positions for the diamidate linkages were located at the 5′, 3′, and 5′/3′ termini of these chimeras. Yields were comparable to those obtained for the N,N-dimethylamino PMO chimeras [15].

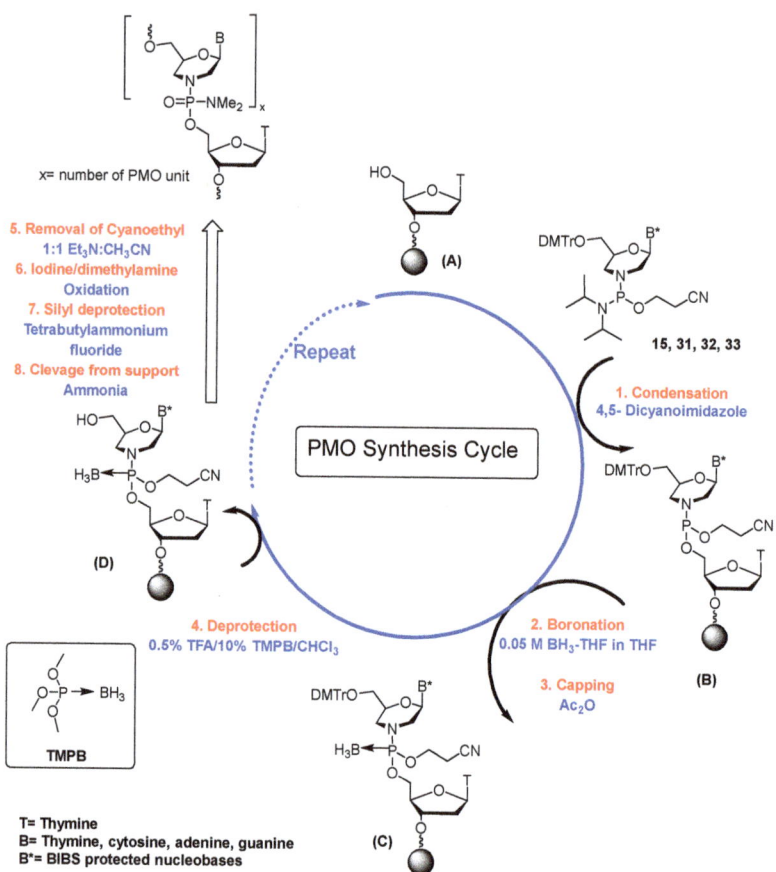

Scheme 4. Synthesis cycle for the preparation of PMOs.

Table 1. Solid-phase synthesis parameters used for PMOs and PMO-DNA chimeras.

Reactions	Wash/Reagents/Solvents	Time (Seconds)
Detritylation	10% TMPB + 0.5% TFA in CHCl$_3$	Flow 45 s
Condensation	0.1 M Morpholinonucleoside phosphoradiamidite (**15, 31, 32, 33**) in ACN + Activator 0.12 M dicyanoimidazole in ACN	Hold 300 s
	0.1 M Deoxynucleoside phosphoradiamidite (**34, 35, 36, 37**) in ACN + Activator 0.25 M 5-ethylthio-1H-tetrazole in ACN	Hold 180 s
Boronation	0.05 M BH$_3$-THF in THF	Flow 15 s, Hold 45 s
Oxidation	0.02 M Iodine in THF/water/pyridine	Flow 8 s, hold 15 s
Capping	Cap A (THF/Pyridine/Ac$_2$O) + Cap B (16% N-methylimidazole in THF)	Flow 10 s, hold 5 s

4. Synthesis of MOs and MO/DNA Chimeras

As outlined earlier in this review, Zhang and collaborators [12,13] synthesized oligonucleotides having up to four thymidine morpholino phosphoramidate internucleotide linkages (MOs, **5** Figure 1). The biochemical results from these investigations were intriguing. These MOs formed duplexes with complementary RNA and were stable toward degradation with snake venom phosphodiesterase. Moreover, when up to four morpholino uridine phosphoramidates were incorporated into siRNA duplexes, the resulting RISC complex showed potent silencing activity in HeLa cells.

Because of these exciting results, MO synthesis was extended to all four standard bases, with these oligonucleotides having 15 to 20 nucleotides [18]. The building blocks used for this research are shown in Scheme 5. Because there is no boronation step, as was the case for the synthesis of PMOs via P(III) chemistry, the standard amide-protecting groups were observed to be appropriate for studies on MO synthesis (Scheme 5; 15, 38, 39, 40). Following an extensive study focused on the activation of these building blocks with 5-ethylthio-1H-tetrazole (ETT), a solid-phase synthesis cycle (Table 2) was developed for preparing MOs [18]. However, despite considerable experimentation, the MOs could not be isolated free of oligonucleotide degradation products. The problem primarily centers on the reactivity of the morpholino phosphoramidate linkage toward aqueous acid. Thus, during the final deprotection step using aqueous acid to remove the 5'-O-(4,4'-dimethoxytrityl) group, considerable degradation of the MO was observed, and the final MO product could not be isolated in pure form. Although the synthesis of MOs with all four bases is now possible, it is clear that a different synthesis strategy must be introduced in order to overcome the inherent instability of the morpholino phosphoramidate linkage toward aqueous acid. As stated in the referenced manuscript [18], "However, due to the inherent acid and base stability of P-N linkages in phosphoramidate ODNs, an improved synthetic methodology that integrates orthogonal protecting group strategies for the morpholino nucleoside building blocks as well as the terminal hydroxyl protecting group of the final ODN may be required to enable their successful isolation for further research."

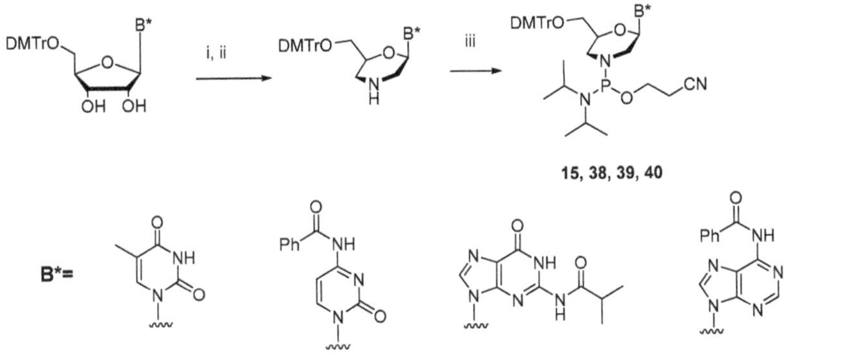

Reagents: (i) NaIO$_4$ (1.2 equiv), anhydrous methanol; (NH$_4$)$_2$B$_2$O$_7$ (1.2 equiv), 6 h. (ii) NaCNBH$_3$ (2.0 equiv); CH$_3$COOH (2.0equiv), 16 h. (iii) (OCH$_2$CH$_2$CN)P[N(iPr)$_2$]$_2$ (1.2 equiv), 5-ethylthio-1H-tetrazole (0.5 equiv), CH$_2$Cl$_2$, 30 min

Scheme 5. Synthesis of phosphorodiamidite morpholino building blocks for MOs and TMOs.

Table 2. Solid-phase synthesis cycle used for MOs and MO-DNA chimeras.

Reactions	Wash/Reagents/Solvents	Time (Second)
Detrylation	3% Trichloroacetic acid in Dichloromethane	95 s
Condensation	0.1M 5'-DMTrO-morpholinonucleoside phosphordiamidite in acetonitrile and 0.12 M 5-ethylthio-1H-tetrazole in acetonitrile	600 s
	0.1M 5'-DMTrO-2'-deoxynucleoside phosphoramidites and 0.12 M 5-ethylthio-1H-tetrazole in acetonitrile	30 s
Capping	Cap A (THF/Pyridine/Ac$_2$O) + Cap B (16% N-methylimidazole in THF)	5 s
Oxidation	0.1 M (1S)-(+)-(10-camphorsulfonyl)-oxaziridine in anhydrous acetonitrile	180 s

5. Synthesis of TMOs and TMO/DNA Chimeras

The solid-phase synthesis of MOs was carried out by oxidizing compound B, as shown in Scheme 4. The resulting morpholino phosphoramidate oligonucleotides were obtained in low overall yields, and their crude reaction mixtures could not be purified using the conventional DMT-On/Off procedure due to their instability to aqueous acid. We conjectured that replacing the nonbridging P(O) linkages in morpholino phosphoramidates with P(S) might improve their hydrolytic stability under acidic conditions and increase nuclease resistance towards intracellular enzymes. This presumption led us to recently focus entirely on thiophosphoramidate morpholino oligonucleotides (TMOs, 4 Figure 1) [15,19]. Morpholino phosphorodiamidites 15, 38, 39, and 40 (Scheme 5) were used as building blocks to prepare TMOs. Because we did not have a boronation step in the TMO synthesis strategy, the standard amide protecting groups were used for the exocyclic amino groups of cytosine, guanine, and adenine. Several activators (ETT; Tetrazole; Dicyanoimidazole (DCI)) at various concentrations were studied, with the best activation of morpholino phosphorodiamidites being ETT at 0.12 M. The first step for synthesizing TMOs was to remove the 5′-DMT group from the 2′-deoxynucleoside attached to the silica support (Scheme 6 and Table 3). This group was removed using 3% trichloroacetic acid in dichloromethane. This 2′-deoxynucleoside was then reacted with a morpholino phosphorodiamidites of mABz, mGiBu, mCBz, or mT (15, 38, 39, and 40) in the presence of 0.12 M ETT in anhydrous acetonitrile (300 s condensation time). The resulting morpholino phosphoramidite (E) was then converted to the thiophosphoramidate dinucleotide (G) using 3-[(dimethylaminomethylene)amino]-3H-1,2,4-dithiazole-5-thione (DDTT). Subsequent detritylation followed by repetitive synthesis cycles generated a TMO oligonucleotide having the desired sequence and length. When using the Universal Support III in order to generate a TMO having only morpholino nucleosides, the first morpholino phosphorodiamidite coupling step was followed by oxidation instead of sulfurization. Thus, during the ammonia cleavage of the TMO from the support, a TMO would be generated having a 3′-morpholine.

The solid-phase synthesis of TMOs and their chimeras resembles the synthesis of phosphorothioate DNA synthesis, with two differences: (i) the ETT concentration was reduced to 0.12 M, and (ii) a 300 s coupling time was used. We noticed that reducing the coupling time of phosphorodiamidite building blocks (41, 42, 43, or 44) to 300 s (0.12 M ETT) resulted in synthesis yields for 20–25 mer TMOs that were approximately 50%–60% the amount for a phosphorothioate DNA on the same support. For the synthesis of TMO–DNA chimeras, a 30 s coupling time was used for commercially available 2′-deoxyribonucleoside 3′-phosphoramidites.

The final steps for the synthesis of TMOs and TMO/DNA chimeras begin with cleavage from the synthesis support and removal of protecting groups using aqueous ammonia at 55 °C for 18 h (for the Universal Support III, 2 M ammonia/methanol, 60 min followed by aqueous ammonia, 55 °C, 16 h). Post-deprotection, the ammonia solutions were removed by evaporation. Product mixtures were dissolved in 3% aqueous methanol, filtered through a 0.2-micron filter, and analyzed via LCMS. The reaction mixtures were redissolved in 3% aqueous acetonitrile and purified using ion-pair RP-HPLC. Post-purification, selected fractions were dried and treated with 500 μL of 50% aqueous acetic acid to remove the DMT group (5 min). The acidic reaction mixture was cooled in an ice bath followed by neutralization with triethylamine. After evaporating to dryness, the TMOs and TMO/DNA chimeras were re-purified using ion-pair RP-HPLC. The pure product fractions were combined and evaporated to dryness. Desalting was carried out using NAP DNA purification columns.

We have successfully developed a chemical synthesis strategy for synthesizing TMOs and TMO/DNA chimeras using phosphoramidite chemistry. In contrast to P(V)-based PMO synthesis, the P(III) approach lends itself to the incorporation of several modifications, such as 2′-OMe, 2′-F, 2′-O-MOE, LNA, and others, that provide a route for generating an array of previously unexplored oligonucleotide therapeutic drug candidates.

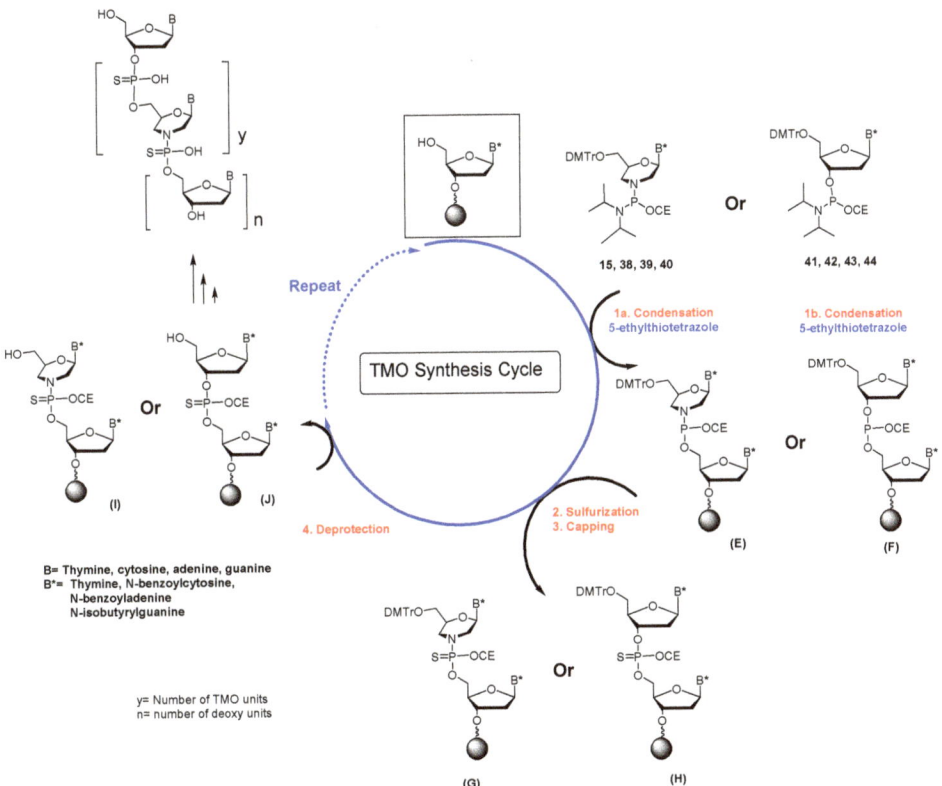

Scheme 6. Synthesis cycle for TMOs and TMO-DNA chimeras.

Table 3. Solid-phase synthesis parameters used for TMOs and TMO-DNA chimeras.

Reactions	Wash/Reagents/Solvents	Time (Seconds)
Detritylation	3% Trichloroacetic acid in Dichloromethane	Flow 45 s
Condensation	0.1 M Morpholino hosphorodiamidite (**15, 38, 39, 40**) in can + Activator 0.12 M 5-ethylthio-1H-tetrazole canACN	Hold 300 s
	0.1 M 2′-Deoxynucleoside 3′-phosphoradiamidite (**41, 42, 43, 44**) in ACN + Activator 0.12 M 5-ethylthio-1H-tetrazocanin ACN	Hold 30 s
Thiolation	3-[(Dimethylaminomethylene)amino]-3H-1,2,4-dithiazole-5-thione (DDTT)	360 s
Capping	Cap A (THF/Pyridine/Ac$_2$O) + Cap B (16% N-methylimidazole in THF)	Hold 5 s

6. Biological and Biochemical Properties of TMOs

Since the binding affinity of an oligonucleotide plays a critical role in determining its therapeutic potential, the biophysical properties of TMOs and TMO/DNA chimeras have been analyzed [19]. Chimeric TMOs with alternating thiophosphoramidatemorpholino/2′-deoxynucleoside 3′-thiophosphate structures showed an increase in binding affinity (+6–10 °C) toward DNA. This could be because of increased flexibility to an otherwise rigid construct, which permits more efficient binding with the complementary strand. A significant Tm loss (>12 °C) was observed for the fully modified thiophosphoramidatemorpholino/2′-deoxynucleoside 3′-thiophosphate oligomer when forming a duplex with complementary DNA. This result suggests that the exclusive presence of rigid morpholino moieties results in a backbone conformation that is not helpful to efficient DNA binding. However, when TMOs form duplexes with complementary RNA, both the chimeric TMO/DNA and the completely thiomorpholino oligonucleotides form duplexes where the Tms are 10 degrees higher than for the RNA/DNA duplex. To gain further insights into their helical structure, RNA heteroduplexes of TMOs with complementary RNA were assessed using circular dichroism (CD) and compared with control duplexes (DNA−pS/RNA and canonical DNA/RNA). The CD spectra of these TMO heteroduplexes revealed structural similarities to the DNA−pS/RNA duplex. Both fully modified TMO/RNA and chimeric TMO/RNA duplexes closely resemble an A-form RNA/RNA duplex which provides a logical explanation for their higher binding affinity towards complementary RNA. The RNase H1 activity of TMOs was assessed by preannealing TMOs with complementary RNA and treatment with Escherichia coli RNase H1. Fully modified TMOs do not activate RNase H1, presumably due to their conformational rigidity. However, gapmers having TMO caps and 2′-deoxynucleoside 3′-thiophosphate gaps are RNase H1 active. This suggests that exclusively TMO-modified oligonucleotides are potential candidates for exon skipping experiments, where they function as steric blockers, whereas gapmers having TMO wings can be used for antisense experiments that involve mRNA degradation through the activation of the RNase H1 pathway.

TMOs were used to address how intron retention influences the spatio-temporal dynamics of transcripts from two clinically relevant genes: TERT (Telomerase Reverse Transcriptase) pre-mRNA and TUG1 (Taurine-Upregulated Gene 1) lncRNA [20]. Data suggest that the splicing of TERT-retained introns occurs during mitosis. In contrast, TUG1 has a bimodal distribution of fully spliced cytoplasmic and intron-retained nuclear transcripts. Upon the addition of TMOs complementary to TERT and TUG 1 preRNA, nuclear TERT and TUG1 intron retentions are correlative and do not show a causality of intron retention driving their subcellular localization. To test this hypothesis, 20mer TMOs were designed against the two TUG1 donor splice sites, each hybridizing to 2 nt of the exon and 18 nt of the intron sequence. These TMOs were tested for the functionality of intron-retention events using RNA targeting to block intron excision. It was observed that TMOs enforcing intron 1 retention decreased the splicing of intron 1 by ~60% compared to the control TMO. Intron 11-containing TERT (assessed by monitoring intron 11 and exon 11 to intron 11 junction) was increased ~32% compared to control TMO. As additional controls, we applied primers at the upstream exon 10 to exon 11 junction, which was not affected with TERT TMO treatment, and exon 10 to exon 12 junction, which was decreased ~50%, in accordance with the decrease in exon 11 to exon 12 junction. These findings demonstrate that TMOs effectively block splicing and change cellular localization and availability of the RNA.

In another study, various TMOs were synthesized and evaluated for their efficacy to induce exon skipping in a Duchenne Muscular Dystrophy (DMD) in vitro model using H2K mdx mouse myotubes [21,22]. Experiments demonstrated that TMOs can efficiently internalize and induce excellent exon-23 skipping potency compared with a conventional PMO control and other widely used nucleotide analogs, such as 2′-O-methyl and 2′-O-methoxyethyl antisense oligonucleotides. TMOs performed well at low concentrations (5–20 nM); hence, the dosages can be minimized, which may improve the drug safety profile. These results in the H2K mdx myotubes demonstrate an opportunity for using TMOs as therapeutic ASOs in the treatment of various genetic diseases.

These experiments strongly support the conclusion that TMO oligonucleotides are quite active in cell nuclei. This is because the excision of introns is a nuclear activity.

7. Conclusions

Chemical methods, focused on the use of nucleotide P(III) building blocks, are used for the synthesis of oligonucleotides that have found major transformative applications in biology, biochemistry, molecular biology, and biophysical chemistry. These building blocks are extremely stable, easily activated, and readily prepared. Recently, we have used P(III) chemistry to prepare several morpholino phosphorodiamidite building blocks to synthesize many novel PMO and TMO analogues. Currently, more than 20 biological studies focused on the use of TMOs as therapeutic drugs for the treatment of various genetic diseases are underway. These collaborations are being carried out with cells in culture, and several have progressed to studies in mouse models. Without exception, TMOs are more active than any other analogue tested. However, major obstacles such as challenging large-scale production and extensive toxicity studies still need to be addressed. TMOs can perhaps be exceptional candidates for future biological applications in the oligonucleotide field.

Author Contributions: Paul was the pioneer of the PMO and TMO project in the Caruthers' lab. All authors have read and agreed to the published version of the manuscript.

Funding: We would like to acknowledge The University of Colorado for the financial assistance.

Institutional Review Board Statement: Not applicable.

Informed Consent Statement: Not applicable.

Data Availability Statement: Not applicable.

Acknowledgments: Over the years, several other members of the Caruthers' laboratory have worked on this morpholino platform to refurbish the chemistry and the synthesis of various morpholino oligonucleotide analogues. Acknowledgment for the research contributions of various colleagues can be found in the references. Marvin H. Caruthers would like to dedicate this review to the memory of Enrique Pedroso. While Enrique was part of my research group, he pioneered our very early attempts to develop new methods for chemically synthesizing RNA. I will always have a warm recollection of my interactions with Enrique and remain saddened that he had to pass way too soon.

Conflicts of Interest: S.P. is an employee of Agilent Technologies.

Sample Availability: Not applicable.

References

1. Lim, K.R.Q.; Maruyama, R.; Yokota, T. Eteplirsen in the treatment of Duchenne muscular dystrophy. *Drug Des. Dev. Ther.* **2017**, *11*, 533–545. [CrossRef] [PubMed]
2. Heo, Y.-A. Golodirsen: First approval. *Drugs* **2020**, *80*, 329–333. [CrossRef] [PubMed]
3. Dhillon, S. Viltolarsen: First approval. *Drugs* **2020**, *80*, 1027–1031. [CrossRef] [PubMed]
4. Shirley, M. Casimersen: First approval. *Drugs* **2021**, *81*, 875–879. [CrossRef] [PubMed]
5. Summerton, J.E. Invention and early history of morpholinos: From pipe dream to practical products. *Morpholino Oligomers Methods Mol. Biol.* **2017**, *1565*, 1–15. [CrossRef] [PubMed]
6. Moulton, H.M.; Moulton, J.D. (Eds.) *Morpholino Oligomers*; Humana Press: Totowa, NJ, USA, 2017. [CrossRef]
7. Summerton, J.; Weller, D. Morpholino antisense oligomers: Design, preparation, and properties. *Antisense Nucleic Acid Drug Dev.* **1997**, *7*, 187–195. [CrossRef] [PubMed]
8. Bhadra, J.; Kundu, J.; Ghosh, K.C.; Sinha, S. Synthesis of phosphorodiamidate morpholino oligonucleotides by H-phosphonate method. *Tetrahedron Lett.* **2015**, *56*, 4565–4568. [CrossRef]
9. Beaucage, S.L.; Caruthers, M.H. Studies on Nucleotide Chemistry V: Deoxynucleoside Phosphoramidites—A New Class of Key Intermediates for Deoxypolynucleotide Synthesis. *Tetrahedron Lett.* **1981**, *22*, 1859–1862. [CrossRef]
10. Beaucage, S.L.; Caruthers, M.H. *Current Protocols in Nucleic Acids Chemistry*; Beaucage, S.L., Bergstrom, D.E., Glick, G.D., Jones, R.A., Eds.; Unit 3.3; John Wiley & Sons, Inc.: New York, NY, USA, 2000; pp. 1–20.
11. Paul, S.; Roy, S.; Monfregola, L.; Shang, S.; Shoemaker, R.; Caruthers, M.H. Oxidative substitution of boranephosphonate diesters as a route to post-synthetically modified DNA. *J. Am. Chem. Soc.* **2015**, *137*, 3253–3264. [CrossRef] [PubMed]
12. Zhang, N.; Tan, C.; Cai, P.; Jiang, Y.; Zhang, P.; Zhao, Y. Synthesis and properties of morpholino chimeric oligonucleotides. *Tetrahedron Lett.* **2008**, *49*, 3570–3573. [CrossRef]
13. Zhang, N.; Tan, C.; Cai, P.; Zhang, P.; Zhao, Y.; Jiang, Y. RNA interference in mammalian cells by siRNAs modified with morpholino nucleoside analogues. *Bioorg. Med. Chem.* **2009**, *17*, 2441–2446. [CrossRef] [PubMed]
14. Roy, S.; Olesiak, M.; Shang, S.; Caruthers, M.H. Silver Nanoassemblies Constructed from Boranephosphonate DNA. *J. Am. Chem. Soc.* **2013**, *135*, 6234–6241. [CrossRef] [PubMed]
15. Caruthers, M.H.; Paul, S. Synthesis of Backbone Modified Morpholino Oligonucleotides and Their Chimeras Using Phosphoramidite Chemistry. U.S. Patent 11,230,565 B2, 25 January 2022.
16. Summerton, J.; Weller, D. Uncharged Morpholino-Based Polymers Having Achiral Intersubunit Linkages. U.S. Patent 5,034,506, 7 July 1991.

17. Paul, S.; Nandi, B.; Pattanayak, S.; Sinha, S. Synthesis of 5-alkynylated uracil–morpholino monomers using Sonogashira coupling. *Tetrahedron Lett.* **2012**, *53*, 4179–4183. [CrossRef]
18. Krishna, H.; Jastrzebska, K.; Caruthers, M. Exploring site-specific activation of bis-N,N'-dialkylaminophosphordiamidites and the synthesis of morpholinophosphoramidate oligonucleotides. *FEBS Lett.* **2019**, *593*, 1459–1467. [CrossRef] [PubMed]
19. Langner, H.K.; Jastrzebska, K.; Caruthers, M.H. Synthesis and Characterization of Thiophosphoramidate Morpholino Oligonucleotides and Chimeras. *J. Am. Chem. Soc.* **2020**, *142*, 16240–16253. [CrossRef] [PubMed]
20. Dumbovic, G.; Braunschweig, U.; Langner, H.K.; Smallegan, M.; Biayna, J.; Hass, E.P.; Jastrzebska, K.; Blencowe, B.; Cech, T.R.; Caruthers, M.H.; et al. Nuclear compartmentalization of TERT mRNA and TUG1 lncRNA is driven by intron retention. *Nat. Commun.* **2021**, *12*, 3308–3326. [CrossRef] [PubMed]
21. Caruthers, M.H.; Paul, S.; Veedu, R.N.; Jastrzebska, K.; Krishna, H. Thiomorpholino Oligonucleotides for the Treatment of Duchene Muscular Dystrophy. U.S. Patent 20230193266 A1, 22 June 2023.
22. Le, B.T.; Paul, S.; Jastrzebska, K.; Langer, H.; Caruthers, M.H.; Veedu, R.N. Thiomorpholino oligonucleotides as a robust class of next generation platforms for alternate mRNA splicing. *Proc. Natl. Acad. Sci. USA* **2022**, *119*, e2207956119. [CrossRef] [PubMed]

Disclaimer/Publisher's Note: The statements, opinions and data contained in all publications are solely those of the individual author(s) and contributor(s) and not of MDPI and/or the editor(s). MDPI and/or the editor(s) disclaim responsibility for any injury to people or property resulting from any ideas, methods, instructions or products referred to in the content.

molecules

Article

Convenient Solid-Phase Attachment of Small-Molecule Ligands to Oligonucleotides via a Biodegradable Acid-Labile P-N-Bond

Nadezhda O. Kropacheva [1,2], Arseniy A. Golyshkin [2], Mariya A. Vorobyeva [1] and Mariya I. Meschaninova [1,*]

1. Institute of Chemical Biology and Fundamental Medicine, Siberian Branch, Russian Academy of Sciences, Novosibirsk 630090, Russia
2. Department of Natural Sciences, Novosibirsk State University, Novosibirsk 630090, Russia
* Correspondence: mesch@niboch.nsc.ru; Tel.: +7-383-363-5129

Abstract: One of the key problems in the design of therapeutic and diagnostic oligonucleotides is the attachment of small-molecule ligands for targeted deliveries in such a manner that provides the controlled release of the oligonucleotide at a certain moment. Here, we propose a novel, convenient approach for attaching ligands to the 5′-end of the oligonucleotide via biodegradable, acid-labile phosphoramide linkage. The method includes the activation of the 5′-terminal phosphate of the fully protected, support-bound oligonucleotide, followed by interaction with a ligand bearing the primary amino group. This technique is simple to perform, allows for forcing the reaction to completion by adding excess soluble reactant, eliminates the problem of the limited solubility of reagents, and affords the possibility of using different solvents, including water/organic media. We demonstrated the advantages of this approach by synthesizing and characterizing a wide variety of oligonucleotide 5′-conjugates with different ligands, such as cholesterol, aliphatic oleylamine, and p-anisic acid. The developed method suits different types of oligonucleotides (deoxyribo-, 2′-O-methylribo-, ribo-, and others).

Keywords: 5′-functionalization; small molecules; conjugates of oligonucleotides; solid-phase synthesis; pH-sensitive phosphoramidate linkage; siRNA

1. Introduction

Functional nucleic acids (FNAs) (catalytic NAs, aptamers, small, interfering RNAs, antisense oligonucleotides, etc.) and FNA-based constructs represent very promising, highly selective research tools for molecular biology, as well as potential therapeutic agents for viral, oncological, and other diseases [1–6]. However, their applications for targeting certain biomolecules inside the cell face the problem of insufficient cell delivery. The negative charge of the sugar-phosphate backbone hinders NA penetration through the negatively charged cell membrane. The cell delivery of FNAs can be improved by conjugation with small, transporting molecules, such as vitamins, dendrimers, lipophilic compounds, peptides, cationic lipids, polymers, etc. (e.g., [7–14]).

Small molecules can be attached to functional NAs either noncovalently or covalently via chemical bonds, which, in turn, can be stable or labile under intracellular conditions. The use of biodegradable bonds allows for the release of an FNA cargo from the carrier molecule under the appropriate intracellular conditions, which, in turn, can enhance its biological activity. In particular, acid-labile linkers are widely used to address this issue: acetals, ketals, beta-thiopropionates, oximes, orthoesters, hydrazones, etc. [15–17]. In our work, we chose a phosphoramidate bond for creating an acid-labile linker.

The literature describes several synthetic approaches to the synthesis of the conjugates of nucleotides or oligonucleotides containing a P-N-bond at the terminal or internucleoside phosphates. One of the most common and frequently used options for obtaining a phosphoramide bond is the Atherton–Todd reaction: the conversion of a dialkyl phosphite into dialkyl chlorophosphate in the presence of carbon tetrachloride, followed by the reaction with a primary amine [18,19]. Alternatively, dialkyl phosphite can be oxidized in the presence of elemental iodine, followed by interaction with amino-containing ligands [20,21]. These reactions allow for the introduction of modification during solid-phase oligonucleotide synthesis. The Staudinger reaction provides another option for obtaining such modifications by solid-phase synthesis. For example, the authors of [22,23] obtained a phosphoramide bond through the Staudinger reaction between azidoalkyl-modified lipophilic molecules and an internucleoside 2-cyanoethylphosphite of the polymer-bound protected oligonucleotide, which is formed at the condensation stage in a standard automated synthesis. Dovydenko et al. [24] proposed a variant of the solid-phase phosphotriester approach: the active 5′-arylphosphodiester derivative reacted with the polymer-bound protected oligonucleotide, with the subsequent replacement of the azole moiety by an alkyl amine.

Otherwise, the ligands can be introduced through the P-N-bond at the 5′-/3′-end or the 2′-position of the fully deblocked oligonucleotide by activating the phosphoric acid residue with carbodiimide (EDC) in the presence of methylimidazole or with the ox/red pair triphenylphosphine/dipyridyl disulfide (PPh_3/$(PyS)_2$) [25–31]. Based on the last option, we proposed a synthetic approach to the 5′-modification of oligonucleotides. Here, we combine the solid-phase approach, with all its advantages, with the activation of the terminal phosphate of the protected polymer-bound oligonucleotide (ribo-, deoxyribo- or 2′-O-methylribo-) via PPh_3/$(PyS)_2$/DMAP, followed by the interaction with amino-containing ligands of various chemical nature.

2. Results and Discussion

As we mentioned above, there are two strategies for the synthesis of oligonucleotide conjugates with small molecules: (1) the introduction of ligands into a fully deblocked oligonucleotide (synthesis "in solution"), and (2) the interaction of ligands with a support-bound protected oligonucleotide (solid-phase conjugation) [32–36]. Each of them has its own advantages and shortcomings. From our point of view, the solid-phase approach is preferable for obtaining conjugates with small transporting molecules. Protected oligonucleotide is covalently bound with the carrier, which allows for using almost any combination of solvents, including water-organic media. This possibility becomes especially important for attaching lipophilic molecules since they require nonpolar solvents that are poorly suited to deblocked oligonucleotides. The presence of standard protecting groups in the oligonucleotide makes it possible to introduce the ligand selectively at a given position, using it in significant excesses when necessary. The easy removal of unreacted components and side products by simple washing of the carrier and high conjugation efficiency make the solid-phase approach handy and versatile.

The common scheme of the proposed solid-phase 5′-functionalization of oligonucleotides (deoxyribo-, ribo-, and 2′-O-methylribo-) includes four steps (Figure 1): (1) introduction of a phosphate group at the 5′-end of the oligonucleotide chain during an automatic synthesis, (2) the deprotection of the terminal phosphate, (3) the activation of 5′-phosphate of the protected support-bound oligonucleotide by PPh_3/$(PyS)_2$, with the formation of the 5′-DMAP-intermediate, (4) the interaction of the intermediate with amino-modified molecules, and (5) the standard deprotection of the subsequent oligonucleotide conjugates. The approach is easy to use, does not require changes in the standard automatic protocol for the synthesis of oligonucleotides, and allows for the obtainment of a set of different conjugates, starting from one oligonucleotide.

Our study includes three steps: the synthesis of amino-containing ligands, the development of a method for the synthesis of oligonucleotide conjugates with an acid-labile bond, and the study of the stability of an acid-labile bond at different pH values.

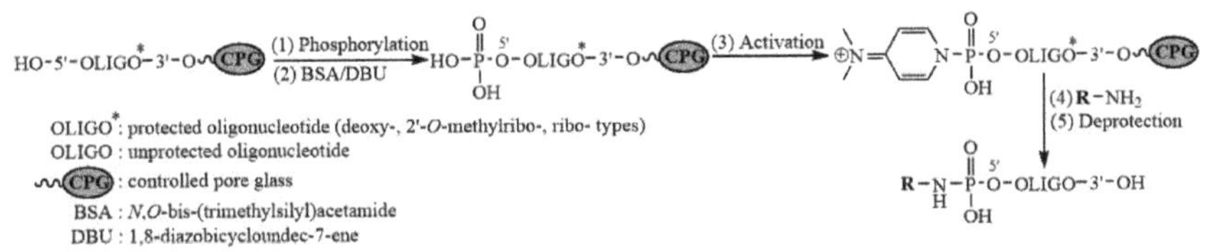

Figure 1. The common scheme for oligonucleotide solid-phase 5′-functionalization. R–NH$_2$ structures: see Supplementary Table S1.

2.1. Synthesis of Amino-containing Ligands

In order to demonstrate the possibilities of the developed approach, we used a set of small molecules bearing a primary amino group: commercially available amino-containing ligands (pyrenemethylamine, aliphatic diamine, oleylamine, propargylamine, 3-aminopropan-1-ol,), as well as home-made amino-modified cholesterol and p-anisic acid. Some of these molecules were reported to be used for cell delivery. In particular, lipophilic oleylamine and cholesterol are capable of interacting with cell membranes [6–8], and anisamide (a derivative of p-anisic acid) possess a high affinity for the sigma receptor 1 (δ1), which is overexpressed on the surface of cancer cells [7–9,37–39]. It is important to note that some small molecule ligands can be sensitive to the conditions of oligonucleotides deprotection. Therefore, one should carefully choose the small molecules for solid-phase conjugation with oligonucleotides and take into account the possibility of their destruction under the conditions of the final release of the oligonucleotide conjugates.

The synthesis of amino-containing cholesterols (**I**) was carried out according to our previous work [40] (Figure 2): 1,6-diaminohexane was used in excess to introduce cholesteryl chloroformate into the process. The carboxylic group of p-anisic acid was activated using N,N′-dicyclohexylcarbodiimide (DCC) in the presence of N-hydroxybenzotriazole (HOBt), analogous with [41]. The activated p-anisic acid reacted with the mono-N-Boc-protected 1,6-diaminohexane, then the Boc-protected group was removed with a formic acid, as described in [42] (Figure 2). NMR was used to confirm the structures of the ligands with amino modifications. For further processing, see Section 3: Materials and Methods.

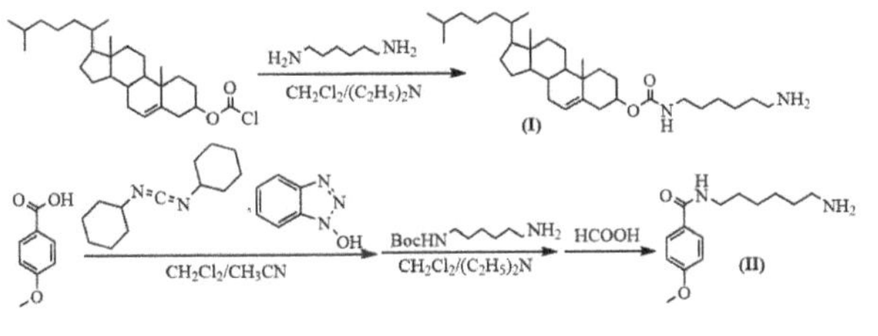

Figure 2. Schemes of the syntheses of amino-modified cholesterol (**I**) and N-(6-aminohexyl)-4-methoxybenzamide (**II**). Boc—*tert*-butyloxycarbonyl.

2.2. Solid-Phase Synthesis of Oligonucleotide Conjugates

We optimized the methodology of the proposed solid-phase approach to the synthesis of 5′-modified oligonucleotides and tested its possibilities by obtaining various conjugates of the model oligodeoxyribonucleotide dT$_7$ (**1–9**) (Figure 3, Table 1).

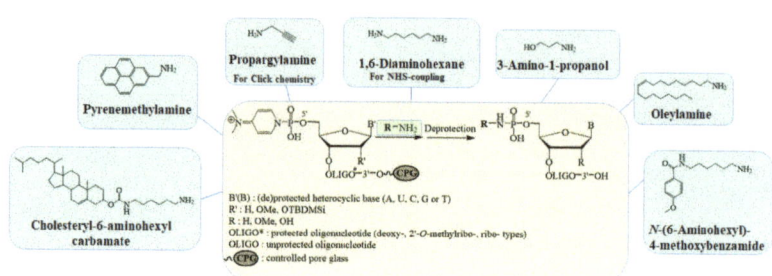

Figure 3. Solid-phase synthesis of 5′-conjugates of oligonucleotides containing an acid-labile P-N-bond (Table 1).

Table 1. Synthesized conjugates of oligonucleotides and their characteristics.

№	Oligonucleotide Conjugate, 5′-3′	RP HPLC Retention Time, min [1]	Molecular Weight Calculated	Molecular Weight Experimental [2]	Yield, % [3]
1	MB-**L₆**-NH-p-d(TTTTTTT)	12.88 (+3.64)	2379.7	2378.5	19
2	Chol-C(O)-**L₆**-NH-p-d(TTTTTTT)	24.05 (+14.81)	2658.2	2656.5	18
3	Oleyl-NH-p-d(TTTTTTT)	23.53 (+14.29)	2396.8	2395.5	24
4	CH≡C-CH₂-NH-p-d(TTTTTTT)	9.93 (+0.69)	2184.4	2183.0	23
5	Pyr-CH₂-NH-p-d(TTTTTTT)	15.21 (+5.97)	2360.3	2358.9	23
6	HO-(CH₂)₃-NH-p-d(TTTTTTT)	9.63 (+0.39)	2204.4	2203.2	22
7	NH₂-(CH₂)₆-NH-p-d(TTTTTTT)	9.43 (+0.19)	2245.6	2244.0	18
8	Biot-NH-(CH₂)₆-NH-p-d(TTTTTTT)	11.88 (2.64)	2471.6	2470.2	17 [5,*]
9	CH₃-NH-p-d(TTTTTTT)	9.55 (+0.31)	2161.1	2159.7	21 [*]
10	Oleyl-NH-p-GᵐGᵐCᵐUᵐUᵐGᵐAᵐCᵐAᵐ	17.12 (+7.01)	3310.3	3309.0	18
11	MB-**L₆**-NH-p-GᵐGᵐCᵐUᵐUᵐGᵐAᵐCᵐAᵐ	12.01 (+1.9)	3293.3	3291.3	18
12	CH≡C-CH₂-NH-p-GᵐGᵐCᵐUᵐUᵐGᵐAᵐCᵐAᵐ	10.39 (+0.28)	3098.1	3096.2	21
13	FAM-click-CH₂-NH-p-GᵐGᵐCᵐUᵐUᵐGᵐAᵐCᵐAᵐ	11.81 (+1.70)	3555.5	3554.5	21 [5,*]
14	MB-**L₆**-NH-p-GGCUUGACAAGUUGUAUAUGGᵐ	n/a [4]	7080.4	7080.28	20
15	Chol-C(O)-**L₆**-NH-p-GGCUUGACAAGUUGUAUAUGGᵐ	n/a [4]	7358.9	7358.98	17
16	Oleyl-NH-p-GGCUUGACAAGUUGUAUAUGGᵐ	n/a [4]	7097.6	7097.2	21
17	CH≡C-CH₂-NH-p-GGCUUGACAAGUUGUAUAUGGᵐ	n/a [4]	6885.1	6886.4	20
18	GalNAc-click-CH₂-NH-p-GGCUUGACAAGUUGUAUAUGGᵐ	n/a [4]	7263.7	7263.87	19 [5,*]

[1] For the RP-HPLC conditions, see Section 3: Materials and Methods. The difference between the retention times for the nonmodified controls 5′-p-d(TTTTTTT) (9.24 min) or 5′-p-GᵐGᵐCᵐUᵐUᵐGᵐAᵐCᵐAᵐ (10.11 min) are given in the brackets. [2] Obtained by ESI or MALDI-TOF mass spectrometry. [3] The yields of conjugates after deblocking and isolation were calculated based on the molar amount of the first support-bound nucleoside. [4] Not available, characterized by PAGE only. [5] After all conjugations and isolation. [*] For the description, see Supplementary material. Chol-C(O)-**L₆**-NH–, cholesteryl-6-aminohexylcarbamate residue; Oleyl-NH–, oleylamine residue; Pyr-CH₂-NH–, pyrenemethylamine residue; MB-**L₆**-NH-p-, N-(6-aminohexyl)-4-methoxybenzamide residue; NH₂-(CH₂)₆-NH-, 1,6-diaminohexane residue; HO-(CH₂)₃-NH–, 3-amino-1-propanole residue; CH≡C-CH₂-NH–, propargylamine residue; Biot-, Biotin residue (Supplementary Figure S2); FAM-click-CH₂-NH-, FAM residue with 1,2,3-triazole linker (Supplementary Figure S3); GalNAc-click, GalNAc residue with 1,2,3-triazole linker (Supplementary Figure S3); -p-, -P(O)(OH)-; **L₆**-, -NH(CH₂)₆-; N, ribonucleotide; Nᵐ, 2′-O-methylribonucleotide; d(N), deoxyribonucleotide.

The approach was developed on the basis of the solution method described in [30,31]. After phosphorylation in an automatic mode, this was followed by the removal of the protective groups from the phosphates of the oligonucleotide. We processed a protected polymer-bound model oligonucleotide 5′-p-dT$_7$ with N,O-bis(trimethylsilyl)acetamide (BSA), with subsequent treatment via 1,8-diazabicyclo[5.4.0]undec-7-ene (DBU), analogous with [31] (Figure 1). It was shown that BSA provides a rearrangement of the linkage between the oligonucleotide and the support, which makes it stable. In turn, DBU gently removes the cyanoethyl-protective groups from the phosphates, leaving the oligonucleotide attached to the support [43]. Since the terminal phosphate is capable of incorporating two ligands, as was shown by the example of pyrene derivatives in [30], we additionally optimized the conditions for obtaining mono conjugates. We thoroughly selected the best solvent for each amino ligand's solid-phase addition (Supplementary Table S1). We also showed that the formation of mono conjugates requires the removal of activating agents (Ph$_3$P/(PyS)$_2$/DMAP) and the three-fold washing of the support with the attached oligonucleotide before coupling with the amino-containing ligand. The treatment by an aqueous solution of methylamine or its mixture with ammonia to remove the protective groups and cleave the oligonucleotide from the polymer carrier did not destroy the P-N-bond [19,20,24]. In our work, we used an aqueous solution of methylamine. During the analyses of the reaction mixtures of the oligonucleotide conjugates after deprotection, we found an unknown product in some cases and assumed it to be the 5′-p-dT$_7$ derivative, with a methylamine at the 5′-phosphate. In order to check this hypothesis, the support-bound 5′-p-dT$_7$ with an activated 5′-phosphate was washed with DMSO and immediately treated with a methylamine solution. In this case, the deblocking and cleavage of the oligonucleotide from the support occur simultaneously, with methylamine attachment at the terminal phosphate (Supplementary Figure S1). The reaction mixtures were analyzed by reversed-phase high-performance liquid chromatography (RP-HPLC) and analytical gel electrophoresis (Supplementary Figure S1). The degree of conversion of 5′-p-dT$_7$ into the corresponding conjugates was 75–95%, according to RP-HPLC data. Figure 4 shows some of the typical examples of chromatogram profiles for conjugates (**1, 4–7**). The nature of the ligand affected the retention time of the conjugate during RP-HPLC: lipophilic conjugates with cholesterol (**2**) and oleylamine (**3**) had the highest retention times (Table 1).

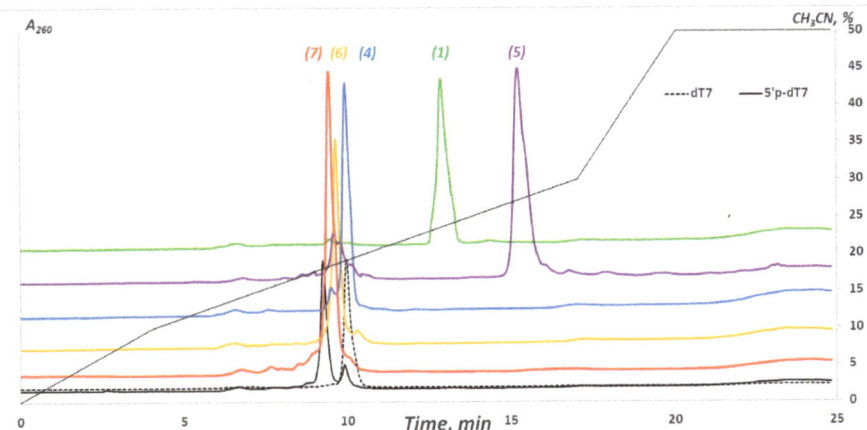

Figure 4. RP-HPLC analyses of reaction mixtures upon the conjugation of 5′-p-dT$_7$ with different amino ligands (Table 1). dT$_7$: 5′-d(TTTTTTT); 5′-p-dT$_7$: 5′-p-d(TTTTTTT); (**1**): MB-L$_6$-NH-p-dT$_7$; (**4**): CH≡C-CH$_2$-NH-p-dT$_7$; (**5**): Pyr-CH$_2$-NH-p-dT$_7$; (**6**): HO-(CH$_2$)$_3$-NH-p-dT$_7$; (**7**): NH$_2$-(CH$_2$)$_6$-NH-p-dT$_7$; MB-L$_6$-NH-p-, N-(6-aminohexyl)-4-methoxybenzamide residue; Pyr-CH$_2$-NH-, pyrenemethylamine residue; -p-, -P(O)(OH)-; d(N), deoxyribonucleotide. See Section 3: Materials and Methods for details.

The developed approach was tested for obtaining RNA and 2′-O-Me-RNA oligonucleotides. As a biologically active synthetic RNA, we chose the siRNA directed to the

557–577 region of mRNA to the MDR1 gene (multiple drug resistance gene). MDR1 encodes the membrane protein P-glycoprotein that is responsible for the transmembrane efflux of such substances as lipids, steroids, peptides, bilirubin, etc., thereby providing the effect of drug resistance [44]. Modifications at the 5′-end of the sense strand of siRNA do not affect the activity of siRNA [44]. It has been shown that the use of siRNAs with the sense strand divided into two fragments is a promising approach to gene silencing [45]. Therefore, our next task was to synthesize the 5′-mono-conjugates of the RNA sense strand and the 5′-mono-conjugates of the 2′-O-Me RNA half of the sense strand. The necessity of 2′-O-protective groups during the RNA chemical synthesis adds one more deblocking step to their removal and requires the optimization of the conjugation protocol. We found that the treatment of the conjugates containing a phosphoramide bond with a standard mixture of NMP/TEA·3HF/TEA for the removal of the 2′-O-TBDMS protective groups partly cleaved this bond, giving the 5′-phosphate-containing oligonucleotide. However, the treatment with a 1 M solution of tetrabutylammonium fluoride in tetrahydrofuran, followed by neutralization with triethylammonium acetate buffer (pH 7.0), and desalting on a C-18 cartridge, preserves the integrity of the P-N-bond. Thus, for the complete deblocking of conjugates of various types of oligonucleotides containing an acid-labile P-N-bond, methylamine treatment suits well, with or without the TBAF treatment and desalting.

The presence of 5′-amino- or 5′-alkyne groups within the oligonucleotide allows their further functionalization via well-known reactions with NHS-activated esters or click-chemistry. We demonstrated these possibilities by the interaction of model aliphatic amino-modified oligonucleotide (**7**) or propargylamine-modified oligonucleotides (**12, 17**) with Biotin-NHS or GalNAc/FAM azide, respectively. The degree of conversion was about 80%, according to the HPLC data (Supplementary Figures S2 and S3, respectively).

All obtained oligonucleotide conjugates were isolated by preparative gel electrophoresis. The yields of the conjugates related to the first support-bound nucleoside were 18–24% (Table 1), which is comparable to the yields of nonconjugated (parent) oligonucleotides of the same length. The homogeneities of the resulting conjugates were confirmed by analytical denaturing PAGE, followed by mass spectrometry. Notably, in some cases, we observed the cleavage of the P-N-bond during mass analysis, especially when recording MALDI spectra with the use of acidic matrices (Supplementary Table S2), similar to the effect that was registered earlier for the oligonucleotide conjugates with acid-labile hydrazone linkage [46].

2.3. Stability of the P-N-Bond within the Oligonucleotide Conjugates at Different pH Values

When small transporting molecules are introduced into the conjugate through a labile bond, upon fulfilling their role, they have to be cleaved from the cargo in the endosomes and lysosomes and leave the therapeutic NA free to perform its function. Therefore, it is important to quantitatively assess the liability of the linkage between the oligo and the transporting ligand. For example, it has been shown that the phosphoramide bond between the oligonucleotide and polyethylene glycol was completely cleaved at 37 °C for 5 h at pH 4.7 [25]. The authors of [47] systematically studied the uptake of lipid nanoparticles loaded with siRNAs and their intracellular transport and endosomal release and found that in the course of these processes, the pH values varied in the range of 4.5–6.5, and the total time was approx. 5 h.

We studied the stability of the phosphoramide bond within the synthesized conjugates at different pH values. The conjugates of oligonucleotide 5′-p-siDmS with N-(6-aminohexyl)-4-methoxybenzamide, cholesteryl-6-aminohexylcarbamate, oleylamine, and GalNAc (**14–16, 18**) were incubated in acetate buffer with pH values of 6.0, 5.2, and 4.5 at 37 °C for 1–24 h, and were then analyzed by the gel electrophoresis. According to the obtained data (Figure 5), at pH 6.0, the P-N-bond is hydrolyzed by no more than 20%, and at pH values lower than 5.2, it becomes significantly less stable. We observed the highest degree of P-N-bond cleavage at pH 4.5 for the conjugate 5′-p-siDmS with the N-(6-aminohexyl)-4-methoxybenzamide (**14**).

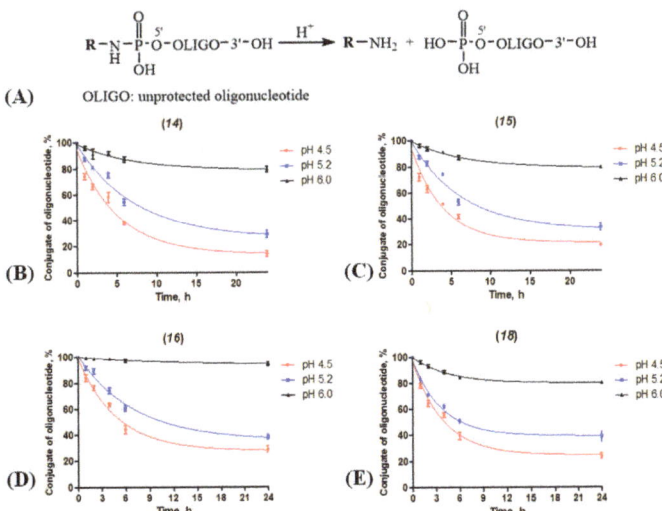

Figure 5. Scheme of hydrolysis (**A**) and kinetic curves of P-N-bond cleavage in conjugates of 5′-p-siDmS (**14–16, 18**, Table 1) with *p*-anisic acid (**B**), cholesterol (**C**), oleyl (**D**) and GalNAc (**E**) at different pH values. Quantification of the full-size conjugate (%, axis Y) in relation to incubation time and pH. The results are presented as mean values (±SD) from three independent experiments. Section 3: Materials and Methods for details.

3. Materials and Methods

3.1. Chemicals and Reagents

A controlled pore glass support (CPG) derivatized with 2′-*O*-methyl-A, 2′-*O*-methyl-G, deoxythymidine, 5′,*N*-protected 2′-*O*-methylribo- (A, C, G, or U), 2′-*O*-TBDMS-ribo (A, C, G, or U) and deoxyribo (dT) phosphoramidites, 2-[2-(4,4′-dimethoxytrityloxy)ethylsulfonyl]ethyl-(2-cyanoethyl)-(*N*,*N*-diisopropyl)-phosphoramidite (CPR, Chemical Phosphorylation Reagent) were purchased from Glen Research Inc. (Sterling, VA, USA). Propargylamine, 1,6-diaminohexane, (pyrene-1-yl-methyl)amine hydrochloride, *p*-anisic (4-methoxybenzoic) acid, *N*,*N*′-dicyclohexylcarbodiimide (DCC), 1-hydroxybenzotriazole hydrate (HOBt), *N*,*N*-diisopropylethylamine (DIPEA), α-GalNac-azide, and 1 M TBAF solution in THF were purchased from Sigma-Aldrich (St. Louis, MO, USA), *N*-Boc-1,6-diaminohexane hydrochloride and ethoxytrimethylsilane were obtained from Alfa Aesar (Heysham, UK); cholesterol chloroformate and oleylamine were obtained from Acros Organics (Geel, Belgium); 3-amino-1-propanol, triphenylphosphine (PPh$_3$), and 2,2′-dipyridyl disulfide ((PyS)$_2$) were obtained from Fluka (St. Louis, MO, USA). FAM-NHS, FAM-azide, 10 mM Cu(II)-TBTA Stock in 55% DMSO, and ascorbic acid were purchased from Lumiprobe (Moscow, Russia). All solvents (THF, DMSO, CH$_3$CN (various vendors)) were dried by 3 Å molecular sieves or by distillation and stored over CaH$_2$. Small molecule ligands were analyzed by thin-layer chromatography (TLC) using DC-Alufolien Kieselgel 60 F$_{254}$ plates (Merck, Darmstadt, Germany) at 254 nm ultraviolet light.

3.2. Physical Measurements

AVANCE III 400 and 300 NMR spectrometers (Bruker Corporation, Billerica, MA, USA) were used to record the ^{1}H-NMR spectra of the small molecule ligands, and CDCl$_3$ was used as the solvent.

A MALDI-TOF Autoflex Speed mass spectrometer (Bruker Corporation, Billerica, MA, USA) or an Agilent G6410A LC-MS/MS Instrument (Agilent Technologies, Santa Clara, CA, USA) was used for the recording of mass spectra.

A NanoDrop 1000 spectrophotometer (Thermo Fisher Scientific, Waltham, MA, USA) was used to measure the oligonucleotide solutions' optical densities.

After analytical gel-electrophoresis, the gels were either stained with a Stains-all dye for qualitative visualization or stained with ethidium bromide and quantified using the E-Box (Vilber, Marne-la-Vallée, France).

3.3. Pr2.3Containing Compounds

Cholesteryl-6-aminohexylcarbamate (**I**) was prepared according to [40]. The yield of (**I**) was 70%, Rf 0.02 (TLC, 10% EtOH in CH_2Cl_2). ^{1}H-NMR (300 MHz, $CDCl_3$, ppm): 2.81 (m, 2H, NH_2CH_2-), 3.13 (t, 2H, -$CONH-CH_2$-), 4.46 (m, 1H, oxycyclohexyl), 5.34 (s, 1H, alkenyl) (Supplementary Table S3).

N-(6-Aminohexyl)-4-methoxybenzamide (**II**)

4-Methoxybenzoic acid (0.3 g, 2.0 mmol) was dissolved in CH_2Cl_2 (12 mL) and simultaneously DCC (0.8 g, 4.0 mmol), previously dissolved in CH_2Cl_2 (10 mL), and HOBt (0.5 g, 4.0 mmol) was added, by analogy with [41]. The reaction was monitored by TLC (5% EtOH in CH_2Cl_2). After 16 h of shaking at room temperature, the reaction mixture was centrifuged and separated from the precipitate. The resulting derivative in solution was added to a solution of mono-N-Boc-protected hexamethylenediamine (1 g, 4.0 mmol) and abs. DIPEA (1 mL) in CH_2Cl_2 (5 mL). After 16 h of shaking at room temperature, the reaction mixture was evaporated, dissolved in 20 mL of CH_2Cl_2, and extracted with water (3 × 20 mL). Anhydrous Na_2SO_4 was used to dry the organic layer, which was then completely evaporated in vacuum to dryness. The substance was separated using column chromatography, then dried using evaporation. The yield was 41%, Rf 0.56. ^{1}H-NMR (400 MHz, $CDCl_3$, ppm): 1.27–1.49 (m, 15H, -CH_2- and -C-(CH_3)$_3$), 1.57 (m, 2H, -CH_2-), 3.09 (dd, 2H, -CH_2-NHBoc), 3.39 (dd, 2H, -CONH-CH_2-), 3.81 (s, 3H, -O-CH_3), 6.89 (m, 2H, -CH-, benzene ring), 7.74 (d, 2H, -CH-, benzene ring) (Supplementary Table S3).

To remove the Boc-protecting group, mono-N-Boc-protected N-(6-aminohexyl)-4-methoxybenzamide was dissolved in 10 mL of CH_2Cl_2 and formic acid (1 mL) was added to the solution, according to [42]. After 2 h of shaking at room temperature, the reaction mixture was evaporated, dissolved in 30 mL of CH_2Cl_2, and washed with 0.1 M NaOH saturated with NaCl (4 × 20 mL). Anhydrous Na_2SO_4 was used to dry the organic layer, which was then completely evaporated in a vacuum until dry. The yield was 83%, Rf 0.02. ^{1}H-NMR (400 MHz, $CDCl_3$, ppm): 1.24–1.51 (m, 6H, -CH_2-), 1.58 (m, 2H, -CH_2-), 2.98 (m, 2H, NH_2CH_2-), 3.41 (m, 2H, -CONH-CH_2-), 3.82 (s, 3H, -O-CH_3), 6.90 (d, 2H, -CH-, benzene ring), 7.71 (d, 2H, -CH-, benzene ring) (Supplementary Table S3).

3.4. Synthesis of Polymer-Bound Oligonucleotides

Oligodeoxyribonucleotides, oligoribonucleotides, and oligo(2′-O-methylribonucleotides) were synthesized, as described in our previous work [40] (Supplementary Experimental Section S1). 5′-Phosphorylation of support-bound oligonucleotides was carried out as a standard automatic phosphoramidite cycle with the use of CPR phosphoramidite (0.1 M in anhydrous CH_3CN); coupling time was 10 min. The subsequent solid-phase conjugation was carried out using polymer-bound DMTr-off oligonucleotides (see Section 3.5).

3.5. Solid-Phase Synthesis of Oligonucleotide Conjugates

BSA/DBU treatment to remove protecting groups from phosphates: removal of protecting groups from the internucleotide and 5′-terminal phosphates was carried out by analogy with [31]. The support-bound oligonucleotide was treated with 400 µL of THF/BSA (1/1, v/v) mixture for 30 min when shaken at room temperature, followed by the addition of 21 µL of DBU and shaking for next 30 min at room temperature. After that, the support was successively washed with THF (3 × 200 µL), CH_3CN (3 × 200 µL), CH_2Cl_2 (3 × 200 µL) and air dried.

Synthesis of monoconjugates (**1–7, 9–12, 14–17**) (see Figure 1): Protecting groups were removed from the internucleotide and 5′-terminal phosphates of oligonucleotides, as described above. Then, Ph$_3$P (7.9 mg, 0.03 mmol), (PyS)$_2$ (6.6 mg, 0.03 mmol), DMAP (5.9 mg, 0.05 mmol), 200 µL abs. DMSO were added to the polymerized oligonucleotide (5–10 mg, 0.15–0.3 µmol) and left to shake for 20 min at 37 °C. The solution was decanted; the support was washed with abs. DMSO (3 × 250 µL). The solution of an amino-containing small molecules (cholesteryl-6-aminohexylcarbamate (**I**) (4.0 mg, 7.5 µmol) in 400 µL of CH$_2$Cl$_2$; 1-pyrenmethylamine hydrochloride (2.0 mg, 7.5 µmol) in 400 µL of DMSO/DIPEA mixture (4/1, *v/v*); *N*-(6-aminohexyl)-4-methoxybenzamide (**II**) (1.9 mg, 7.5 µmol) in 400 µL CH$_2$Cl$_2$; oleylamine (5.0 µL, 15 µmol) in 400 µL CH$_2$Cl$_2$; 1,6-diaminohexane (1.7 mg, 15 µmol) in 400 µL CH$_2$Cl$_2$; propargylamine (1.0 µL, 15 µmol) in 400 µL THF; 3-amino-1-propanol (1.1 µL, 15 µmol) in 400 µL of THF) was added to a 5′-phosphate activated oligonucleotide on a polymer carrier and left stirring for 16 h at 37 °C. At the end of the reaction, the solutions were decanted, and the polymer was washed with THF, CH$_2$Cl$_2$ or DMSO (3 × 300 µL), acetone (2 × 200 µL), and air dried. Next, unblocking, analysis of the reaction mixture by analytical gel electrophoresis and RP-HPLC and isolation by preparative gel electrophoresis were carried out (see Section 3.6).

3.6. Deprotection and Isolation of the Oligonucleotides and Their Conjugates

The oligonucleotide conjugates were cleaved from the support and deprotected by 40% aq.CH$_3$NH$_2$ (300 µL) for 2 h at room temperature, followed by Speedvac concentration. 2′-O-TBDMS groups were removed upon treatment with 1 M TBAF in THF (200 µL) overnight at room temperature, followed by the addition of 1 M TEAAc (pH 7.0) (600 µL), removed THF by Speedvac concentrator, and desalted with C18-cartridge or Amicon Ultra 3K (Millipore, Burlington, MA, USA). Unmodified control oligonucleotides were cleaved from the support and deprotected in the same way. 2′-O-TBDMS groups were removed upon treatment with a mixture of NMP/TEA·3HF/TEA (150/100/75, *v/v/v*) for 1.5 h at 65 °C, followed by treatment with ethoxytrimethylsilane. Deprotected oligonucleotides and their conjugates were isolated by 12% denaturing polyacrylamide gel electrophoresis (PAGE), followed by elution from the gel with 0.3 M NaClO$_4$ solution, desalted with Amicon Ultra 3K, and precipitated as sodium salts. The total yields per the first nucleotide base are shown in Table 1. The purified oligonucleotide conjugates were characterized by RP-HPLC, PAGE and mass spectrometry (Table 1, Supplementary Table S2, Figures S1–S4).

3.7. Synthesis of Biotin Conjugate (8) Using NHS Esters

Biotin derivative (**8**) was obtained according to the NHS protocol of the reagent supplier (Lumiprobe, Moscow, Russia). The solution of the Biotin-NHS (0.5 mg, 1.1 mmol) in DMSO (80 µL) was added to the amino-modified oligonucleotide (**7**) (150 nmol) in 0.5 M Tris-HCl, pH 8.3 (20 µL). The mixture was incubated for 2 h at room temperature, precipitated with 2% solution of NaClO$_4$ in acetone, washed with acetone, and air dried. The reaction mixture was dissolved and analyzed using PAGE and RP-HPLC (Supplementary Figure S2). Isolation was carried out as described above. The total yield per the first nucleotide base is shown in Table 1.

3.8. Synthesis of Conjugates (13, 18) Using Click-Chemistry

Triethylammonium acetate buffer (pH 7.0), 10 mM FAM-azide, or GalNac-azide in DMSO (20 µL), 5 mM ascorbic acid solution in water and 10 mM Cu(II)-TBTA stock in 55% DMSO were added to the water solution of 5′-alkyne-modified oligonucleotides (**12** or **17**) (100 nmol), according to the protocol of the click reagent supplier (Lumiprobe, Moscow, Russia). The reaction mixtures were incubated overnight at room temperature. The oligonucleotide conjugates were precipitated with 2% NaClO$_4$ in acetone and washed with acetone. The pellet was air-dried, dissolved in water, and analyzed by RP-HPLC and/or PAGE (Supplementary Figure S3). The conversion of the oligonucleotide to the conjugate was almost quantitative, according to the PAGE and RP-HPLC (Supplementary Figure S3). Isolation was carried out, as described above. The total yields per the first nucleotide base are shown in Table 1.

3.9. RP-HPLC Analysis of the Oligonucleotide and Their Conjugates

Reversed-phase HPLC analysis of the oligonucleotides and their conjugates was performed on an Alphachrom A-02 high-performance liquid chromatograph (EcoNova, Novosibirsk, Russia) with the use of a ProntoSil-120-5-C18 AQ (75 × 2.0 mm, 5.0 µm) column, applying a gradient elution from 0% to 50% (25 min) of CH_3CN in 0.02 M triethylammonium acetate buffer, pH 7.0, at a flow rate 100 µL per min.

3.10. Stability of the P-N-Bond within the Oligonucleotide Conjugates (14–16, 18) at Different pH Values

Conjugates of 5'-p-siDmS oligonucleotide (14–16, 18) with N-(6-aminohexyl)-4-methoxybenzamide, cholesterol-6-aminohexylcarbamate, oleylamine and GalNAc (27 nmol) were kept in a NaOAc-buffer (0.05M, 90 µL) with pH values 6.0, 5.2, and 4.5 at 37 °C. After 1, 2, 4, 6, and 24 h, 15 µL aliquots were taken, and the oligonucleotides were precipitated with 2% $NaClO_4$ in acetone and washed with acetone. The pellet was air-dried, dissolved in water, analyzed by denatured gel electrophoresis, and stained with ethidium bromide. The resulting electropherograms were digitized and processed using the Quantity One program (BioRad, Hercules, CA, USA). Each experiment was repeated at least three times. The statistical analyses were performed using GraphPad Prism 6.01 (GraphPad Software, San Diego, CA, USA). The outcome variables are expressed as means ± standard deviations (SDs).

4. Conclusions

We proposed a new, convenient solid-phase approach for attaching various transporting small molecules to the 5'-end of an oligonucleotide via the biodegradable, acid-labile phosphoramide bond. The method is simple and efficient and allows for the fine-tuning of the ratio of different solvents for a desired ligand over a wide range. Moreover, the unreacted reaction components can easily be removed at any step by washing since the conjugate is attached to the support during the whole synthesis. The method is based on the activation of the 5'-terminal phosphate of a protected support-bound oligonucleotide, followed by interaction with a small molecule bearing a primary amino group. We demonstrated the advantages of this approach in the synthesis of a series of oligonucleotide 5'-conjugates with different ligands, such as cholesterol, aliphatic amine, N-acetylgalactosamine (GalNAc), and p-anisic acid (anisamide). The obtained conjugates were characterized by HPLC, analytical PAGE, and mass-spectrometry. The effective release of the oligonucleotide from the small molecules was shown under mildly acidic conditions that are close to the pH value within endosomes/lysosomes. Our subsequent studies will include a series of in vitro experiments to examine the influence of the small molecules themselves and also the type of the linker (stable/labile) on the biological activity of functional nucleic acids bearing transport ligands. The developed method is compatible with various types of oligonucleotides (deoxyribo-, 2'-O-methylribo-, ribo- and others) and can be further used for obtaining the conjugates of antisense oligonucleotides, siRNAs, miRNAs, or aptamers with transporting ligands to improve their cell delivery and cargo release inside the cell.

Supplementary Materials: The following supporting information can be downloaded at: https://www.mdpi.com/article/10.3390/molecules28041904/s1, Table S1: The amino ligands used for solid-phase attachment to oligonucleotides and selected optimal solvents for this reaction; Figure S1: Electrophoretic analysis of reaction mixtures upon solid-phase conjugation; Figure S2: Functionalization of the 5'-amino-modified oligonucleotide (7) with Biotin N-hydroxysuccinimide ester; Figure S3: Attachment of FAM or α-GalNAc azides to the 5'-alkyne-modified oligonucleotide (12) or (17) using "click"-chemistry reaction; Table S2: Representative ESI or MALDI-TOF mass spectra of the 5'-conjugates of oligonucleotides; Table S3: ^{1}H-NMR spectra of amino-containing ligands; Figure S4: Full-size images of electropherograms after PAGE analysis and Stains-all staining for 5'-phosphorylated oligonucleotides and their conjugates (1–18); Experimental Section S1: Automated synthesis of polymer-bound oligonucleotides; Table S4: Stability of the P-N-bond within the oligonucleotide conjugates (14–16, 18) at different pH values.

Author Contributions: Conceptualization, M.I.M. and M.A.V.; methodology, M.I.M.; investigation, N.O.K., A.A.G. and M.I.M.; visualization, N.O.K. and M.I.M.; writing—original draft preparation, N.O.K. and M.I.M.; writing—review and editing, M.I.M. and M.A.V. All authors have read and agreed to the published version of the manuscript.

Funding: This research was funded by Russian Scientific Foundation: 19-14-00251.

Institutional Review Board Statement: Not applicable.

Informed Consent Statement: Not applicable.

Data Availability Statement: Not applicable.

Acknowledgments: We are grateful for the support of the Centre of Spectral Investigations at NIOCh SB RAS and the Core Facility for Mass Spectrometry at ICBFM SB RAS for NMR and mass spectra analyses, respectively.

Conflicts of Interest: The authors declare no conflict of interest.

Dedication: Dedicated to the memory of our colleague, Alya G. Venyaminova (ICBFM, Novosibirsk, Russia), an expert in RNA chemistry and applications.

Sample Availability: Samples of the compounds are not available from the authors.

References

1. Laganà, A.; Shasha, D.; Croce, C.M. Synthetic RNAs for Gene Regulation: Design Principles and Computational Tools. *Front. Bioeng. Biotechnol.* **2014**, *2*, 65. [CrossRef] [PubMed]
2. Sridharan, K.; Gogtay, N.J. Therapeutic Nucleic Acids: Current Clinical Status. *Br. J. Clin. Pharmacol.* **2016**, *82*, 659–672. [CrossRef] [PubMed]
3. Panigaj, M.; Johnson, M.B.; Ke, W.; McMillan, J.; Goncharova, E.A.; Chandler, M.; Afonin, K.A. Aptamers as Modular Components of Therapeutic Nucleic Acid Nanotechnology. *ACS Nano* **2019**, *13*, 12301–12321. [CrossRef] [PubMed]
4. Smith, C.I.E.; Zain, R. Therapeutic Oligonucleotides: State of the Art. *Annu. Rev. Pharmacol. Toxicol.* **2019**, *59*, 605–630. [CrossRef]
5. Jani, S.; Ramirez, M.S.; Tolmasky, M.E. Silencing Antibiotic Resistance with Antisense Oligonucleotides. *Biomedicines* **2021**, *9*, 416. [CrossRef]
6. Bajan, S.; Hutvagner, G. RNA-Based Therapeutics: From Antisense Oligonucleotides to MiRNAs. *Cells* **2020**, *9*, 137. [CrossRef]
7. Winkler, J. Oligonucleotide Conjugates for Therapeutic Applications. *Ther. Deliv.* **2013**, *4*, 791–809. [CrossRef]
8. Juliano, R.L. The Delivery of Therapeutic Oligonucleotides. *Nucleic Acids Res.* **2016**, *44*, 6518–6548. [CrossRef]
9. Nakagawa, O.; Ming, X.; Huang, L.; Juliano, R.L. Targeted Intracellular Delivery of Antisense Oligonucleotides via Conjugation with Small-Molecule Ligands. *J. Am. Chem. Soc.* **2010**, *132*, 8848–8849. [CrossRef]
10. Springer, A.D.; Dowdy, S.F. GalNAc-SiRNA Conjugates: Leading the Way for Delivery of RNAi Therapeutics. *Nucleic Acid Ther.* **2018**, *28*, 109–118. [CrossRef]
11. Dong, Y.; Siegwart, D.J.; Anderson, D.G. Strategies, Design, and Chemistry in SiRNA Delivery Systems. *Adv. Drug Deliv. Rev.* **2019**, *144*, 133–147. [CrossRef] [PubMed]
12. Benizri, S.; Gissot, A.; Martin, A.; Vialet, B.; Grinstaff, M.W.; Barthélémy, P. Bioconjugated Oligonucleotides: Recent Developments and Therapeutic Applications. *Bioconjug. Chem.* **2019**, *30*, 366–383. [CrossRef] [PubMed]
13. Hawner, M.; Ducho, C. Cellular Targeting of Oligonucleotides by Conjugation with Small Molecules. *Molecules* **2020**, *25*, 5963. [CrossRef] [PubMed]
14. Zhang, Y.; Sun, C.; Wang, C.; Jankovic, K.E.; Dong, Y. Lipids and Lipid Derivatives for RNA Delivery. *Chem. Rev.* **2021**, *121*, 12181–12277. [CrossRef]
15. Wolff, J.A.; Rozema, D.B. Breaking the Bonds: Non-Viral Vectors Become Chemically Dynamic. *Mol. Ther.* **2008**, *16*, 8–15. [CrossRef]
16. Leriche, G.; Chisholm, L.; Wagner, A. Cleavable Linkers in Chemical Biology. *Bioorg. Med. Chem.* **2012**, *20*, 571–582. [CrossRef]
17. Choy, C.J.; Ley, C.R.; Davis, A.L.; Backer, B.S.; Geruntho, J.J.; Clowers, B.H.; Berkman, C.E. Second-Generation Tunable PH-Sensitive Phosphoramidate-Based Linkers for Controlled Release. *Bioconjug. Chem.* **2016**, *27*, 2206–2213. [CrossRef]
18. Le Corre, S.S.; Berchel, M.; Couthon-Gourvès, H.; Haelters, J.-P.; Jaffrès, P.-A. Atherton–Todd Reaction: Mechanism, Scope and Applications. *Beilstein J. Org. Chem.* **2014**, *10*, 1166–1196. [CrossRef]
19. Vlaho, D.; Fakhoury, J.F.; Damha, M.J. Structural Studies and Gene Silencing Activity of SiRNAs Containing Cationic Phosphoramidate Linkages. *Nucleic Acid Ther.* **2018**, *28*, 34–43. [CrossRef]
20. Cooke, L.A.; Frauendorf, C.; Gîlea, M.A.; Holmes, S.C.; Vyle, J.S. Solid-Phase Synthesis of Terminal Oligonucleotide–Phosphoramidate Conjugates. *Tetrahedron Lett.* **2006**, *47*, 719–722. [CrossRef]

21. Gołębiewska, J.; Rachwalak, M.; Jakubowski, T.; Romanowska, J.; Stawinski, J. Reaction of Boranephosphonate Diesters with Amines in the Presence of Iodine: The Case for the Intermediacy of H-Phosphonate Derivatives. *J. Org. Chem.* **2018**, *83*, 5496–5505. [CrossRef] [PubMed]
22. Kupryushkin, M.S.; Apukhtina, V.S.; Vasilyeva, S.V.; Pyshnyi, D.V.; Stetsenko, D.A. A New Simple and Convenient Method for Preparation of Oligonucleotides Containing a Pyrene or a Cholesterol Moiety. *Russ. Chem. Bull.* **2015**, *64*, 1678–1681. [CrossRef]
23. Derzhalova, A.; Markov, O.; Fokina, A.; Shiohama, Y.; Zatsepin, T.; Fujii, M.; Zenkova, M.; Stetsenko, D. Novel Lipid-Oligonucleotide Conjugates Containing Long-Chain Sulfonyl Phosphoramidate Groups: Synthesis and Biological Properties. *Appl. Sci.* **2021**, *11*, 1174. [CrossRef]
24. Dovydenko, I.S.; Kupryushkin, M.S.; Pyshnyi, D.V.; Apartsin, E.K. A Convenient Solid Phase Approach to Obtain Lipophilic 5′-Phosphoramidate Derivatives of DNA and RNA Oligonucleotides. *Nucleosides Nucleotides Nucleic Acids* **2018**, *37*, 102–111. [CrossRef] [PubMed]
25. Jeong, J.H.; Kim, S.W.; Park, T.G. Novel Intracellular Delivery System of Antisense Oligonucleotide by Self-Assembled Hybrid Micelles Composed of DNA/PEG Conjugate and Cationic Fusogenic Peptide. *Bioconjug. Chem.* **2003**, *14*, 473–479. [CrossRef] [PubMed]
26. Wang, T.-P.; Ko, N.C.; Su, Y.-C.; Wang, E.-C.; Severance, S.; Hwang, C.-C.; Shih, Y.T.; Wu, M.H.; Chen, Y.-H. Advanced Aqueous-Phase Phosphoramidation Reactions for Effectively Synthesizing Peptide–Oligonucleotide Conjugates Trafficked into a Human Cell Line. *Bioconjug. Chem.* **2012**, *23*, 2417–2433. [CrossRef]
27. Mukaiyama, T.; Hashimoto, M. Synthesis of Oligothymidylates and Nucleoside Cyclic Phosphates by Oxidation-Reduction Condensation. *J. Am. Chem. Soc.* **1972**, *94*, 8528–8532. [CrossRef]
28. Zarytova, V.; Ivanova, E.; Venyaminova, A. Design of Functional Diversity in Oligonucleotides via Zwitter-Ionic Derivatives of Deprotected Oligonucleotides. *Nucleosides Nucleotides* **1998**, *17*, 649–662. [CrossRef]
29. Grimm, G.N.; Boutorine, A.S.; Helene, C. Rapid Routes of Synthesis of Oligonucleotide Conjugates from Non-Protected Oligonucleotides and Ligands Possessing Different Nucleophilic or Electrophilic Functional Groups. *Nucleosides Nucleotides Nucleic Acids* **2000**, *19*, 1943–1965. [CrossRef]
30. Novopashina, D.S.; Totskaya, O.S.; Kholodar', S.A.; Meshchaninova, M.I.; Ven'yaminova, A.G. Oligo(2′-O-Methylribonucleotides) and Their Derivatives: III. 5′-Mono- and 5′-Bispyrenyl Derivatives of Oligo(2′-O- Methylribonucleotides) and Their 3′-Modified Analogues: Synthesis and Properties. *Russ. J. Bioorganic Chem.* **2008**, *34*, 602–612. [CrossRef]
31. Krasheninina, O.A.; Novopashina, D.S.; Lomzov, A.A.; Venyaminova, A.G. 2′-Bispyrene-Modified 2′-O-Methyl RNA Probes as Useful Tools for the Detection of RNA: Synthesis, Fluorescent Properties, and Duplex Stability. *ChemBioChem* **2014**, *15*, 1939–1946. [CrossRef] [PubMed]
32. Lönnberg, H. Solid-Phase Synthesis of Oligonucleotide Conjugates Useful for Delivery and Targeting of Potential Nucleic Acid Therapeutics. *Bioconjug. Chem.* **2009**, *20*, 1065–1094. [CrossRef] [PubMed]
33. Cedillo, I.; Chreng, D.; Engle, E.; Chen, L.; McPherson, A.; Rodriguez, A. Synthesis of 5′-GalNAc-Conjugated Oligonucleotides: A Comparison of Solid and Solution-Phase Conjugation Strategies. *Molecules* **2017**, *22*, 1356. [CrossRef] [PubMed]
34. Singh, Y.; Murat, P.; Defrancq, E. Recent Developments in Oligonucleotide Conjugation. *Chem. Soc. Rev.* **2010**, *39*, 2054–2070. [CrossRef]
35. Raouane, M.; Desmaële, D.; Urbinati, G.; Massaad-Massade, L.; Couvreur, P. Lipid Conjugated Oligonucleotides: A Useful Strategy for Delivery. *Bioconjug. Chem.* **2012**, *23*, 1091–1104. [CrossRef]
36. Gooding, M.; Malhotra, M.; Evans, J.C.; Darcy, R.; O'Driscoll, C.M. Oligonucleotide Conjugates—Candidates for Gene Silencing Therapeutics. *Eur. J. Pharm. Biopharm.* **2016**, *107*, 321–340. [CrossRef]
37. Dasargyri, A.; Kümin, C.D.; Leroux, J.-C. Targeting Nanocarriers with Anisamide: Fact or Artifact? *Adv. Mater.* **2017**, *29*, 1603451. [CrossRef]
38. Hayashi, T.; Su, T. Sigma-1 Receptors at Galactosylceramide-Enriched Lipid Microdomains Regulate Oligodendrocyte Differentiation. *Proc. Natl. Acad. Sci. USA* **2004**, *101*, 14949–14954. [CrossRef]
39. Qu, D.; Jiao, M.; Lin, H.; Tian, C.; Qu, G.; Xue, J.; Xue, L.; Ju, C.; Zhang, C. Anisamide-Functionalized PH-Responsive Amphiphilic Chitosan-Based Paclitaxel Micelles for Sigma-1 Receptor Targeted Prostate Cancer Treatment. *Carbohydr. Polym.* **2020**, *229*, 115498. [CrossRef]
40. Meschaninova, M.I.; Novopashina, D.S.; Semikolenova, O.A.; Silnikov, V.N.; Venyaminova, A.G. Novel Convenient Approach to the Solid-Phase Synthesis of Oligonucleotide Conjugates. *Molecules* **2019**, *24*, 4266. [CrossRef]
41. Jain, P.K.; Friedman, S.H. The ULTIMATE Reagent: A Universal Photocleavable and Clickable Reagent for the Regiospecific and Reversible End Labeling of Any Nucleic Acid. *ChemBioChem* **2018**, *19*, 1264–1270. [CrossRef] [PubMed]
42. Coffey, D.S.; McDonald, A.I.; Overman, L.E.; Rabinowitz, M.H.; Renhowe, P.A. A Practical Entry to the Crambescidin Family of Guanidine Alkaloids. Enantioselective Total Syntheses of Ptilomycalin A, Crambescidin 657 and Its Methyl Ester (Neofolitispates 2), and Crambescidin 800. *J. Am. Chem. Soc.* **2000**, *122*, 4893–4903. [CrossRef]
43. Sekine, M.; Tsuruoka, H.; Iimura, S.; Kusuoku, H.; Wada, T.; Furusawa, K. Studies on Steric and Electronic Control of 2′–3′ Phosphoryl Migration in 2′-Phosphorylated Uridine Derivatives and Its Application to the Synthesis of 2′-Phosphorylated Oligouridylates. *J. Org. Chem.* **1996**, *61*, 4087–4100. [CrossRef] [PubMed]

44. Petrova, N.S.; Chernikov, I.V.; Meschaninova, M.I.; Dovydenko, I.S.; Venyaminova, A.G.; Zenkova, M.A.; Vlassov, V.V.; Chernolovskaya, E.L. Carrier-Free Cellular Uptake and the Gene-Silencing Activity of the Lipophilic SiRNAs Is Strongly Affected by the Length of the Linker between SiRNA and Lipophilic Group. *Nucleic Acids Res.* **2012**, *40*, 2330–2344. [CrossRef]
45. Bramsen, J.B.; Laursen, M.B.; Damgaard, C.K.; Lena, S.W.; Babu, B.R.; Wengel, J.; Kjems, J. Improved Silencing Properties Using Small Internally Segmented Interfering RNAs. *Nucleic Acids Res.* **2007**, *35*, 5886–5897. [CrossRef] [PubMed]
46. Meschaninova, M.I.; Entelis, N.S.; Chernolovskaya, E.L.; Venyaminova, A.G. A Versatile Solid-Phase Approach to the Synthesis of Oligonucleotide Conjugates with Biodegradable Hydrazone Linker. *Molecules* **2021**, *26*, 2119. [CrossRef]
47. Gilleron, J.; Querbes, W.; Zeigerer, A.; Borodovsky, A.; Marsico, G.; Schubert, U.; Manygoats, K.; Seifert, S.; Andree, C.; Stöter, M.; et al. Image-Based Analysis of Lipid Nanoparticle–Mediated SiRNA Delivery, Intracellular Trafficking and Endosomal Escape. *Nat. Biotechnol.* **2013**, *31*, 638–646. [CrossRef]

Disclaimer/Publisher's Note: The statements, opinions and data contained in all publications are solely those of the individual author(s) and contributor(s) and not of MDPI and/or the editor(s). MDPI and/or the editor(s) disclaim responsibility for any injury to people or property resulting from any ideas, methods, instructions or products referred to in the content.

Article

DNA Base Excision Repair Intermediates Influence Duplex–Quadruplex Equilibrium

Mark L. Sowers [1,2], James W. Conrad [1], Bruce Chang-Gu [1,2], Ellie Cherryhomes [1], Linda C. Hackfeld [1] and Lawrence C. Sowers [1,3,*]

1. Department of Pharmacology and Toxicology, University of Texas Medical Branch, 301 University Boulevard, Galveston, TX 77555, USA
2. MD-PhD Combined Degree Program, University of Texas Medical Branch, 301 University Boulevard, Galveston, TX 77555, USA
3. Department of Internal Medicine, University of Texas Medical Branch, 301 University Boulevard, Galveston, TX 77555, USA
* Correspondence: lasowers@utmb.edu; Tel.: +1-409-772-9678

Citation: Sowers, M.L.; Conrad, J.W.; Chang-Gu, B.; Cherryhomes, E.; Hackfeld, L.C.; Sowers, L.C. DNA Base Excision Repair Intermediates Influence Duplex–Quadruplex Equilibrium. *Molecules* 2023, *28*, 970. https://doi.org/10.3390/molecules28030970

Academic Editors: Ramon Eritja, Daniela Montesarchio and Montserrat Terrazas

Received: 30 December 2022
Revised: 12 January 2023
Accepted: 13 January 2023
Published: 18 January 2023

Copyright: © 2023 by the authors. Licensee MDPI, Basel, Switzerland. This article is an open access article distributed under the terms and conditions of the Creative Commons Attribution (CC BY) license (https://creativecommons.org/licenses/by/4.0/).

Abstract: Although genomic DNA is predominantly duplex under physiological conditions, particular sequence motifs can favor the formation of alternative secondary structures, including the G-quadruplex. These structures can exist within gene promoters, telomeric DNA, and regions of the genome frequently found altered in human cancers. DNA is also subject to hydrolytic and oxidative damage, and its local structure can influence the type of damage and its magnitude. Although the repair of endogenous DNA damage by the base excision repair (BER) pathway has been extensively studied in duplex DNA, substantially less is known about repair in non-duplex DNA structures. Therefore, we wanted to better understand the effect of DNA damage and repair on quadruplex structure. We first examined the effect of placing pyrimidine damage products uracil, 5-hydroxymethyluracil, the chemotherapy agent 5-fluorouracil, and an abasic site into the loop region of a 22-base telomeric repeat sequence known to form a G-quadruplex. Quadruplex formation was unaffected by these analogs. However, the activity of the BER enzymes were negatively impacted. Uracil DNA glycosylase (UDG) and single-strand selective monofunctional uracil DNA glycosylase (SMUG1) were inhibited, and apurinic/apyrimidinic endonuclease 1 (APE1) activity was completely blocked. Interestingly, when we performed studies placing DNA repair intermediates into the strand opposite the quadruplex, we found that they destabilized the duplex and promoted quadruplex formation. We propose that while duplex is the preferred configuration, there is kinetic conversion between duplex and quadruplex. This is supported by our studies using a quadruplex stabilizing molecule, pyridostatin, that is able to promote quadruplex formation starting from duplex DNA. Our results suggest how DNA damage and repair intermediates can alter duplex-quadruplex equilibrium.

Keywords: DNA quadruplex; duplex-quadruplex equilibrium; base excision repair; pyridostatin; glycosylase; telomere

1. Introduction

DNA constantly undergoes persistent damage and repair. If this damage goes unrepaired, it can result in genomic instability and mutations that predispose towards the development of cancer. The primary DNA repair pathway for repairing endogenous deamination and oxidative damage is the base excision repair (BER) pathway [1,2]. Multiple groups have made substantial efforts to explore mechanisms of DNA repair. However, most of these studies have been conducted with canonical duplex DNA and not with the unusual secondary structures also found in our genome.

Substantially less is known about non-canonical DNA secondary structures and their potential roles. In DNA, specific sequence motifs are known to adopt non-duplex structures, including Z-DNA, triplexes, quadruplexes, hairpins, and cruciforms [3,4]. Deviating from normal Watson–Crick base pairing, and forming Hoogsteen pairing, protonated bases, or some combination of the two, allows for the formation of various alternative secondary structures. Sequence motifs known to form these different structures are found throughout the genome, including promoter regions and the telomeric ends of chromosomes. The biological functions of unusual DNA structures are incompletely understood but are thought to be involved in cellular processes, including DNA replication, recombination, and gene expression [5].

Among the possible non-canonical structures, the G-quadruplex has been studied in the greatest detail. Some investigators have estimated that there may be as many as ~700,000 quadruplex-forming regions in the human genome [6]. Many are present in the promoter regions of oncogenes and in telomeric repeats that are aberrantly extended in most human cancers. Under appropriate salt conditions, quadruplexes can form from DNA sequences containing four or more runs of three guanines (G), each separated by a few nucleotides (loops). Notably, various quadruplex configurations have been reported depending on the sequence and type of cationic salt conditions, including antiparallel, parallel, hybrid, and others. Furthermore, at different time scales, interconversions between different possible configurations can also occur [7–9]. Consequently, quadruplex structures can be dynamic and undergo folding, unfolding, and adopt different configurations.

How DNA damage might influence quadruplex structure, or how its unique structure impacts BER is complex. The formation of alternative secondary structures, including the quadruplex, potentially interfering with DNA repair has been discussed previously [3,10,11]. Importantly, nearly all studies of DNA damage in quadruplex DNA have only focused on the impacts of oxidative damage to guanine [12–17]. Guanine is typically directly involved in the secondary structure of the quadruplex. Depending on the position of the oxidized guanine, quadruplex folding can be influenced [14,16,17]. Additionally, the type of guanine oxidation product, its position, and which DNA glycosylase is present, can all influence BER in quadruplex DNA.

These previous studies have revealed how some DNA glycosylases can or cannot act on guanine oxidation damage products but have overlooked two additional aspects of DNA damage and repair in quadruplexes. Firstly, it is valuable to examine DNA damage other than guanine oxidation. DNA bases other than guanine can also undergo endogenous DNA damage, including hydrolytic deamination and oxidation [18–20]. In order to investigate the role of pyrimidine DNA damage on quadruplex formation and subsequent repair, we inserted uracil (U), 5-hydroxymethyluracil (5hmU), and 5-fluorouracil (5FU) into the loop regions, in the place of thymine (T), of the human telomere sequence A(GGGTTA)$_3$GGG (Tel22) [21]. The pyrimidine analogs examined herein represent hydrolytic deamination of cytosine (C) to U, oxidation of T to 5hmU, and the incorporation of the chemotherapy agent 5FU into DNA (Figure 1).

The second unexplored factor is that previous investigations of DNA quadruplexes have been conducted primarily with the G-rich strand alone. While this ensures quadruplex formation, under most circumstances the complementary strand is present in vivo. When the complementary strand is present, an equilibrium is established between the duplex and quadruplex, with the duplex being the predominate configuration [22–24]. To date, few studies have examined how DNA damage and repair might influence the equilibrium that exists between duplex and quadruplex structures.

To address these two key factors, we first examined the influence of these analogs on the capacity of Tel22 to form a quadruplex structure using circular dichroism (CD) and fluorescence resonance energy transfer (FRET). We then probed the quadruplex structures with glycosylases and an AP endonuclease, which comprise the first two steps of BER, to determine how such structures influence DNA repair. Finally, we examined how the BER intermediates alter duplex–quadruplex equilibrium.

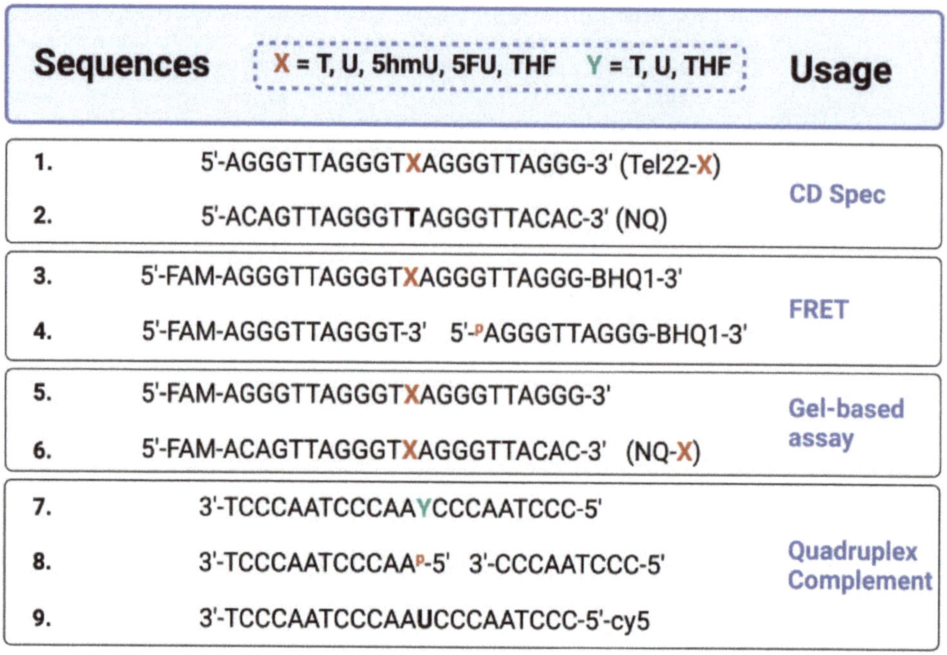

Sequences	X = T, U, 5hmU, 5FU, THF Y = T, U, THF	Usage
1.	5'-AGGGTTAGGGTXAGGGTTAGGG-3' (Tel22-X)	CD Spec
2.	5'-ACAGTTAGGGTTAGGGTTACAC-3' (NQ)	
3.	5'-FAM-AGGGTTAGGGTXAGGGTTAGGG-BHQ1-3'	FRET
4.	5'-FAM-AGGGTTAGGGT-3' 5'-ᵖAGGGTTAGGG-BHQ1-3'	
5.	5'-FAM-AGGGTTAGGGTXAGGGTTAGGG-3'	Gel-based assay
6.	5'-FAM-ACAGTTAGGGTXAGGGTTACAC-3' (NQ-X)	
7.	3'-TCCCAATCCCAAYCCCAATCCC-5'	Quadruplex Complement
8.	3'-TCCCAATCCCAAᴾ-5' 3'-CCCAATCCC-5'	
9.	3'-TCCCAATCCCAAUCCCAATCCC-5'-cy5	

Figure 1. Sequences of oligonucleotides used in this study. Synthetic oligonucleotides of the telomere repeat sequence (Tel22), known to form a quadruplex structure, were prepared with various modifications. CD studies were performed on oligos without a fluorophore or quencher [Oligos 1–2]. FRET studies used a 6-carboxyfluorescein (FAM) and a 3′-non-fluorescent quencher (BHQ1) [Oligos 3–4], containing thymine (T) or its analogs uracil (U), 5-hydroxymethyluracil (5hmU), 5-fluorouracil (5FU), and a stable synthetic abasic site (THF). For gel-based studies, FAM-only oligos were used [Oligos 5–6]. A non-quadruplex (NQ) sequence of the same length [Oligos 2 and 6] was also prepared. In certain experiments, the quadruplex strand was annealed to a complementary strand [Oligos 7–9] to form a duplex where 'Y' is either T, U, THF, or two shorter oligos that simulate a one-base gap. Created with Biorender.com.

2. Results

A series of oligonucleotides were synthesized for this study and the sequences and modifications are shown in Figure 1. Non-fluorescent quadruplex-forming oligonucleotides were prepared for CD spectroscopy studies as well as complementary oligonucleotides to form a duplex with the Tel22 quadruplex sequence. We used 5′-6-carboxyfluorescein (FAM) fluorophore-labeled oligonucleotides for gel-based studies. For FRET studies, we added a non-fluorescent quencher, 3′-BHQ1, in addition to the 5′-FAM fluorophore.

2.1. Pyrimidine Analogs in Loops Do Not Affect Quadruplex Formation

CD spectra were acquired for oligonucleotides containing the wild-type 22-mer telomeric repeat sequence (Tel22-T) or those containing a U, 5hmU, 5FU, or a stable abasic site, THF, (Tel22-X) as shown in Figure 1. Unless otherwise indicated, a physiologic-like buffer containing 150 mM KCl and 15 mM NaCl in 20 mM Tris buffer at pH 7.4 was used, and data were acquired at 37 °C (Figure 2). As a negative control, we used a non-quadruplex (NQ) forming 22-mer (Figure 1). Upon quadruplex formation, a positive band was seen near 295 nm, in accordance with a previous study demonstrating this telomeric repeat sequence forms a hybrid type quadruplex in the presence of K$^+$ [21]. Spectra of the quadruplexes substituted with U, 5hmU, 5FU, and a tetrahydrofuran (THF) abasic site were nearly indistinguishable. The quadruplex containing a 5hmU substitution (red) showed a lower band intensity at 295 nm and increased band intensity at 250 nm. This could potentially be attributed to both intra- and intermolecular hydrogen bonding of 5hmU with the N7 of an adjacent guanine [25].

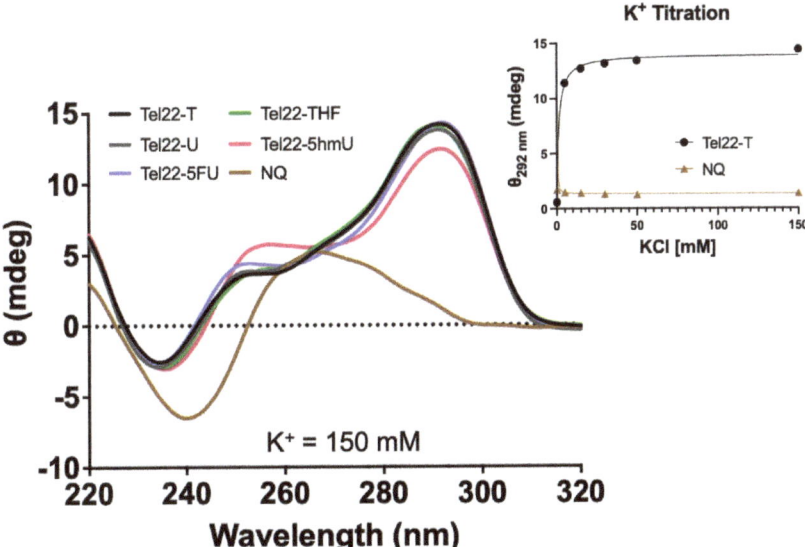

Figure 2. Pyrimidine analogs in the loop region did not disrupt G-quadruplex formation. CD spectra were acquired for quadruplex Tel22-X oligonucleotides or a non-quadruplex-forming (NQ) oligonucleotide. A 4 µM solution was prepared in 20 mM Tris buffer, pH 7.4, 150 mM KCl, and 15 mM NaCl and CD spectra were obtained from 320 to 220 nm at 37 °C. The spectra here were consistent with the formation of a hybrid-type G-quadruplex [21]. The 5hmU-containing oligonucleotide appeared slightly different from the others likely due to both intra- and intermolecular hydrogen bonding of 5hmU and N7 of adjacent guanines [25]. Tel22-T and the NQ were titrated from 0 to 150 mM KCl in a 20 mM Tris buffer at pH 7.4. Using a one-site specific binding model in the PRISM software, we estimated K_d of 1.2 ± 0.2 mM for quadruplex binding while no changes in CD spectra were seen with the NQ oligonucleotide. Titration experiments were done in triplicate, error bars of S.D. were smaller than the data points. Oligonucleotides were equilibrated in buffer at RT for a minimum of 30 min.

The CD spectra confirmed the formation of a quadruplex and that pyrimidine substitutions in the central loop did not interfere or prevent quadruplex formation. Quadruplex formation is highly dependent upon the environmental conditions. In particular, the presence of Na^+ or K^+ can greatly influence quadruplex formation and the type of quadruplex structure. We obtained the CD spectrum of the Tel22-T oligonucleotide as a function of KCl concentration (Figure 2). The approximate K_d was 1.2 ± 0.2 mM and complete quadruplex formation was seen by 30 mM KCl while the NQ sequence showed no peak at 295 nm or any changes as a function of K^+.

2.2. The Activity of DNA Glycosylases and AP Endonuclease 1 (APE1) Are Impaired by Quadruplex Structure

Because the pyrimidine modifications U, 5hmU, and 5FU did not inhibit quadruplex formation, we then sought to understand how these structures would be substrates for DNA repair enzymes. In the studies shown in Figure 3, the targets for repair were placed into the quadruplex-forming strand and incubated with either DNA glycosylases or APE1, and enzymatic activity was quantified by gel [26]. When the quadruplex-forming oligonucleotide containing a U was paired with a complementary strand (U:A), the U was rapidly excised by UDG at a rate similar to the repair of U from a non-quad-forming single strand.

However, when in a quadruplex, U excision by UDG was significantly inhibited (Figure 3A). UDG can also excise 5FU (Figure 3B), albeit more slowly. As with U repair, removal of 5FU from a quadruplex was slower than in duplex or single-stranded DNA.

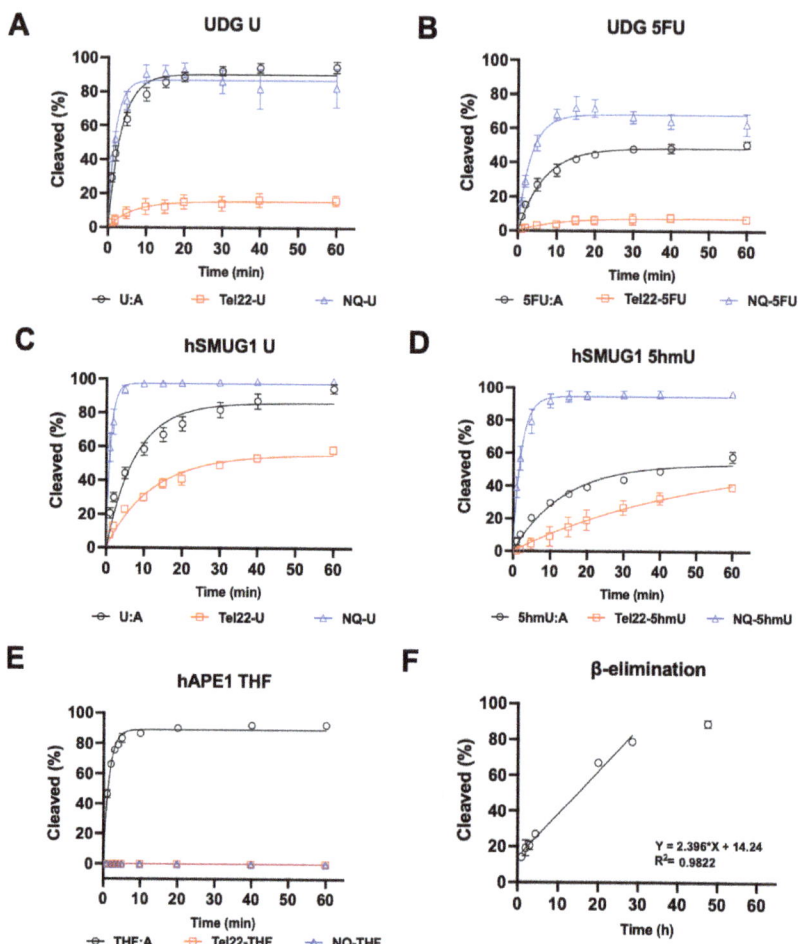

Figure 3. Quadruplex inhibited DNA glycosylase and APE1 activity. UDG, hSMUG1, and APE1 activity on quadruplex oligonucleotides (Tel22-X) containing U, 5hmU, 5FU, or THF was compared to duplex, or single-stranded DNA using the NQ-X oligonucleotides. (**A**) UDG (0.613 nM, 300:1 DNA to enzyme ratio) was fastest on single-stranded U (NQ-U) and duplex (U:A) followed by quadruplex (Tel22-U). (**B**) UDG (6.13 nM, 30:1 DNA to enzyme ratio) was overall less efficient on 5FU but the trend was the same, NQ-5FU > 5FU:A > Tel22-5FU. (**C,D**) hSMUG1 (6.2 nM, 30:1 DNA:enzyme ratio) followed a similar trend with either U or 5hmU. (**E**) Under the conditions tested, APE1 (1.31 nM, 140:1 DNA:enzyme ratio) was only active on duplex (THF:A) and no activity on Tel22-THF or NQ-THF was observed. (**F**) Because of the poor activity of APE1 on THF in a quadruplex, we wanted to estimate the rate of spontaneous β-elimination. We generated an abasic site in situ by using an excess of UDG to remove U from Tel22-U containing oligonucleotide. We then continued incubating the quadruplex containing abasic site for up to 50 h at 37 °C and measured spontaneous β-elimination over time using gel electrophoresis. The rate of elimination was determined by a linear fit to be 0.024 h^{-1} with a half-life of 21 h. Gels scans are shown in Figures S1–S16 Supplementary material.

We then examined hSMUG1 activity under similar conditions but reduced the K$^+$ concentration from 150 mM to 50 mM for optimal enzymatic activity [27]. hSMUG1 can excise U and 5hmU (Figure 3C,D). The trend in repair rates for U and for 5hmU was single-strand > duplex > quadruplex. hSMUG1 was not as inhibited as UDG by the quadruplex structure. Our results are in accord with previous studies that showed quadruplex formation inhibiting glycosylase excision [13,15,28].

Glycosylase excision of a target base is the initiating event for BER. The resulting abasic site is then cleaved by a lyase or AP endonuclease. In this experiment, an abasic site was simulated with the THF-abasic site. Substrates were incubated with repair enzymes in buffer at 37 °C, and the resulting fluorescently tagged oligonucleotides were separated by gel electrophoresis and quantified. When in a duplex (Figure 3E), the abasic site was cleaved by APE1. However, when an abasic site was in a quadruplex or single-stranded context, APE1 could not cleave its substrate or was greatly inhibited. This is consistent with previous reports of APE1 inhibition by quadruplex formation [29,30].

The inability of APE1 to cleave an abasic site when not in a duplex would halt the BER pathway. Furthermore, abasic sites are labile to attack by water resulting in β-elimination and strand cleavage, which could be cytotoxic. Here, we generated an abasic site in a U-containing quadruplex with UDG and then incubated the oligonucleotide at 37 °C for several days. Aliquots were taken as a function of time, and oligonucleotides were separated by gel electrophoresis. The apparent half-life for the cleavage of an abasic site in a quadruplex structure was approximately 21 h (Figure 3F). Previously, the rate of abasic site cleavage in duplex DNA was reported to be 190 h at pH 7.4 at 37 °C in a buffer containing Mg^{2+} [31].

2.3. Quadruplex–Duplex Equilibrium Can Be Monitored by FRET

Quadruplex formation can also be monitored by FRET as described previously [32]. In this system, a fluorophore (FAM) was placed on the 5′-end of the oligonucleotide, and a non-fluorescent quencher (BHQ1) was placed at the 3′-end. When paired with a complementary sequence, the quadruplex-forming strand formed a duplex, separating the fluorophore and quencher. This yielded a maximum fluorescence intensity as seen for the wild-type Tel22 sequence (Figure 4, Tel22-T:A). However, in the absence of the complementary strand, the Tel22 sequence may fold into a quadruplex structure. This brings the fluorophore and quencher in proximity, resulting in a minimum fluorescent intensity (Figure 4). Neither the pyrimidine analogs nor the abasic site (Tel22-THF) oligonucleotides altered quadruplex formation. Therefore, quadruplex formation was not affected by these modifications when placed in the loop regions.

The overlaid fluorescence spectra of the Tel22-X sequences alone and annealed to its complement (i.e., duplex) are shown in Figure S17, respectively. The presence of the complementary strand mostly inhibited quadruplex formation regardless of the pyrimidine analog. We also examined how a gap in the Tel22 sequence (Figure 1, Oligo 4) would affect quadruplex formation as the presence of an abasic site BER intermediate surprisingly did not inhibit quadruplex formation (Figures 2 and 4, Tel22-THF). However, formation of a gap generates two fragments of the quadruplex where the FAM and BHQ1 ends are now separate strands. As expected, the quadruplex fragments (Oligo 4) abolished quadruplex formation and a similar fluorescence intensity to the duplex oligonucleotides was observed (Quadruplex Fragments, Figure S17).

Upon addition of the complementary strand, unwinding of the quadruplex with formation of a duplex occurred with all systems studied here; however, the transition was not instantaneous. Fluorescence intensity as a function of time was monitored with a qPCR instrument. Experimental fluorescence intensity from the Tel22-T as it unfolded and formed a duplex with its complement (T:A) is shown in the inset of Figure 4. The data from the three separate reactions were best fit by a double exponential equation as shown (red). The initial half-life was 1.9 ± 0.5 min while the second half-life was 33 ± 11 min. We propose that the quadruplex and complement initially bind at one end in a fast step and the remaining quadruplex is then unzipped, forming the duplex in a slow step. Our study is in accord with previous studies, which show that quadruplex–duplex interconversion can be slow and may occur through one or more stable intermediates [33,34]. The significance of this finding is that the kinetics of interconversion might become important when trying to interpret the biological consequences of quadruplex-forming sequences [35].

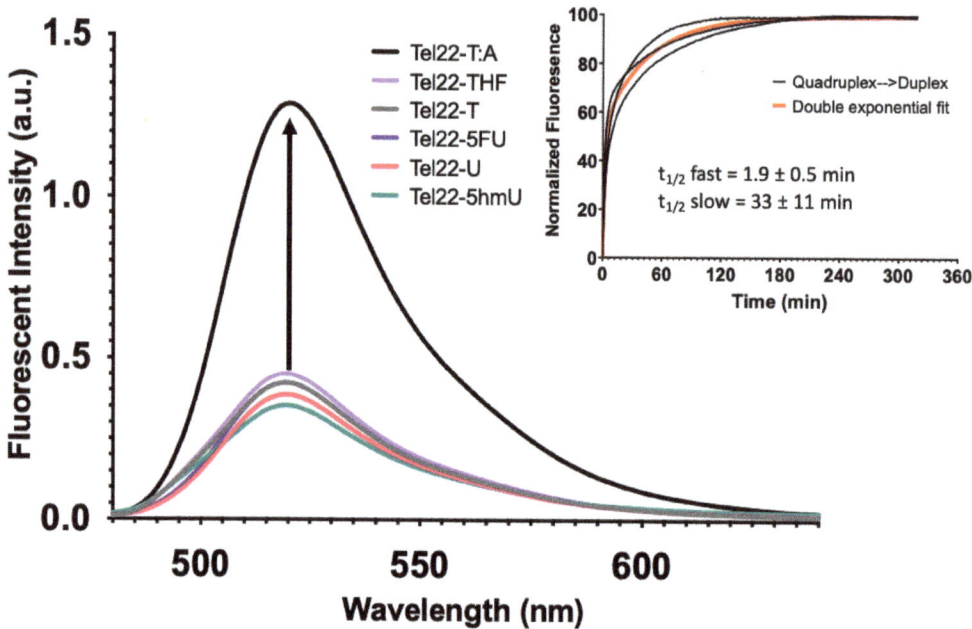

Figure 4. G-quadruplex formation quenched fluorescence and the addition of a complementary strand promoted duplex formation. While quadruplexes are highly stable secondary DNA structures, in the presence of the complementary strand, they are in equilibrium with the corresponding duplex. A FRET strategy was employed to monitor quadruplex formation at 37 °C. Relatively low fluorescent excitation was observed when oligonucleotides formed a G-quadruplex as the fluorophore (FAM) and quencher (BHQ1) were in close contact. However, fluorescence increased when the quadruplex-containing sequence was annealed to its complementary strand to form a duplex (Inset). We then performed a time course experiment to examine the kinetics of quadruplex unfolding as it transitioned to a duplex. Black curves represent three independent experiments and the exponential fit in red. A single exponential fit poorly; therefore, we applied a double-exponential fit, which suggested that there was a relatively fast phase of quadruplex unfolding to one or more intermediate structures, followed by a slower complete unfolding to duplex. The first half-life was 1.9 ± 0.5 min and the second was 33 ± 11 min. The equation for the curve was $y = A_1\left(1 - e^{-k_1 t}\right) + A_2\left(1 - e^{-k_2 t}\right)$ where k_1 and k_2 are 0.37 min^{-1} and 0.02 min^{-1}, respectively. The first and second amplitudes were estimated to be 54.5 and 99.3, respectively.

2.4. Base Excision Repair and Quadruplex Stabilizing Ligands Shift the Duplex–Quadruplex Equilibrium towards Quadruplex

We then examined the fluorescence intensity of oligonucleotide complexes that recapitulate the steps in BER (Figure 5A). The fluorescence spectra for the Tel22-T quadruplex annealed to a complement containing uracil (A:U) were nearly identical to the wild-type duplex from Figure 4 (Tel22-T:A). When the U in the complementary strand was replaced with a THF abasic site, the fluorescence intensity at equilibrium decreased (magenta), reflecting a shift in the duplex–quadruplex equilibrium toward the quadruplex. A one-base gap was simulated by adding two oligonucleotides complementary to the 5′-side and the 3′-side of the quadruplex-forming sequence (Figure 1, Oligo 8). The fluorescence intensity dropped further (green) to a level approximately midway between the duplex and quadruplex. We sought to further examine the duplex–quadruplex equilibrium by non-denaturing gel electrophoresis (Figure 5B). The fluorescently labeled A:T, A:U, and A:THF duplexes were observed as a single band with the same relative gel migration. The quadruplex, in the absence of a complementary strand, was observed as a single band with lower gel migration. However, the quadruplex sequence with a one-base gap in the complementary strand did not show resolution towards either a single-stranded quadruplex or duplex (Figure 5B). The fluorescence data as well as the gel data suggest this system is best described as an ensemble of interconverting structures, including both duplex and quadruplex configurations, in approximately equal proportions.

To further explore this hypothesis, we used a quadruplex-stabilizing small molecule, pyridostatin, to examine if the equilibrium could be shifted from duplex towards quadruplex. We titrated A:U, A:THF, and A:Gap oligonucleotides with increasing amounts of pyridostatin and monitored changes in fluorescence intensity using a qPCR instrument (Figure 5C). In the absence of pyridostatin, the A:U and A:THF oligonucleotides had relatively high fluorescence while the A:Gap oligonucleotide had roughly ~50% of the fluorescence intensity. Surprisingly, 5–10 µM of pyridostatin was sufficient to drop the fluorescent intensity to that of the A:Gap oligonucleotide. With all the oligonucleotides examined, increasing amounts of pyridostatin further quenched the fluorescence. This suggests that duplex and quadruplex may interconvert and pyridostatin traps the oligonucleotide in the quadruplex state. In all cases, 50 µM of pyridostatin drove all complexes to quadruplex. These results were consistent with previous reports that demonstrated a small molecule could promote quadruplex formation in the presence of its complementary C-rich strand [36].

In the study shown in Figure 5, repair intermediates were generated by combining various oligonucleotides and allowing them to form duplexes prior to the fluorescence measurements. Having examined the fluorescence intensity changes associated with simulated BER intermediates, we sought to determine if similar results would be obtained in a reconstituted DNA repair system (Figure 6). In this study, we measured the fluorescence spectra of the same A:U duplex (Figure 6, black). Upon incubation with uracil DNA glycosylase (UDG), an abasic site was formed and fluorescence dropped (Figure 6, magenta). The simultaneous addition of UDG and APE1 generated a repair gap with a further drop in fluorescence (Figure 6, green). These data revealed that the fluorescence changes for the simulated and enzyme-generated intermediates were essentially the same, and therefore measuring fluorescence could be used to monitor enzymatic DNA repair reactions. Our earlier studies showed that some time was required for these systems to reach equilibrium (Figure 4, Inset). We therefore measured the change in fluorescence intensity of this system as a function of time following the addition of UDG and APE1. Data from three independent experiments were averaged (Figure 6, Inset). The apparent half-life for reaching equilibrium was approximately 3.2 ± 0.4 min. It appears that as a gap was formed in the complementary strand (Figure S18), simultaneously the quadruplex-forming strand underwent changes in its configuration such that it brought the 5' and 3' ends closer into contact. Regardless, the apparent rate at which a quadruplex-like structure began to form from a damaged duplex was substantially faster than the rate of quadruplex unfolding during duplex formation (Figure 4, Inset).

Previous studies on the biological consequences of quadruplex-forming sequences have examined the systems once equilibrium has been achieved. Our results suggest that achieving structural equilibrium in such systems may be slower than biochemical reactions including base excision and AP endonuclease cleavage. Therefore, inferring biological consequences of quadruplex structures when at equilibrium might need to be reconsidered when such structures are generated as part of a biochemical pathway.

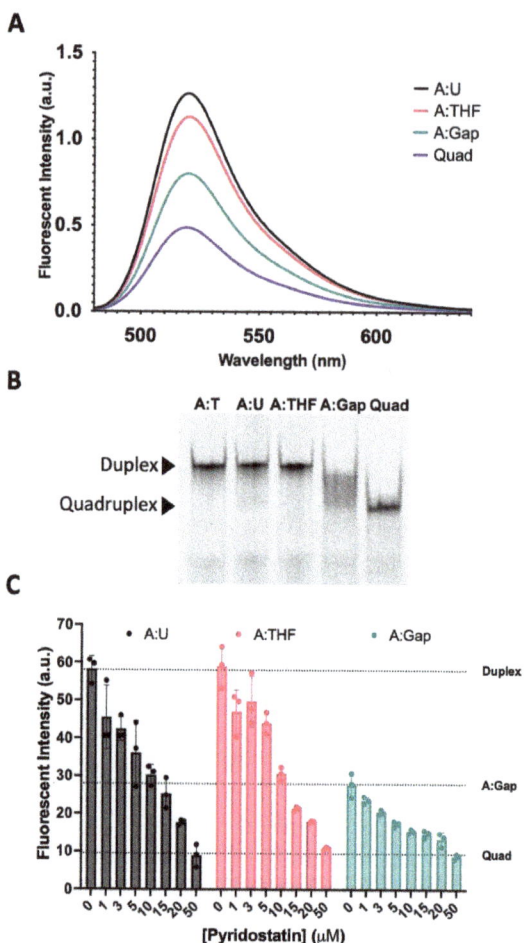

Figure 5. Reconstitution of repair intermediates promoted quadruplex formation. (**A**) Annealing Tel22-T with a complementary strand containing U (A:U) promoted duplex formation as indicated by an increase in fluorescence relative to the quadruplex-only control (Quad). When a complementary strand containing a stable abasic site (A:THF) was used instead, fluorescence was reduced relative to the A:U duplex, suggesting the destabilization of the duplex and promoting quadruplex. Interestingly, generating a gap in the complementary strand (A:Gap) further destabilized the duplex and shifted the equilibrium towards quadruplex. (**B**) Using a 20% native polyacrylamide gel, we demonstrated that A:U and A:THF were duplex as they migrated the same as the A:T control (Lanes 1–3). On the other hand, a gapped complement (A:Gap) showed some intermediate between duplex and quadruplex (Lane 4). Lane 5 was the quadruplex-only control. (**C**) We saw a concentration-dependent decrease in fluorescence when we titrated oligonucleotides that could form a duplex with a quadruplex-stabilizing small molecule, pyridostatin. Consistent with A and B, the A:Gap oligonucleotide had ~50% of the fluorescence as either A:U or A:THF. Oligonucleotides were prepared as a 1 µM solution in 20 mM Tris, pH 7.4, 150 mM KCl, and 15 mM NaCl and equilibrated in buffer for a minimum of 30 min. 1.2 equivalents of the corresponding complementary strands were annealed at 90 °C for 5 min and cooled at RT. Gel samples were prepared identically. Pyridostatin was allowed to equilibrate with oligonucleotides for 30 min at 37 °C and fluorescence emission of FAM was acquired using a qPCR instrument.

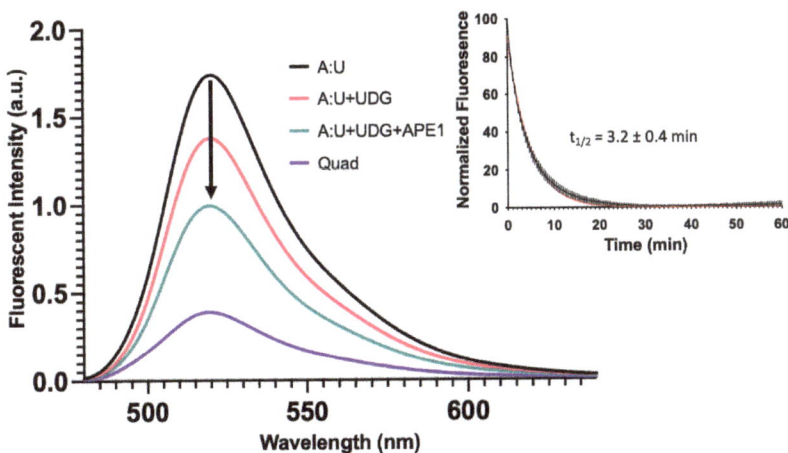

Figure 6. Base excision repair promotes quadruplex formation. In Figure 5, we demonstrated with synthetic oligonucleotides that the presence of a complementary strand containing an abasic site or gap destabilized the duplex and promoted quadruplex formation. Here, we simulated the same process but enzymatically prepared the abasic site and gapped DNA duplex using UDG or UDG and APE1, respectively. The fluorescent emission spectrum was taken for the quadruplex-only strand, duplex oligonucleotide containing a U in the C-rich strand (A:U), duplex [1 μM] incubated with 10 U of UDG (3.39 pmol, 34 nM), and duplex incubated with 10 U of UDG (3.39 pmol, 34 nM) and 20 U of APE1 (0.71 pmol, 7.1 nM) for 2 h at 37 °C in 100 μL total volume. Following UDG treatment, fluorescence decreased ~25% followed by a ~50% decrease in fluorescence with UDG and APE1. This suggested that the equilibrium between duplex and quadruplex could shift following DNA repair. We then wanted to estimate the time scale for secondary structure changes in the quadruplex strand, as a gap formed in the opposing strand. The duplex A:U containing oligonucleotide [1 μM] was treated with 1.25 U of UDG (0.42 pmol, 16.8 nM) and 2.5 U of APE1 (0.09 pmol, 3.6 nM) and fluorescence was monitored over time at 37 °C in a qPCR instrument in 25 μL total volume. As a gap was introduced into the opposing strand, fluorescence of the quadruplex strand decreased with time. This suggested that as the gap was introduced, the quadruplex-forming strand underwent relatively fast changes in configuration that brought the FAM and BHQ1 quencher closer together. The half-life was estimated to be 3.2 ± 0.4 min using a single exponential decay (Inset). This contrasted with the much slower unfolding of the quadruplex (Figure 4, Inset). The black curve represents the average of three replicates with the vertical lines as the S.D. The red curve shows a single exponential fit. In an identical experiment where the complementary oligonucleotide containing U was labeled with Cy5, we monitored the gap formation by gel (Figure S18).

3. Discussion

3.1. DNA Damage and Repair in Quadruplex DNA

The DNA of all organisms is persistently damaged by endogenous reactive molecules as well as exogenous agents [18]. Most of the single-base lesions can be repaired by the BER pathway. Previous BER studies have focused primarily upon the repair of normal duplex DNA and not quadruplex structures, which can affect DNA repair. Furthermore, the studies that have examined base lesions and their repair in DNA quadruplexes have primarily only examined the oxidative damage and repair of guanine adducts [12–17]. In addition to DNA damage of guanine, other forms of DNA damage can and do occur, such as hydrolytic deamination of C to U, oxidation of T to 5hmU, abasic sites, and strand breaks. These forms of DNA damage and their repair have been understudied in the context of DNA quadruplexes.

Pyrimidine analogs can also be misincorporated by DNA polymerases, in place of T, under physiological conditions. This is exploited by chemotherapy drugs including methotrexate, which impairs folate metabolism. This compromises thymidylate synthase resulting in increased formation of dUTP, which can compete with dTTP during DNA replication, placing U into DNA [18,37]. Similarly, pyrimidine analogs 5hmU and 5FU can be converted to the corresponding 5'-triphosphates and compete for incorporation into DNA during replication [38–43]. The cytotoxicity resulting from the presence of U, 5hmU, and 5FU incorporation into DNA is incompletely understood. If unrepaired, the presence of these analogs can interfere with DNA–protein interactions. Furthermore, the initiation of the BER pathway can be cytotoxic when the total number of strand breaks exceeds the threshold for repair capacity [37–39,41].

Pyrimidine analogs in the loops of quadruplexes did not disrupt folding. U, 5hmU, and 5FU did not inhibit quadruplex formation by either CD spectroscopy or FRET-based techniques (Figures 2 and 4). Similarly U and likely 5FU and 5hmU, when located in the C-rich complementary strand of a corresponding duplex, did not substantially diminish duplex stability or alter the equilibrium between duplex and quadruplex configurations (Figures 5 and 6). Surprisingly, the introduction of an abasic site, mimicking a BER intermediate, into the quadruplex sequence did not inhibit its formation (Figures 2 and 4). However, if a single-base gap was introduced into the loop region, generating quadruplex fragments (Figure 1, Oligo 4), this abolished quadruplex formation (Figure S17, Quadruplex Fragments).

While these pyrimidine analogs did not inhibit quadruplex formation, their repair was significantly hampered. The three analogs studied here (U, 5hmU, and 5FU) are substrates for the monofunctional glycosylases UDG and SMUG1 when located in duplex structures but are excised much more slowly when in the loop regions of quadruplex structures (Figure 3). Importantly, APE1 was unable to cleave an abasic site when it was present in either single-stranded or quadruplex DNA. These results are in accord with previous studies [10,28–30]. Abasic sites are unable to be repaired by APE1 and could undergo spontaneous β-elimination, generating a strand break. We estimated the apparent half-life for spontaneous β-elimination to be 21 h, which was significantly faster than the rate previously reported for duplex DNA, 190 h [31].

The challenges with repairing these analogs in quadruplexes could represent a therapeutic approach for cancer treatment. For example, incorporation of some pyrimidine analogs, and their subsequent repair by DNA glycosylases, in quadruplexes could be poorly repaired and potentially lethal. If the glycosylases can remove the analogs, this would result in an abasic site. Because of the very poor APE1 activity on quadruplex forming DNA, this would result in spontaneous β-elimination of the abasic site and formation of a single-strand break. Closely spaced single-strand breaks could result in potentially lethal double-strand breaks [44].

3.2. Base Excision Repair Promotes Quadruplex Formation

Most quadruplex-forming sequences are in equilibrium with duplex DNA. Although the thermal stability of a quadruplex is similar to that of the corresponding duplex, the duplex is the preferred configuration under physiological conditions of salt concentration and temperature [22–24]. Given sufficient time, we observed that the quadruplex converted to duplex at physiologic conditions (Figure 4, Inset). The addition of a single strand complement to a preformed quadruplex could unwind the quadruplex with formation of a duplex that had a fast and slow half-life of approximately 2 and 33 min, respectively. These results were similar in timescale to quadruplex unfolding seen previously [34].

Duplex–quadruplex equilibrium can be shunted towards quadruplex by destabilizing the duplex. The BER intermediates, introduced either synthetically or enzymatically in the complementary strand of a quadruplex sequence, altered the duplex–quadruplex equilibrium (Figures 5 and 6). DNA glycosylase mediated excision of an analog from the strand complementary to the quadruplex strand, shifted the equilibrium slightly toward the quadruplex (Figure 6). Cleavage of the resulting abasic site with APE1, generating a one-base gap, resulted in roughly equal populations of duplex and quadruplex configurations. Using a FRET approach, we observed that the formation of a quadruplex from a duplex was not instantaneous and proceeded with a half-life of approximately 3 min (Figure 6, Inset).

3.3. Duplex to Quadruplex Transition Using Pyridostatin

Previous studies have suggested that the toxicity of quadruplex-stabilizing small molecules, such as pyridostatin, might be attributed to the stabilization of quadruplex structures, which could inhibit DNA replication [45]. We examined the capacity of pyridostatin to drive the duplex DNA towards quadruplex formation (Figure 5C). Unexpectedly with increasing concentrations of pyridostatin, a duplex formed from a quadruplex-forming 22-mer (Tel22) and its complementary strand (A:U) was shifted towards quadruplex formation. Despite introducing an abasic site into the C-rich complementary strand, a similar concentration of pyridostatin was required to generate predominantly quadruplex DNA. Upon introduction of a gap, a similar amount of pyridostatin was needed to form quadruplex, although this destabilized the duplex configuration. This indicates that the binding affinity of pyridostatin to the quadruplex was unchanged since the quadruplex sequence itself was otherwise unaltered. Therefore, we propose that despite duplex being favored under physiologic conditions, duplex and quadruplex can rapidly interconvert and pyridostatin serves to trap the quadruplex configuration—preventing its conversion back to duplex (Figure 7). This was demonstrated when we added a 50-fold excess of pyridostatin relative to DNA and trapped all available quadruplex-forming oligonucleotides (Figure 5C). Thus, the cytotoxicity of pyridostatin may be attributed to stabilizing existing quadruplexes but also trapping the quadruplex configuration in the apparently hundreds of thousands of quadruplex-forming regions that are in equilibrium with duplex [6]. Furthermore, the duplex–quadruplex equilibrium may similarly be influenced and regulated by the presence of quadruplex-binding enzymes [29,36,45–47].

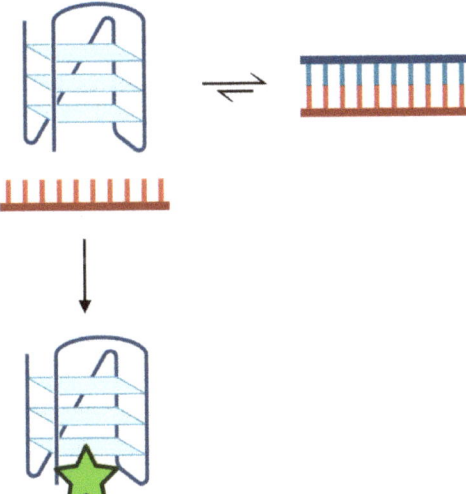

Figure 7. Duplex–quadruplex equilibrium scheme. Potentially quadruplex-forming regions of the genome are in equilibrium with duplex. Duplex is the preferred configuration in the presence of the complementary strand [22–24]. Transiently, a quadruplex may form that can be 'trapped' by quadruplex-binding ligands, such as pyridostatin (green star). The ligand stabilizes the quadruplex structure such that it prevents its unfolding and reforming duplex. Created with Biorender.com.

3.4. Conclusions and Limitations of This Study

The biological functions of unusual, non-duplex structures in DNA are not well understood. However, the work presented here demonstrates that the introduction of some modified bases into DNA, and the subsequent conversion to intermediates of the BER pathway, can alter duplex–quadruplex equilibrium under physiological conditions. The intersection of BER repair intermediates generated by the repair of pyrimidine analogs and unusual DNA structures results in complexities that may contribute to the cytotoxicity of these analogs that has implications for the development of chemotherapies. Within the context of antimicrobial or antitumor chemotherapy, various pyrimidine analogs could potentially be used in combinations where specific glycosylases are known to be differentially expressed. It remains unclear if overexpression and high activity of DNA glycosylases, which would promote strand-breaks, would be more cytotoxic than a lack of repair activity, resulting in an accumulation of the nucleoside analog and potential disrupt ion of DNA-protein interaction.

This study focused on the human telomeric repeat sequence, which is one of the most well-studied DNA quadruplex-forming sequences [12,16,21]. The similarity among quadruplex-forming oligonucleotides is the core of hydrogen-bonding guanine bases, and variability is created by the length and base composition of the loop regions. The results of our study might not be generally extendible to other sequences with varying loop lengths, such as c-MYC and hTERT promoters, although duplex to quadruplex transition with small molecules has been shown with the c-MYC promoter [36,48].

Structural restraints induced by the quadruplex might interfere with base extrusion and therefore glycosylase excision [28]. Monofunctional glycosylases recognize a damaged or modified base, generally in a duplex, and extrude that damaged base from the helix into a binding pocket. Short loops like those found in the human telomere appear to interfere with base extrusion and therefore glycosylase excision. However, if the enzyme is active on single-strand DNA and a loop is of sufficient length, target bases might be more readily excised by monofunctional glycosylase and the abasic site cleaved by an endonuclease or lyase [28].

The modified bases examined here, U, 5hmU, and 5FU are targets of the monofunctional glycosylases of the BER pathway, but not of the bifunctional glycosylases. Therefore, the implications of this study are limited to targets of the monofunctional glycosylases. Bifunctional glycosylases, including FPG and OGG1, are also components of BER. The bifunctional glycosylases excise target bases such as 8-oxoguanine and cleave the DNA backbone resulting in simultaneous strand cleavage. Further studies are in progress to examine the activity of the bifunctional glycosylases on DNA sequences that can form alternative structures.

4. Materials and Methods

4.1. Oligonucleotide Synthesis

Oligonucleotides were synthesized by solid-phase phosphoramidite methods and purified using HPLC. All phosphoramidites, including those for the modified bases U, 5hmU, 5FU, THF, 6FAM, and BHQ1, were obtained from Glen Research. Oligonucleotide composition was verified using MALDI-MS and enzymatic digestion, followed by HPLC analysis (Supplementary Section S1). Oligonucleotide purity was verified using gel electrophoresis.

4.2. Enzymes

E. coli uracil DNA glycosylase (UDG, ca# M0280S), human AP endonuclease 1 (APE1, ca# M0282S), and human single-strand selective monofunctional uracil DNA glycosylase (hSMUG1, ca# M0336S) were purchased from New England Biolabs (NEB).

4.3. Buffers and Reagents

UDG reactions were conducted in a buffer containing 15 mM NaCl, 150 mM KCl, 20 mM Tris, and pH 7.4. APE1 reactions were conducted in buffer containing 15 mM NaCl, 150 mM KCl, 10 mM magnesium acetate (Mg-Ac), 20 mM Tris, pH 7.4. hSMUG1 reactions were conducted in a buffer containing 15 mM NaCl, 50 mM KCl, 20 mM Tris, and pH 7.4. Unless otherwise stated, all chemicals were purchased from Sigma-Aldrich. Pyridostatin was purchased from APExBIO (ca# A3742) and dissolved in double-deionized water and stored at 4 °C or −20 °C for longer term storage.

4.4. CD Spectroscopy Studies

CD spectra were obtained on a Jasco J-815 CD spectrometer. Spectra were obtained at 37 °C from 320 to 220 nm. Oligonucleotides (4 µM) were typically prepared in a buffer containing 20 mM Tris, pH 7.4, 15 mM NaCl, and 150 mM KCl. For the K^+ titration experiments, NaCl was excluded, and the KCl concentration was varied from 0 to 150 mM KCl. Oligonucleotides were equilibrated at room temperature (RT) for at least 30 min before spectra were acquired. Spectra were acquired in a 100 µL cuvette with a 1 cm path length (Starna, ca#26.100LHS-Q-10/Z15). Pyridostatin was added to some oligonucleotides as indicated in the figure legends to measure its effect.

4.5. Fluorescence Studies

Solutions containing oligonucleotides with a 5'-6FAM fluorophore and a 3'-BHQ1 quencher were prepared as a 1 µM solution (100 pmol in 100 µL) in 20 mM Tris buffer, pH 7.4, 15 mM NaCl, and 150 mM KCl unless otherwise indicated using the Jasco J-815. For duplex formation, 1.2 equivalents (120 pmol) of the C-rich complementary strand were annealed at 90 °C for 5 min and cooled to RT. All samples were equilibrated for 30 min at RT prior to the acquisition of fluorescence spectra.

When comparing fluorescence spectra from different oligonucleotides, absolute fluorescence varied among the samples. Fluorescence intensity for each oligonucleotide was therefore normalized to the fluorescence intensity measured in the absence of added K^+ or Na^+ cations, in 20 mM Tris buffer at pH 7.4.

Fluorescence emission spectra were acquired at 37 °C from 480 to 640 nm in 1 nm intervals with an excitation of 495 nm using a Jasco J-815 CD spectrometer equipped with a fluorimeter. Sensitivity was set to a default 600 V, D.I.T 0.125 s, and bandwidth set to 10 nm. A 100 µL fluorometer cuvette was used (Starna, ca# 16.100F-Q-10/Z15).

To examine the effect of BER intermediates on the duplex–quadruplex equilibrium, the labeled quadruplex strand was annealed with 1.2 equivalents (120 pmol) of each complementary strand containing C, U, a THF abasic site or a one-base gap in 20 mM Tris buffer pH 7.4 with 150 mM KCl and 15 mM NaCl, heated to 90 °C for 5 min and cooled to RT. Spectra were then acquired as described above.

To simulate BER intermediates prepared enzymatically, we used UDG (abasic site), or UDG and APE1 (one-base gap). The fluorescent emission spectrum was taken for the quadruplex-only strand and a duplex oligonucleotide containing a U in the C-rich strand (A:U). In addition, spectra were also acquired for the duplex incubated with 10 U of UDG (3.39 pmol, 34 nM), and duplex incubated with 10 U of UDG (3.39 pmol, 34 nM) and 20 U of APE1 (0.71 pmol, 7.1 nM) for 2 h at 37 °C. For all samples, 10 mM Mg-Ac was supplemented to each reaction.

4.6. Quadruplex Fluorescence-Based Kinetic Studies and Pyridostatin Titration

Kinetic measurements were performed in 96-well plates on a Roche 480 Lightcycler II qPCR instrument using the default excitation and emission filters for fluorescein (FAM). Pyridostatin was added to some oligonucleotides, as indicated in the figure legends, and fluorescence was measured after oligonucleotides were incubated with increasing amounts of pyridostatin for 30 min at 37 °C.

The quadruplex unfolding and duplex formation time-course experiments were conducted on a Roche 480 qPCR instrument at 37 °C. Quadruplex-forming oligonucleotides (25 pmol, 1 µM) in 20 mM Tris buffer, pH 7.4, 15 mM NaCl, and 150 mM KCl were first equilibrated for 30 min at RT, at a final volume of 25 µL. Each reaction was then placed on ice for ~20 min and an equimolar amount (25 pmol) of unlabeled complementary strand was added. Fluorescence emission intensity at ~520 nm was then measured every 20 s for 35 min and every 60 s for the remaining 285 min. The data were fit to a two-phase exponential and plotted with GraphPad PRISM 9.4.1. These experiments were performed in triplicate.

We monitored structural changes in a quadruplex-forming strand of a duplex as U in the complementary C-rich strand was removed. The U-containing duplex was incubated simultaneously with 1.25 U of UDG (0.42 pmol, 16.8 nM) and 2.5 U of APE1 (0.09 pmol, 3.6 nM) in the buffer described above supplemented with 10 mM Mg-Ac. Fluorescence intensity was measured every 20 s at 37 °C. A single exponential decay was used to fit the data, and all studies were performed in triplicate.

4.7. Native Gel Electrophoresis of DNA Repair Intermediates

Oligonucleotides used in the above studies were examined using 20% polyacrylamide native gel electrophoresis (PAGE) in 1× TBE buffer (Fisher Scientific, Hampton, NH, USA). Oligonucleotide mixtures (2.5 pmol) were mixed with an equal volume of 20% glycerol and loaded onto the gels and electrophoresed for 70 min at 180 V, 4 °C.

4.8. DNA Glycosylase and AP Endonuclease Studies by Denaturing Gel Electrophoresis

Time points were taken from 2.5 pmol aliquots (12.5 µL, 0.2 µM) of a single reaction, per analog, that consisted of 28.75 pmol (0.2 µM) of oligonucleotide in 144 µL. Reactions were prepared in the appropriate buffer for each enzyme and equilibrated at RT for 30 min. After 30 min, the negative control aliquot containing no enzyme was taken. Enzyme was then added, and the reaction incubated at 37 °C. Aliquots were then taken at each indicated time point up to 60 min. Reactions were quenched by mixing an aliquot (12.5 µL) with 3 µL of 1 M NaOH and placing on ice. All reactions were performed with 3–4 replicates.

UDG reactions with uracil-containing DNA had 0.26 U of UDG (0.0882 pmol, 0.613 nM), which is a 300:1 DNA:enzyme ratio. UDG reactions with 5FU-containing DNA had 2.6 U of UDG (0.882 pmol, 6.13 nM), which is a 30:1 ratio. hSMUG1 reactions with U and 5hmU-containing DNA used 0.8925 pmol (6.2 nM) of hSMUG1 (13.1 U), which is a 30:1 ratio. APE1 reactions used 0.189 pmol (1.31 nM) of APE1 (5.3 U), which is a 140:1 ratio.

Reactions containing UDG were in a buffer containing 15 mM NaCl, 150 mM KCl, 20 mM Tris, pH 7.4. APE1 reactions were prepared in an identical buffer but supplemented with 10 mM Mg-Ac. Reactions with hSMUG1 contained 15 mM NaCl, 50 mM KCl, 20 mM Tris, pH 7.4. The KCl concentration was reduced to 50 mM as hSMUG1 activity is salt-dependent [27].

Following glycosylase removal of a target pyrimidine, the DNA backbone was cleaved by addition of 3 µL 1 M NaOH and heating at 95 °C for 10 min. Each reaction was then neutralized with 3 µL of 1 M acetic acid and an equal volume of formamide was added. Before loading gels, all reactions were heated to 95 °C for 1 min. Samples were then loaded onto a 6 M urea, 20% PAGE gel and run at 180 V for 70 min. A single exponential was fit to the data and rate constants were obtained using PRISM 9.4.1. The rate constants are reported in Table S1. The raw unnormalized gels corresponding to these experiments are shown in the supplementary information (Figures S1–S15).

4.9. Spontaneous β-Elimination Time Course

To measure spontaneous β-elimination of an abasic site at 37 °C, a quadruplex-forming oligonucleotide containing a U (100 pmol in 100 µL, 1 µM) was first incubated in 15 mM NaCl, 150 mM KCl, 20 mM Tris, pH 7.4 at 37 °C with 3.36 pmol (33.6 nM) UDG (10 U). Aliquots were taken at selected times and stored at −20 °C. Samples were then loaded and run on a denaturing gel as described above. The amount of cleaved product was quantified, and the rate of β-elimination was determined by linear regression of data points obtained from 1 h to 28.5 h. Reactions were performed in triplicate (Figure S16). The percentage of β-elimination at each time point was normalized to the total amount of cleavage observed in a sample incubated for 3 h with UDG at 37 °C followed by NaOH-induced cleavage.

4.10. Gel Quantification and Statistical Analysis

Gels images were analyzed in ImageJ as previously described [26]. Error bars represent the standard deviation (S.D.) of three independent experiments. Error bars not seen indicate a S.D. smaller than the data point.

Supplementary Materials: The following supporting information can be downloaded at: https://www.mdpi.com/article/10.3390/molecules28030970/s1, Table S1: DNA glycosylase and APE1 rate constants for Figure 3; Figure S1: Unedited gel scans of UDG time course with U:A substrate in Figure 3A; Figure S2: Unedited gel scans of UDG time course with Tel22-U substrate in Figure 3A; Figure S3: Unedited gel scans of UDG time course with NQ-U substrate in Figure 3A; Figure S4: Unedited gel scans of UDG time course with 5FU:A substrate in Figure 3B; Figure S5: Unedited gel scans of UDG time course with Tel22-5FU substrate in Figure 3B; Figure S6: Unedited gel scans of UDG time course with NQ-5FU substrate in Figure 3B; Figure S7: Unedited gel scans of hSMUG1 time course with substrate U:A in Figure 3C; Figure S8: Unedited gel scans of hSMUG1 time course with substrate Tel22-U in Figure 3C; Figure S9: Unedited gel scans of hSMUG1 time course with substrate NQ-U in Figure 3C; Figure S10: Unedited gel scans of hSMUG1 time course with substrate 5hmU:A in Figure 3D; Figure S11: Unedited gel scans of hSMUG1 time course with substrate Tel22-5hmU in Figure 3D; Figure S12: Unedited gel scans of hSMUG1 time course with substrate NQ-5hmU in Figure 3D; Figure S13: Unedited gel scans of APE1 time course with substrate THF:A in Figure 3E; Figure S14: Unedited gel scans of APE1 time course with substrate Tel22-THF in Figure 3E; Figure S15: Unedited gel scans of APE1 time course with substrate NQ-THF in Figure 3E; Figure S16: Unedited gel scans of the β-elimination time course with UDG treated substrate Tel22-U in Figure 3F; Figure S17: Quadruplexes quench fluorescence compared to duplex. In addition, when the quadruplex oligonucleotide is separated in half (quadruplex fragment), simulating a DNA repair gap, fluorescence is also no longer quenched and similar to the values seen for quadruplex strands annealed to their complement; Figure S18: Rate of gap formation in Figure 6 time course; Supplementary Section S1: This document contains the detailed synthesis and characterization for the oligonucleotides prepared and used in this study.

Author Contributions: Conceptualization, M.L.S. and L.C.S.; methodology, M.L.S., J.W.C., B.C.-G. and L.C.H.; validation, M.L.S. and J.W.C.; formal analysis, M.L.S., J.W.C. and B.C.-G.; investigation, M.L.S., J.W.C. and B.C.-G.; resources, L.C.H. and L.C.S.; data curation, M.L.S., J.W.C., B.C.-G. and L.C.H.; writing—original draft preparation, M.L.S. and L.C.S.; writing—review and editing, M.L.S., J.W.C., E.C. and L.C.S.; visualization, M.L.S., E.C.; supervision, L.C.S.; project administration, L.C.S.; funding acquisition, L.C.S. All authors have read and agreed to the published version of the manuscript.

Funding: This work was funded in part by a grant from the NIH NCI R01CA228085, the John Sealy Distinguished Chair in Cancer Biology, and the NSF EFRI1933321. M.L.S. and B.C.G. were supported in part by the UTMB physician–scientist training program. J.W.C. was funded in part by the CCBTP training fellowship from CPRIT Grant No. RP170593.

Institutional Review Board Statement: Not applicable.

Informed Consent Statement: Not applicable.

Data Availability Statement: All data used to prepare this manuscript are available in the supporting information and Supplementary Section S1.

Conflicts of Interest: The authors declare no conflict of interest.

References

1. Wallace, S.S. Base Excision Repair: A Critical Player in Many Games. *DNA Repair* **2014**, *19*, 14–26. [CrossRef] [PubMed]
2. Krokan, H.E.; Bjørås, M. Base Excision Repair. *Cold Spring Harb. Perspect. Biol.* **2013**, *5*, a012583. [CrossRef] [PubMed]
3. Wang, G.; Vasquez, K.M. Impact of Alternative DNA Structures on DNA Damage, DNA Repair, and Genetic Instability. *DNA Repair* **2014**, *19*, 143–151. [CrossRef]
4. Satange, R.; Chang, C.K.; Hou, M.H. A Survey of Recent Unusual High-Resolution DNA Structures Provoked by Mismatches, Repeats and Ligand Binding. *Nucleic Acids Res.* **2018**, *46*, 6416–6434. [CrossRef] [PubMed]
5. Bansal, A.; Kaushik, S.; Kukreti, S. Non-Canonical DNA Structures: Diversity and Disease Association. *Front. Genet.* **2022**, *13*, 959258. [CrossRef] [PubMed]

6. Chambers, V.S.; Marsico, G.; Boutell, J.M.; di Antonio, M.; Smith, G.P.; Balasubramanian, S. High-Throughput Sequencing of DNA G-Quadruplex Structures in the Human Genome. *Nat. Biotechnol.* **2015**, *33*, 877–881. [CrossRef]
7. Gellert, M.; Lipsett, M.N.; Davies, D.R. Helix Formation by Guanylic Acid. *Proc. Natl. Acad. Sci. USA* **1962**, *48*, 2013–2018. [CrossRef]
8. Guschlbauer, W.; Chantot, J.F.; Thiele, D. Four-Stranded Nucleic Acid Structures 25 Years Later: From Guanosine Gels to Telomer Dna. *J. Biomol. Struct. Dyn.* **1990**, *8*, 491–511. [CrossRef]
9. Chaires, J.B. Human Telomeric G-Quadruplex: Thermodynamic and Kinetic Studies of Telomeric Quadruplex Stability. *FEBS J.* **2010**, *277*, 1098–1106. [CrossRef]
10. Pavlova, A.V.; Kubareva, E.A.; Monakhova, M.V.; Zvereva, M.I.; Dolinnaya, N.G. Impact of G-Quadruplexes on the Regulation of Genome Integrity, Dna Damage and Repair. *Biomolecules* **2021**, *11*, 1284. [CrossRef]
11. Linke, R.; Limmer, M.; Juranek, S.A.; Heine, A.; Paeschke, K. The Relevance of G-quadruplexes for Dna Repair. *Int. J. Mol. Sci.* **2021**, *22*, 12599. [CrossRef] [PubMed]
12. Zhou, J.; Liu, M.; Fleming, A.M.; Burrows, C.J.; Wallace, S.S. Neil3 and NEIL1 DNA Glycosylases Remove Oxidative Damages from Quadruplex DNA and Exhibit Preferences for Lesions in the Telomeric Sequence Context. *J. Biol. Chem.* **2013**, *288*, 27263–27272. [CrossRef] [PubMed]
13. Zhou, J.; Fleming, A.M.; Averill, A.M.; Burrows, C.J.; Wallace, S.S. The NEIL Glycosylases Remove Oxidized Guanine Lesions from Telomeric and Promoter Quadruplex DNA Structures. *Nucleic Acids Res.* **2015**, *43*, 4039–4054. [CrossRef] [PubMed]
14. Lech, C.J.; Cheow Lim, J.K.; Wen Lim, J.M.; Amrane, S.; Heddi, B.; Phan, A.T. Effects of Site-Specific Guanine C8-Modifications on an Intramolecular DNA G-Quadruplex. *Biophys. J.* **2011**, *101*, 1987–1998. [CrossRef]
15. Kuznetsova, A.A.; Fedorova, O.S.; Kuznetsov, N.A. Lesion Recognition and Cleavage of Damage-Containing Quadruplexes and Bulged Structures by DNA Glycosylases. *Front. Cell Dev. Biol.* **2020**, *8*, 595687. [CrossRef]
16. Bielskute, S.; Plavec, J.; Podbevšek, P. Impact of Oxidative Lesions on the Human Telomeric G-Quadruplex. *J. Am. Chem. Soc.* **2019**, *141*, 2594–2603. [CrossRef]
17. Bielskute, S.; Plavec, J.; Podbevšek, P. Oxidative Lesions Modulate G-Quadruplex Stability and Structure in the Human BCL2 Promoter. *Nucleic Acids Res.* **2021**, *49*, 2346–2356. [CrossRef]
18. Lindahl, T. Instability and Decay of the Primary Structure of DNA. *Nature* **1993**, *362*, 709–715. [CrossRef]
19. Mullaart, E.; Lohman, P.H.M.; Berends, F.; Vijg, J. DNA Damage Metabolism and Aging. *Mutat. Res.* **1990**, *237*, 189–210. [CrossRef]
20. Bordin, D.L.; Lirussi, L.; Nilsen, H. Cellular Response to Endogenous DNA Damage: DNA Base Modifications in Gene Expression Regulation. *DNA Repair* **2021**, *99*, 103051. [CrossRef]
21. Ambrus, A.; Chen, D.; Dai, J.; Bialis, T.; Jones, R.A.; Yang, D. Human Telomeric Sequence Forms a Hybrid-Type Intramolecular G-Quadruplex Structure with Mixed Parallel / Antiparallel Strands in Potassium Solution. *Nucleic Acids Res.* **2006**, *34*, 2723–2735. [CrossRef] [PubMed]
22. Phan, A.T.; Mergny, J.-L. Human Telomeric DNA: G-Quadruplex, i-Motif and Watson-Crick Double Helix. *Nucleic Acids Res.* **2002**, *30*, 4618–4625. [CrossRef] [PubMed]
23. Risitano, A.; Fox, K.R. Stability of Intramolecular DNA Quadruplexes: Comparison with DNA Duplexes. *Biochemistry* **2003**, *42*, 6507–6513. [CrossRef] [PubMed]
24. Jaumot, J.; Eritja, R.; Tauler, R.; Gargallo, R. Resolution of a Structural Competition Involving Dimeric G-Quadruplex and Its C-Rich Complementary Strand. *Nucleic Acids Res.* **2006**, *34*, 206–216. [CrossRef]
25. Mellac, S.; Fazakerley, G.V.; Sowers, L.C. Structures of Base Pairs with 5-(Hydroxymethyl)-2′-Deoxyuridine in DNA Determined by NMR Spectroscopy. *Biochemistry* **1993**, *32*, 7779–7786. [CrossRef]
26. Hsu, C.W.; Conrad, J.W.; Sowers, M.L.; Baljinnyam, T.; Herring, J.L.; Hackfeld, L.C.; Hatch, S.S.; Sowers, L.C. A Combinatorial System to Examine the Enzymatic Repair of Multiply Damaged DNA Substrates. *Nucleic Acids Res.* **2022**, *50*, 7406–7419. [CrossRef]
27. Masaoka, A.; Matsubara, M.; Hasegawa, R.; Tanaka, T.; Kurisu, S.; Terato, H.; Ohyama, Y.; Karino, N.; Matsuda, A.; Ide, H. Mammalian 5-Formyluracil-DNA Glycosylase. 2. Role of SMUG1 Uracil-DNA Glycosylase in Repair of 5-Formyluracil and Other Oxidized And. *Biochemistry* **2003**, *42*, 5003–5012. [CrossRef] [PubMed]
28. Holton, N.W.; Larson, E.D. G-Quadruplex DNA Structures Can Interfere with Uracil Glycosylase Activity in Vitro. *Mutagenesis* **2016**, *31*, 385–392. [CrossRef]
29. Fleming, A.M.; Howpay Manage, S.A.; Burrows, C.J. Binding of AP Endonuclease-1 to G-Quadruplex DNA Depends on the N-Terminal Domain, Mg 2+, and Ionic Strength. *ACS Bio. Med. Chem. Au.* **2021**, *1*, 44–56. [CrossRef]
30. Theruvathu, J.A.; Darwanto, A.; Hsu, C.W.; Sowers, L.C. The Effect of Pot1 Binding on the Repair of Thymine Analogs in a Telomeric DNA Sequence. *Nucleic Acids Res.* **2014**, *42*, 9063–9073. [CrossRef]
31. Lindahl, T.; Andersson, A. Rate of Chain Breakage at Apurinic Sites Double-Stranded Deoxyribonucleic Acid. *Biochemistry* **1972**, *11*, 3618–3623. [CrossRef] [PubMed]
32. Simonsson, T.; Sjöback, R. DNA Tetraplex Formation Studied with Fluorescence Resonance Energy Transfer. *J. Biol. Chem.* **1999**, *274*, 17379–17383. [CrossRef]
33. Gray, R.D.; Trent, J.O.; Chaires, J.B. Folding and Unfolding Pathways of the Human Telomeric G-Quadruplex. *J. Mol. Biol.* **2014**, *426*, 1629–1650. [CrossRef] [PubMed]
34. Kumar, N.; Maiti, S. The Effect of Osmolytes and Small Molecule on Quadruplex-WC Duplex Equilibrium: A Fluorescence Resonance Energy Transfer Study. *Nucleic Acids Res.* **2005**, *33*, 6723–6732. [CrossRef]

35. Li, M.H.; Wang, Z.F.; Kuo, M.H.J.; Hsu, S.T.D.; Chang, T.C. Unfolding Kinetics of Human Telomeric G-Quadruplexes Studied by NMR Spectroscopy. *J. Phys. Chem. B* **2014**, *118*, 931–936. [CrossRef] [PubMed]
36. Rangan, A.; Fedoroff, O.Y.; Hurley, L.H. Induction of Duplex to G-Quadruplex Transition in the c-Myc Promoter Region by a Small Molecule. *J. Biol. Chem.* **2001**, *276*, 4640–4646. [CrossRef] [PubMed]
37. Goulian, M.; Bleile, B.; Tseng, B.Y. Methotrexate-Induced Misincorporation of Uracil into DNA. *Proc. Natl. Acad. Sci. USA* **1980**, *77*, 1956–1960. [CrossRef]
38. Kunz, C.; Focke, F.; Saito, Y.; Schuermann, D.; Lettieri, T.; Selfridge, J.; Schär, P. Base Excision by Thymine DNA Glycosylase Mediates DNA-Directed Cytotoxicity of 5-Fluorouracil. *PLoS Biol.* **2009**, *7*, e1000091. [CrossRef]
39. An, Q.; Robins, P.; Lindahl, T.; Barnes, D.E. 5-Fluorouracil Incorporated into DNA Is Excised by the Smug1 DNA Glycosylase to Reduce Drug Cytotoxicity. *Cancer Res.* **2007**, *67*, 940–945. [CrossRef]
40. Pettersen, H.S.; Visnes, T.; Vågbø, C.B.; Svaasand, E.K.; Doseth, B.; Slupphaug, G.; Kavli, B.; Krokan, H.E. UNG-Initiated Base Excision Repair Is the Major Repair Route for 5-Fluorouracil in DNA, but 5-Fluorouracil Cytotoxicity Depends Mainly on RNA Incorporation. *Nucleic Acids Res.* **2011**, *39*, 8430–8444. [CrossRef]
41. Rogstad, D.K.; Darwanto, A.; Herring, J.L.; Rogstad, K.N.; Burdzy, A.; Hadley, S.R.; Neidigh, J.W.; Sowers, L.C. Measurement of the Incorporation and Repair of Exogenous 5-Hydroxymethyl-2′-Deoxyuridine in Human Cells in Culture Using Gas Chromatography-Negative Chemical Ionization-Mass Spectrometry. *Chem. Res. Toxicol.* **2007**, *20*, 1787–1796. [CrossRef]
42. Pospíšil, Š.; Panattoni, A.; Gracias, F.; Sýkorová, V.; Hausnerová, V.V.; Vítovská, D.; Šanderová, H.; Krásný, L.; Hocek, M. Epigenetic Pyrimidine Nucleotides in Competition with Natural dNTPs as Substrates for Diverse DNA Polymerases. *ACS Chem. Biol.* **2022**, *17*, 2781–2788. [CrossRef] [PubMed]
43. Boorstein, R.J.; Chiu, L.; Teebor, G.W. A Mammalian Cell Line Deficient in Activity of the DNA Repair Enzyme 5-Hydroxymethyluracil-DNA Glycosylase Is Resistant to the Toxic Effects of the Thymidine Analog 5-Hydroxymethyl-2′-Deoxyuridine. *Mol. Cell. Biol.* **1992**, *12*, 5536–5540. [PubMed]
44. Dianov, G.L.; Timehenko, T.V.; Sinitsina, O.I.; Kuzminov, A.V.; Medvedev, O.A.; Salganik, R.I. Repair of Uracil Residues Closely Spaced on the Opposite Strands of Plasmid DNA Results in Double-Strand Break and Deletion Formation. *Mol. Gen. Genet.* **1991**, *225*, 448–452. [CrossRef] [PubMed]
45. Rodriguez, R.; Miller, K.M.; Forment, J.V.; Bradshaw, C.R.; Nikan, M.; Britton, S.; Oelschlaegel, T.; Xhemalce, B.; Balasubramanian, S.; Jackson, S.P. Small-Molecule-Induced DNA Damage Identifies Alternative DNA Structures in Human Genes. *Nat. Chem. Biol.* **2012**, *8*, 301–310. [CrossRef]
46. Roychoudhury, S.; Pramanik, S.; Harris, H.L.; Tarpley, M.; Sarkar, A.; Spagnol, G.; Sorgen, P.L.; Chowdhury, D.; Band, V.; Klinkebiel, D.; et al. Endogenous Oxidized DNA Bases and APE1 Regulate the Formation of G-Quadruplex Structures in the Genome. *Proc. Natl. Acad. Sci. USA* **2020**, *117*, 11409–11420. [CrossRef]
47. Meier-Stephenson, V. G4-Quadruplex-Binding Proteins: Review and Insights into Selectivity. *Biophys. Rev.* **2022**, *14*, 635–654. [CrossRef]
48. Palumbo, S.M.L.; Ebbinghaus, S.W.; Hurley, L.H. Formation of a Unique End-to-End Stacked Pair of G-Quadruplexes in the HTERT Core Promoter with Implications for Inhibition of Telomerase by G-Quadruplex-Interactive Ligands. *J. Am. Chem. Soc.* **2009**, *131*, 10878–10891. [CrossRef]

Disclaimer/Publisher's Note: The statements, opinions and data contained in all publications are solely those of the individual author(s) and contributor(s) and not of MDPI and/or the editor(s). MDPI and/or the editor(s) disclaim responsibility for any injury to people or property resulting from any ideas, methods, instructions or products referred to in the content.

Correction

Correction: Clua et al. Properties of Parallel Tetramolecular G-Quadruplex Carrying N-Acetylgalactosamine as Potential Enhancer for Oligonucleotide Delivery to Hepatocytes. *Molecules* 2022, 27, 3944

Anna Clua [1,2], Santiago Grijalvo [1,2], Namrata Erande [3], Swati Gupta [3], Kristina Yucius [3], Raimundo Gargallo [4], Stefania Mazzini [5], Muthiah Manoharan [3] and Ramon Eritja [1,2,*]

1. Institute for Advanced Chemistry of Catalonia (IQAC-CSIC), Jordi Girona 18-26, E-08034 Barcelona, Spain
2. Networking Center on Bioengineering, Biomaterials and Nanomedicine (CIBER-BBN), E-08034 Barcelona, Spain
3. Alnylam Pharmaceuticals, 300 Third Street, Cambridge, MA 02142, USA
4. Department of Chemical Engineering and Analytical Chemistry, University of Barcelona, Martí i Franquès 1-11, E-08028 Barcelona, Spain
5. DEFENS-Dipartimento Di Scienze per Gli Alimenti, la Nutrizione e l'Ambiente, Università degli Studi di Milano, Via Celoria, 2, 20133 Milan, Italy
* Correspondence: recgma@cid.csic.es; Tel.: +34-934-006-145

In the original article [1], there was a mistake in the position of one of the OH of N-acetylgalactosamine. In the original article [1], the OH was in an equatorial position and it should be in an axial position. This mistake affected Scheme 1 and Scheme 2. Below, we include the figures from the original manuscript and the figures that are corrected. We apologize for this unintentional mistake.

Original Scheme 1:

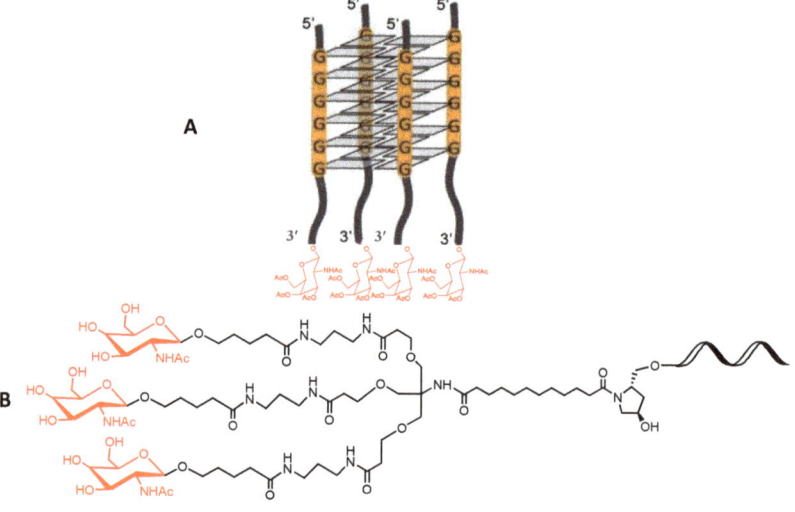

Scheme 1. Chemical schemes of potential multifunctional GalNAc derivatives including (**A**) tetrameric G-quadruplexes studied in this work and (**B**) the triantennary GalNAc.

This should be replaced with the following:

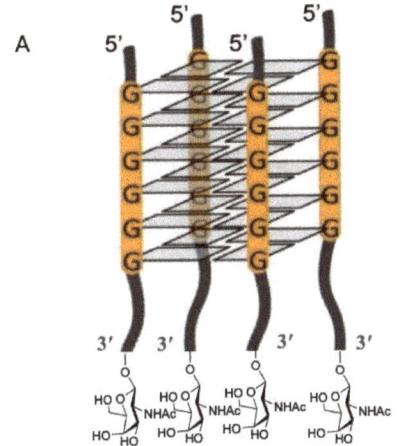

Scheme 1. Chemical schemes of potential multifunctional GalNAc derivatives including (**A**) tetrameric G-quadruplexes studied in this work and (**B**) the triantennary GalNAc.

Original Scheme 2:

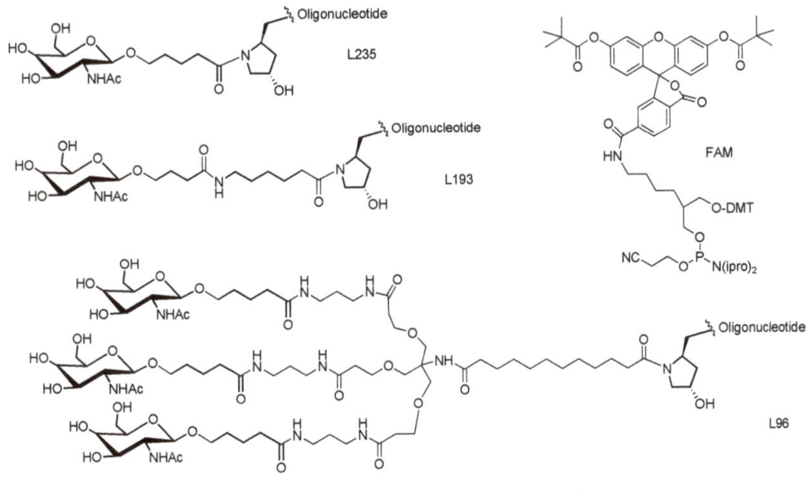

Scheme 2. Chemical structures of the GalNAc and FAM ligands used in this work.

This should be replaced with the following:

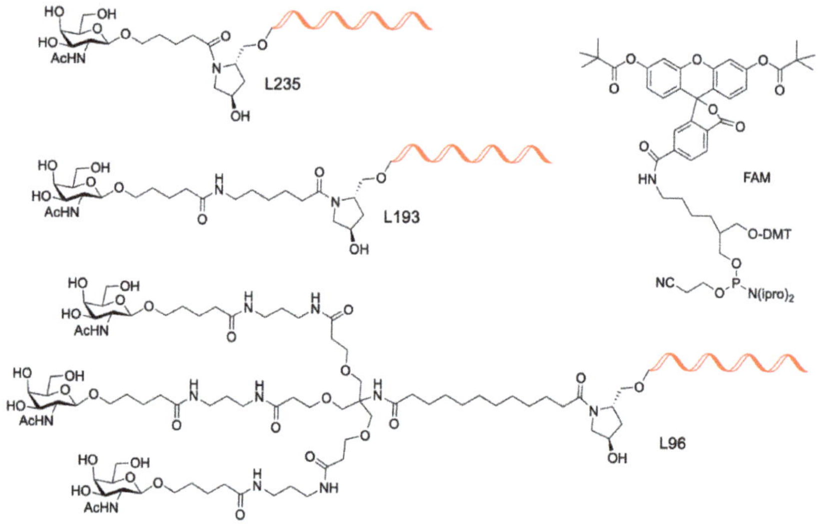

Scheme 2. Chemical structures of the GalNAc and FAM ligands used in this work.

Reference

1. Clua, A.; Grijalvo, S.; Erande, N.; Gupta, S.; Yucius, K.; Gargallo, R.; Mazzini, S.; Manoharan, M.; Eritja, R. Properties of Parallel Tetramolecular G-Quadruplex Carrying N-Acetylgalactosamine as Potential Enhancer for Oligonucleotide Delivery to Hepatocytes. *Molecules* **2022**, *27*, 3944. [CrossRef]

Disclaimer/Publisher's Note: The statements, opinions and data contained in all publications are solely those of the individual author(s) and contributor(s) and not of MDPI and/or the editor(s). MDPI and/or the editor(s) disclaim responsibility for any injury to people or property resulting from any ideas, methods, instructions or products referred to in the content.

Article

Factors Impacting Invader-Mediated Recognition of Double-Stranded DNA

Caroline P. Shepard, Raymond G. Emehiser, Saswata Karmakar and Patrick J. Hrdlicka *

Department of Chemistry, University of Idaho, Moscow, ID 83844-2343, USA
* Correspondence: hrdlicka@uidaho.edu

Abstract: The development of chemically modified oligonucleotides enabling robust, sequence-unrestricted recognition of complementary chromosomal DNA regions has been an aspirational goal for scientists for many decades. While several groove-binding or strand-invading probes have been developed towards this end, most enable recognition of DNA only under limited conditions (e.g., homopurine or short mixed-sequence targets, low ionic strength, fully modified probe strands). Invader probes, i.e., DNA duplexes modified with +1 interstrand zippers of intercalator-functionalized nucleotides, are predisposed to recognize DNA targets due to their labile nature and high affinity towards complementary DNA. Here, we set out to gain further insight into the design parameters that impact the thermal denaturation properties and binding affinities of Invader probes. Towards this end, ten Invader probes were designed, and their biophysical properties and binding to model DNA hairpins and chromosomal DNA targets were studied. A Spearman's rank-order correlation analysis of various parameters was then performed. Densely modified Invader probes were found to result in efficient recognition of chromosomal DNA targets with excellent binding specificity in the context of denaturing or non-denaturing fluorescence in situ hybridization (FISH) experiments. The insight gained from the initial phase of this study informed subsequent probe optimization, which yielded constructs displaying improved recognition of chromosomal DNA targets. The findings from this study will facilitate the design of efficient Invader probes for applications in the life sciences.

Keywords: oligonucleotides; DNA recognition; chromosomes; DNA; pyrene; fluorescence; FISH; karyotyping; SNP; strand invasion

Citation: Shepard, C.P.; Emehiser, R.G.; Karmakar, S.; Hrdlicka, P.J. Factors Impacting Invader-Mediated Recognition of Double-Stranded DNA. *Molecules* 2023, 28, 127. https://doi.org/10.3390/molecules28010127

Academic Editor: Ramon Eritja

Received: 29 October 2022
Revised: 16 December 2022
Accepted: 20 December 2022
Published: 23 December 2022

Copyright: © 2022 by the authors. Licensee MDPI, Basel, Switzerland. This article is an open access article distributed under the terms and conditions of the Creative Commons Attribution (CC BY) license (https://creativecommons.org/licenses/by/4.0/).

1. Introduction

Over the past several decades, numerous chemically modified oligonucleotides and nucleic acid mimics have been designed to target specific sequences of double-stranded DNA (dsDNA) and identify, regulate, and manipulate genes. For example, traditional peptide nucleic acids (PNAs) [1,2] and triplex-forming oligonucleotides (TFOs) [3,4] bind in the major groove of double-stranded DNA (dsDNA), forming Hoogsteen base pairs (bps), which require the presence of extended purine tracts. Pyrrole-imidazole polyamides, on the other hand, have been designed to target complementary sites through binding via the minor groove. However, it has proven challenging to design polyamides that target sufficiently long sequences, as shape complementarity in the minor groove gradually vanishes with increasing probe length [5,6].

Strand-invading approaches—i.e., chemically modified oligonucleotides and nucleic acid mimics capable of unzipping Watson-Crick base pairs of dsDNA targets and forming new, more stable Watson-Crick base pairs between probe strands and the complementary DNA (cDNA) regions—have been explored to overcome the limitations of groove-binding approaches. A key advantage of strand-invading strategies is the prospect of the sequence-unrestricted recognition of dsDNA. Progress towards this end has been realized with various modified single-stranded PNAs [7–12], double-stranded probes such as pseudo-complementary (pc) PNAs [13–15], and related approaches [16–23].

We focused on the development of dsDNA-targeting Invader probes [24], i.e., short DNA duplexes featuring one or more +1 interstrand zipper arrangements [25] of intercalator-functionalized nucleotides such as 2′-O-(pyren-1-yl)methyl-RNA (Figure 1). This monomer arrangement—coined an *energetic hotspot* for brevity—forces pairs of intercalators between the π-stacks of neighboring base pairs in the double-stranded probe, resulting in a violation of the neighbor exclusion principle [26]. The principle asserts that local intercalator densities exceeding one intercalator per two base pairs are unfavorable in DNA duplexes due to limitations in local helix expandability (each intercalation event expands the duplex by ~3.4 Å), and because stabilizing stacking interactions between neighboring base pairs and the first intercalating moiety are perturbed [27–29]. Accordingly, double-stranded Invader probes, featuring two intercalators between the two base pairs of the hotspot region, are partially unwound and labile (Figure 1) [30,31]. The two Invader probe strands, in turn, display high affinity towards cDNA, as duplex formation results in strongly stabilizing stacking interactions between the intercalator and flanking base pairs (the neighbor exclusion principle is no longer violated, as the local intercalator density is one intercalator per two base pairs or less). The difference in stability between the probe-target duplexes, vis-à-vis the double-stranded Invader probe and the dsDNA target region, generates the driving force for dsDNA recognition via double-duplex strand invasion (Figure 1) [24].

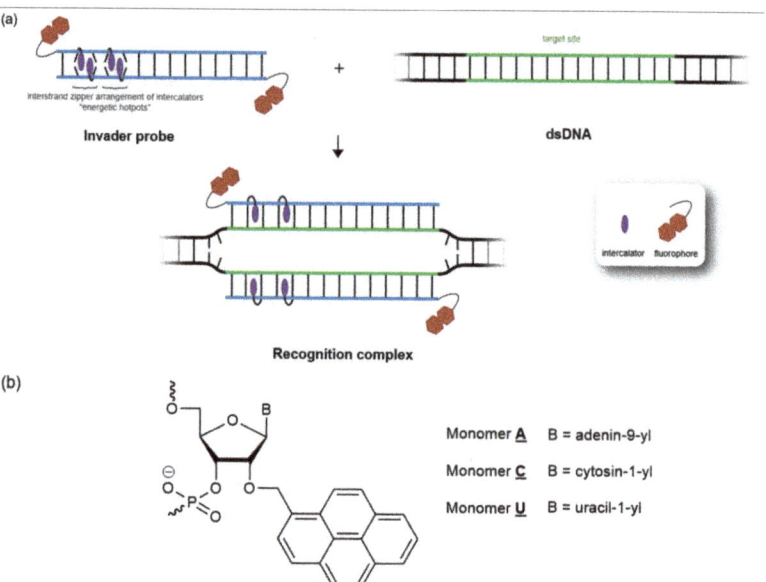

Figure 1. (**a**) Illustration of Invader-mediated recognition of dsDNA via a double-duplex invasion process. (**b**) Structures of Invader monomers used herein.

The sequence-unrestricted recognition of dsDNA targets using Invader probes has been demonstrated, enabling the detection of (i) DNA fragments from specific food pathogens using sandwich assays [32], (ii) telomeric DNA of individual chromosomes in metaphasic spreads [33], and (iii) sex chromosome-specific targets in interphase and metaphase nuclei under non-denaturing conditions [23,24].

In addition to our efforts aimed at optimizing Invader probes through the refinement of the monomer and probe architectures [34–36], early foundational studies provided some insight into the design parameters that impact the dsDNA-recognition efficiency of Invader probes [24,37]. For example, the use of intercalator-functionalized pyrimidine monomers (and avoidance of the corresponding guanine monomers) was found to be preferable for the construction of energetic hotspots. This is because the resulting probe-target duplexes are particularly stabilized when the intercalator-modified monomers are flanked by 3′-purines, thus, increasing the thermodynamic driving force for dsDNA-recognition [37].

In the present study, we set out to gain further insight into the design parameters that impact denaturation properties, driving forces for target recognition, and recognition of chromosomal DNA targets. Towards this end, a library of Invader probes was constructed; their denaturation, thermodynamic and dsDNA-targeting properties were studied; and a Spearman's rank-order correlation analysis of different parameter pairs was performed. The insights gained from the initial phase of this study informed the subsequent optimization of probes, which displayed improved recognition of chromosomal DNA targets.

2. Results and Discussion

2.1. Invader Probe Design

Initially, ten 5′-Cy3-labeled oligodeoxyribonucleotide (ON)-based Invader probes (**INV1-INV10**, Table 1), varying in length (14–16 base pairs) and GC-content (GC%) (30–70%), were designed to target complementary sequences within the DYZ-1 satellite gene (~6 × 10^4 tandem repeats of a ~1175 bp region) located on the bovine (Bos taurus) Y chromosome [38] (NCBI code: M26067, Figure S1). The probes were designed to have modification densities (mod%) of ~20–30%, as earlier studies suggested this level of modification to strike a favorable balance between binding affinity and binding specificity [24,33]. Individual probe strands were obtained using established machine-assisted solid-phase ON synthesis protocols [37].

Table 1. Thermal denaturation temperatures (T_ms) of Invader probes and duplexes between individual probe strands and cDNA. Also shown are the length, thermal advantage values (TA), available free energy for recognition of isosequential dsDNA targets at 310 K (ΔG_{rec}^{310}), modification densities (mod%), and GC-content (GC%) of Invader probes [a].

Probe (Length)	Sequence	T_m [ΔT_m] (°C)			TA (°C)	ΔG_{rec}^{310} (kJ/mol)	mod%	GC%
		Probe Duplex	5′-ON: cDNA	3′-ON: cDNA				
INV1 (15)	5′-Cy3-T**U**ATCA GCAC**U**GUGC-3′ 3′-AA**U**AGTC GTGA**C**ACG-Cy3-5′	52.0 [b] [−4.0]	65.5 [+9.5]	66.0 [+10.0]	+23.5	ND	20.0%	46.7%
INV2 (16)	5′-Cy3-A**U**ACUGG**TT**TGUGUTC-3′ 3′-**T**AUGAC CAAA**C**AC**A**AG-Cy3-5′	34.5 [b] [−18.5]	66.0 [+13.0]	66.0 [+13.0]	+44.5	−50	25.0%	37.5%
INV3 (15)	5′-Cy3-**TU**GUGCC CTGGCAAC-3′ 3′-AACACGG ACCGT**U**G-Cy3-5′	NT	64.0 [+5.5]	62.0 [+3.5]	ND	ND	20.0%	60.0%
INV4 (14)	5′-Cy3-AGCCCUGT GCCCTG-3′ 3′-TCGGGA**C** GGG**A**C-Cy3-5′	61.5 [+1.0]	69.5 [+9.0]	75.5 [+15.0]	+23.0	−7	21.4%	71.4%
INV5 (16)	5′-Cy3-GA**TT**TC AGCCAUGUGC-3′ 3′-C**T**AAAGT CGGTA**C**A**C**G-Cy3-5′	45.0 [−12.0]	63.0 [+6.0]	69.5 [+12.5]	+30.5	−29	18.8%	50.0%
INV6 (16)	5′-Cy3-CUGUGCA ACTCG**T**U**T**G-3′ 3′-GACACGT**T**G ACCAA**A**C-Cy3-5′	63.0 [+5.0]	65.5 [+7.5]	69.0 [+11.0]	+13.5	−22	18.8%	50.0%
INV7 (16)	5′-Cy3-CUGUGC AAUA**TT**TUGT-3′ 3′-GA**C**ACGTTAUAA AA**C**A-Cy3-5′	55.0 [−4.0]	73.0 [+22.0]	71.0 [+20.0]	+38.0	−46	25.0%	31.3%
INV8 (15)	5′-Cy3-**TT**CACA GCCCUGUGC-3′ 3′-AAGUGTCG GGA**C**A**C**G-Cy3-5′	58.5 [b] [−1.5]	70.5 [+10.5]	74.5 [+14.5]	+26.5	−52	20.0%	60.0%
INV9 (15)	5′-Cy3-**TU**A**U**ATG CTGUTCTC-3′ 3′-AA**U**A**U**ACGA CAAGAG-Cy3-5′	55.0 [+9.5]	58.0 [+12.5]	64.0 [+18.5]	+21.5	−19	20.0%	33.3%
INV10 (14)	5′-Cy3-GUGUAG **T**CUAUA**T**G-3′ 3′-**C**ACA**U**CAC A**U**A**U**AC-Cy3-5′	45.5 [+2.0]	65.0 [+21.5]	64.5 [+21.0]	+40.5	−56	28.6%	35.7%
OPT6 (16)	5′-Cy3-CUCU**G**CAACUG GT**U**TG-3′ 3′-GA**C**AC**G**U**T**GAC**C**A AAC-Cy3-5′	49.0 [−9.0]	75.0 [+17.0]	75.0 [+17.0]	+43.0	−84	31.3%	50.0%
OPT8 (15)	5′-Cy3-**TT**CA**C**AG CCCUGUGC-3′ 3′-AAC**G**U**C**G GGA**C**A**C**G-Cy3-5′	38.0 [b] [−22.0]	77.0 [+17.0]	76.5 [+16.5]	+55.5	−93	26.7%	60.0%
OPT9 (15)	5′-Cy3-**T**A**UA**U A**U**GCU GU**T**CTC-3′ 3′-AA**U**A**U**ACGACAAAG AG-Cy3-5′	29.0 [b] [−16.5]	65.0 [+19.5]	64.0 [+18.5]	+54.5	−59	33.3%	33.3%

[a] ΔT_m = change in T_m value relative to corresponding unmodified duplex. T_ms for the corresponding unmodified DNA duplexes are: **DNA1** = 56.0 °C, **DNA2** = 53.0 °C, **DNA3** = 58.5 °C, **DNA4** = 60.5 °C [24], **DNA5** = 57.0 °C, **DNA6** = 58.0 °C, **DNA7** = 51.0 °C, **DNA8** = 60.0 °C, **DNA9** = 45.5 °C, and **DNA10** = 43.5 °C. Thermal denaturation curves were recorded in medium salt buffer ([Na⁺] = 110 mM, [Cl⁻] = 100 mM, pH 7.0 (NaH$_2$PO$_4$/Na$_2$HPO$_4$), [EDTA] = 0.2 mM) and each [ON] = 1.0 µM; see main text for definitions of TA and ΔG_{rec}^{310}. For structures of A, C, and U, see Figure 1. [b] Broad transition. ND = not determined. NT = no clear transition observed in $A_{230-280}$ range.

2.2. Thermal Denaturation Properties of Invader Probes

Thermal denaturation temperatures (T_ms) were determined for the double-stranded Invader probes and the corresponding duplexes between individual probe strands and complementary DNA (Table 1). With the exception of **INV2**, **INV5**, and **INV9**, the Invader probes display substantially similar T_ms as the corresponding unmodified DNA duplexes (see ΔT_ms for probe duplexes, Table 1). Conversely, duplexes between individual Invader probe strands and cDNA display T_ms that, on average, are ~13 °C higher than the corresponding unmodified DNA duplexes with ΔT_ms ranging between +3.5 and +22.0 °C (see ΔT_m values for 5'-ON:cDNA and 3'-ON:cDNA, Table 1). The observed differences in T_m values are in agreement with prior results [24] and reflect that the neighbor exclusion principle is violated in the double-stranded Invader probes (high local intercalator density) but not in duplexes between individual Invader strands and cDNA (lower local intercalator density).

Our Spearman's rank-order correlation analysis of select parameter pairs (full dataset in Supplementary Materials) indicates that there is a lack of significant correlation between the ΔT_m values of the Invader probes and any of the following parameters: length, GC-content, modification density, number of modifications (#mod), or longest unmodified stretch (stretch), at least within the pre-selected design restrictions of the test set ($p \gg 0.05$, entries 1–5, Table 2). In contrast, a significant positive correlation between the ΔT_m values of the probe–cDNA duplexes and modification density was observed ($p < 0.05$, $r_s \gg 0$, entries 6 and 7, Table 2). This, along with a significant negative correlation with the longest unmodified stretch metric ($p < 0.05$, $r_s \ll 0$, entries 8 and 9, Table 2), suggests that densely modified Invader probes with short unmodified stretches yield probe-target duplexes displaying the most prominent increases in T_m values relative to the corresponding unmodified DNA duplexes. Accordingly, average ΔT_ms of ~17 °C and ~10 °C were observed for probe:cDNA duplexes entailing probe strands with modification densities of >20% and ≤20%, respectively, Table 1). Moreover, negative correlations approaching significance were observed between the ΔT_m values of probe-target duplexes and the GC-content or T_m of the corresponding unmodified DNA duplexes (entries 10–13, Table 2). This suggests that the stabilizing impact of the 2'-O-(pyren-1-yl)methyl-RNA monomers in probe:cDNA duplexes is more pronounced when Invader probes are designed to target lower melting AT-rich regions. Accordingly, probe–cDNA duplexes with lower GC-content display greater relative increases (ΔT_m ~16 °C and ~9.5 °C for duplexes with GC% of <50% and ≥50%, respectively, Table 1).

Table 2. Selected data pertaining to denaturation properties and dsDNA-recognition potential from Spearman's rank-order correlation analysis of parameter pairs [a].

Entry	Parameter Pair	Correlation Coefficient r_s	p-Value
1	probe duplex ΔT_m × length	−0.169	0.664
2	probe duplex ΔT_m × GC%	−0.328	0.388
3	probe duplex ΔT_m × mod%	−0.037	0.924
4	probe duplex ΔT_m × #mod	−0.091	0.815
5	probe duplex ΔT_m × stretch	−0.244	0.526
6	5'-ON:cDNA ΔT_m × mod%	0.774	0.009
7	3'-ON:cDNA ΔT_m × mod%	0.661	0.037
8	5'-ON:cDNA ΔT_m × stretch	−0.810	0.005
9	3'-ON:cDNA ΔT_m × stretch	−0.853	0.002
10	5'-ON:cDNA ΔT_m × GC%	−0.762	0.010
11	3'-ON:cDNA ΔT_m × GC%	−0.518	0.125
12	5'-ON:cDNA ΔT_m × dsDNA T_m	−0.697	0.025
13	3'-ON:cDNA ΔT_m × dsDNA T_m	−0.515	0.128
14	TA × probe duplex ΔT_m	−0.567	0.112
15	TA × 5'-ON:cDNA ΔT_m	0.583	0.099
16	TA × 3'-ON:cDNA ΔT_m	0.333	0.381
17	TA × probe duplex T_m	−0.778	0.014
18	TA × GC%	−0.359	0.343
19	TA × mod%	0.662	0.052

Table 2. Cont.

Entry	Parameter Pair	Correlation Coefficient r_s	p-Value
20	TA × #mod	0.822	0.007
21	ΔG_{rec}^{310} × TA	−0.814	0.014
22	ΔG_{rec}^{310} × mod%	−0.583	0.129
23	ΔG_{rec}^{310} × #mod	−0.620	0.101

[a] For the complete dataset, see the Supplementary Materials.

2.3. Driving Force for Recognition of dsDNA Targets

The driving force for the Invader-mediated recognition of isosequential dsDNA targets—i.e., complementary DNA duplexes of identical length and sequence—can be assessed using T_m- or ΔG-based terms. Concerning the former, we define the thermal advantage as TA = 5'-ON:cDNA ΔT_m + 3'-ON:cDNA ΔT_m—probe duplex ΔT_m. Thus, prominently positive TA values are expected for double-stranded probes that are activated for the recognition of isosequential dsDNA targets. Indeed, eight of the ten Invader probes display TA values greater than 20 °C, indicating that these probes are activated for the recognition of complementary dsDNA regions (Table 1).

The Spearman rank-order correlation analysis of the dataset indicates that there are correlations approaching significance between TA values and the metrics used to calculate the term (entries 14—16, Table 2). Moreover, a strongly negative correlation was observed between TA and Invader T_m values (entry 17, Table 2), indicating that low-melting Invader probes exhibit the most pronounced driving forces for the recognition of dsDNA. However, there is no correlation between TA values and GC-content (entry 18, Table 2). Importantly, a significant positive correlation between TA values and modification density or number of modifications was observed (entries 19 and 20, Table 2). Accordingly, the quadruply and most densely modified Invader probes display the most prominent TA values (TAs between 38.0 and 44.5 °C for **INV2**, **INV7** and **INV10**, Table 1).

Alternatively, the available free energy for the recognition of an isosequential dsDNA target at 310 K can be determined as ΔG_{rec}^{310} = ΔG^{310} (5'-ON:cDNA) + ΔG^{310} (3'-ON:cDNA)−ΔG^{310} (probe duplex)−ΔG^{310} (dsDNA) (Table 1 and Table S2). Thermodynamic parameters were derived from thermal denaturation curves via the baseline-fitting method (Tables S2–S4) [39]. Prominently negative values indicate a probe with a strong thermodynamic driving force for the recognition of isosequential dsDNA targets. In agreement with the TA-based conclusions, Invader probes are prominently activated for dsDNA-recognition (ΔG_{rec}^{310} between −7 and −56 kJ/mol, Table 1). This is due to the labile nature of the Invader probes ($\Delta\Delta G^{310}$ values, calculated relative to the corresponding unmodified dsDNA target, range between −1 kJ/mol and +25 kJ/mol; averaging +12 kJ/mol, Table S2) and the prominent stability of the probe-target duplexes ($\Delta\Delta G^{310}$ values range between +19 kJ/mol and −33 kJ/mol, averaging −7 kJ/mol, Table S2). The driving force for dsDNA-recognition is generally due to favorable changes in enthalpy (ΔH_{rec} << 0 kJ/mol, Table S3) [40]. This reflects that the formation of probe duplexes is considerably less enthalpically favorable than the corresponding probe:cDNA duplexes ($\Delta\Delta H$ values range between +174 kJ/mol and +372 kJ/mol, Table S3), which, again, is due to the energetic hotspots and the ensuing violation of the neighbor exclusion principle.

The Spearman rank-order correlation analysis of the dataset confirmed the expected negative correlation between ΔG_{rec}^{310} and TA values (entry 21, Table 2), i.e., negative ΔG_{rec}^{310} values correlate with positive TA values. Negative correlations approaching significance between ΔG_{rec}^{310} values and modification density or number of modifications were also observed (entries 22 and 23, Table 2). Hence, both the ΔG_{rec}^{310} and TA parameters indicate that the thermodynamic gradient for dsDNA recognition is maximized when densely modified Invader probes are used.

2.4. Recognition of Mixed-Sequence Model DNA Hairpin Targets

The dsDNA-recognition characteristics of the initial set of Invader probes were first evaluated using an electrophoretic mobility shift assay (EMSA), in which the probes were incubated with 3′-digoxigenin (DIG)-labeled DNA hairpin (DH) model targets (Figure 2). Each DNA hairpin (**DH1-DH10**) comprises a double-stranded stem that is complementary to the corresponding Invader probe, and in which one end is linked by a decameric thymidine (T_{10}) loop. The resulting hairpins are high-melting (T_ms for **DH1-DH10** between 62 and 82 °C, Table S6). This and the unimolecular nature of the DNA hairpins ensures that both target strands are present in equimolar amounts and unlikely to fray. Invader-mediated recognition of the double-stranded stem region is expected to result in the formation of a ternary recognition complex (RC) that manifests itself as a slower-moving band relative to the DNA hairpin when mixtures are resolved by non-denaturing polyacrylamide gel electrophoresis (nd-PAGE).

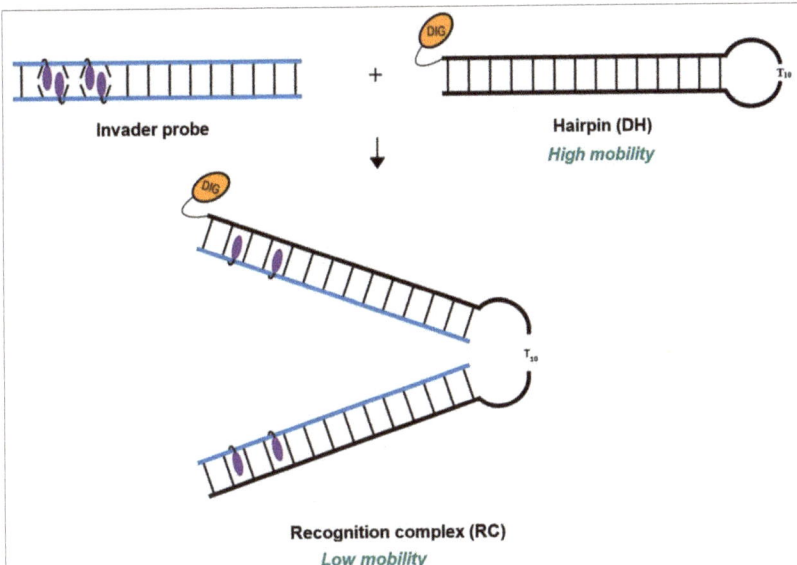

Figure 2. Illustration of EMSA assay used to evaluate dsDNA-recognition of Invader probes.

In the initial screen, a 100-fold molar excess of each Invader probe was incubated with the corresponding DNA hairpin target for 15 h at 37 °C in a HEPES buffer containing 100 mM of NaCl and 5 mM of $MgCl_2$ (Figure 3) [41]. Essentially complete dsDNA-recognition was observed for four of the ten probes (**INV2, INV7, INV8,** and **INV10**), while five probes resulted in moderate recognition (40–70%, **INV1, INV3, INV5, INV6,** and **INV9**) (Figure 3 and Table 3). No dsDNA recognition was observed for **INV4**; this was a surprising result considering that this probe has been used to detect chromosomal DNA targets under non-denaturing FISH conditions [24].

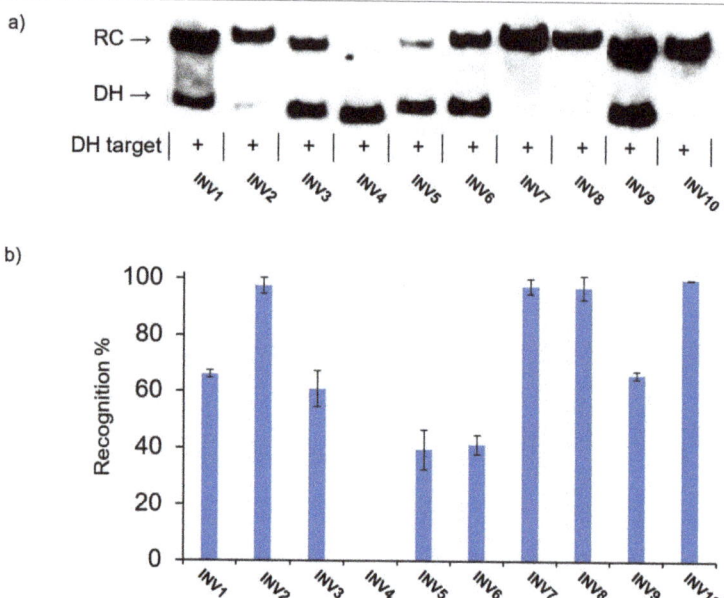

Figure 3. (**a**) Representative electrophoretograms from recognition experiments in which a 100-fold molar excess of Invader probes **INV1-INV10** was incubated with their respective DNA hairpin targets **DH1-DH10**. (**b**) Histograms depict averaged results from at least three recognition experiments with error bars denoting standard deviation. RC = recognition complex. DH = DNA hairpin. DIG-labeled DNA hairpins **DH1-DH10** (34.4 nM, sequences shown in Table S6) were incubated with the corresponding Invader probe in HEPES buffer (50 mM of HEPES, 100 mM of NaCl, 5 mM of MgCl$_2$, pH 7.2, 10% sucrose, 1.44 mM of spermine tetrahyrdochloride) at 37 °C for 15 h. Incubation mixtures were resolved on 12% non-denaturing TBE-PAGE slabs (~70 V, ~4 °C, ~1.5 h).

Table 3. Rec$_{100\times}$ and C$_{50}$ values for recognition of model DNA hairpin targets when using the corresponding Invader probes [a].

Probe	Rec$_{100\times}$ (%)	C$_{50}$ (µM)
INV1	66 ± 1.3	1.3
INV2	97 ± 2.8	0.2
INV3	60 ± 6.2	2.9
INV4	<5	ND
INV5	39 ± 7.0	4.1
INV6	41 ± 3.4	>10
INV7	97 ± 2.6	0.7
INV8	96 ± 4.2	0.6
INV9	66 ± 1.4	1.5
INV10	99 ± 0.0	0.2
OPT6	42 ± 3.4	>10
OPT8	99 ± 0.4	0.2
OPT9	94 ± 8.9	0.6

[a] Rec$_{100\times}$ = level of DNA hairpin recognition using 100-fold molar probe excess (Figure 3). C$_{50}$ values for **INV1-INV10** and **OPT6/8/9** were determined from dose–response curves shown in Figure 4 and Figure S24, respectively. "±" = standard deviation. ND = not determined due to low levels of recognition in preliminary screen.

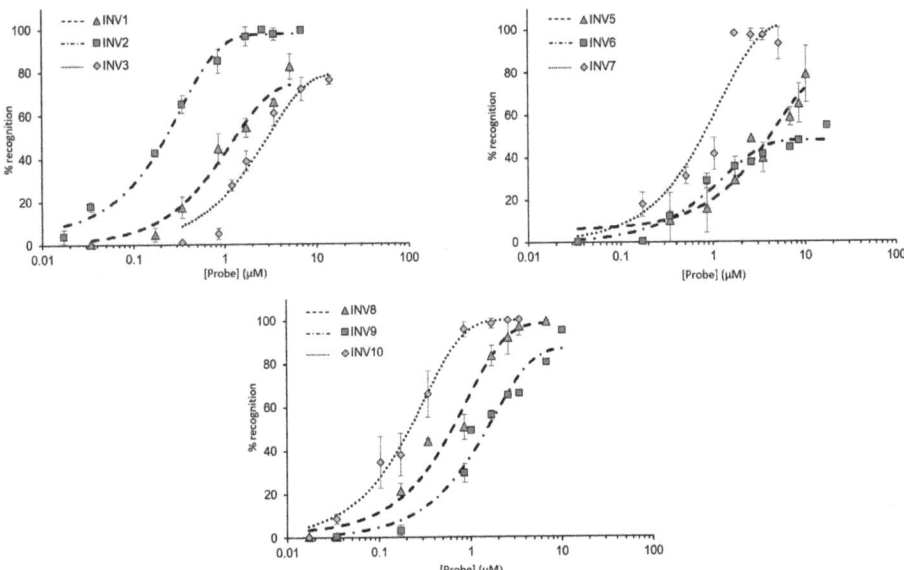

Figure 4. Dose–response curves for recognition of DNA hairpins using **INV1-INV3** (upper left panel), **INV5-INV7** (upper right panel), and **INV8-INV10** (lower panel). Probes were incubated with their respective DNA hairpin targets for 15 h at 37 °C. Experimental conditions are as described in Figure 3, except for variable probe concentrations. Bars denote standard deviations. For the corresponding electrophoretograms, see Figures S11–S13.

Next, dose–response relationships were established to determine C_{50} values, i.e., the probe concentrations resulting in 50% recognition of a corresponding DNA hairpin target (Figure 4). **INV2** and **INV10** displayed the most efficient recognition ($C_{50} \sim 0.2$ μM, Table 3), followed by **INV7** and **INV8** (C_{50} = 0.6–0.7 μM, Table 3). Moderately efficient recognition was observed for **INV1**, **INV3**, **INV5**, and **INV9** (C_{50} = 1.3–4.1 μM, Table 3), whilst **INV6** only displayed marginal recognition of **DH6** ($C_{50} \geq 10$ μM, Table 3).

The Spearman's rank-order correlation analysis of the dataset indicates the presence of significant correlations between the observed C_{50} values and the modification density, number of modifications, or longest unmodified stretch of the Invader probes (entries 1–3, Table 4). Accordingly, the quadruply and most densely modified **INV2**, **INV7**, and **INV10** probes display the lowest C_{50} values, while all but one of the less densely modified probes (mod% < 21.5%) display moderate or no recognition of DNA hairpin targets (Table 3).

Table 4. Selected data from our Spearman's rank-order correlation analysis pertaining to Invader-mediated recognition of DNA hairpins and chromosomal DNA in FISH assays [a].

Entry	Parameter Pair	Correlation Coefficient r_s	p-Value
1	$C_{50} \times$ mod%	−0.850	0.008
2	$C_{50} \times$ #mod	−0.732	0.039
3	$C_{50} \times$ stretch	0.735	0.038
4	$C_{50} \times TA$	−0.598	0.156
5	$C_{50} \times \Delta G_{rec}^{310}$	0.772	0.072
6	$C_{50} \times$ 5'-ON:cDNA ΔT_m	−0.782	0.022
7	$C_{50} \times$ 3'-ON:cDNA ΔT_m	−0.566	0.144
8	$C_{50} \times$ 5'-ON:cDNA ΔG^{310}	0.604	0.113
9	$C_{50} \times$ 3'-ON:cDNA ΔG^{310}	0.749	0.032
10	$C_{50} \times$ probe duplex T_m	0.032	0.945
11	$C_{50} \times$ probe duplex ΔT_m	0.010	0.983
12	$C_{50} \times$ probe duplex ΔG^{310}	−0.187	0.723

Table 4. Cont.

Entry	Parameter Pair	Correlation Coefficient r_s	p-Value
13	C_{50} × probe duplex $\Delta\Delta G^{310}$	0.138	0.795
14	C_{50} × GC%	0.323	0.435
15	d-FISH × mod%	0.713	0.021
16	d-FISH × C_{50}	−0.853	0.007
17	nd-FISH × mod%	0.738	0.015
18	nd-FISH × C_{50}	−0.710	0.049
19	d-FISH × #mod	0.558	0.094
20	d-FISH × stretch	−0.711	0.021
21	nd-FISH × #mod	0.547	0.102
22	nd-FISH × stretch	−0.590	0.073
23	nd-FISH × TA	0.505	0.165
24	nd-FISH × ΔG^{310}_{rec}	−0.583	0.129
25	d-FISH × TA	0.274	0.476
26	d-FISH × ΔG^{310}_{rec}	−0.319	0.441
27	d-FISH × GC%	−0.099	0.785
28	nd-FISH × GC%	0.191	0.597

[a] For the complete dataset, see the Supplementary Materials.

The level of recognition observed for **INV8** is surprising given that it is only ~20% modified (Table 3). However, it should be noted that **INV8** displays favorable TA and ΔG^{310}_{rec} values (TA = 26.5 °C and ΔG^{310}_{rec} = −52 kJ/mol, Table 1). This is relevant since correlations approaching significance were also observed between C_{50} values and TA or ΔG^{310}_{rec} values (entries 4 and 5, Table 4).[r] Further along these lines, correlations approaching significance were observed between C_{50} values and measures of probe:cDNA duplex stability (entries 6–9, Table 4), indicating that the formation of stable probe-target duplexes is an important driver of DNA hairpin recognition. The relatively high levels of hairpin recognition observed with **INV8** may, therefore, be linked to the high stability of the corresponding probe-target duplexes (ΔT_m average of +12.5 °C, $\Delta\Delta G^{310}$ averaging −12.5 kJ/mol, Table 1 and Table S2, respectively).

Somewhat surprisingly, no correlation was observed between the C_{50} values and the T_m, ΔT_m, ΔG^{310}, $\Delta\Delta G^{310}$ values, or the GC-content of the Invader probes (entries 10–14, Table 4). This indicates that the absolute or relative stability of Invader probes—at least within the design constraints of the test set—does not impact hairpin recognition.

The binding specificities of high-affinity Invader probes were evaluated by incubating a 100-fold molar excess of **INV2** and **INV10** with DNA hairpins featuring stems that differ in sequence at one or two positions relative to the probes (sequences shown in Table S6). Both probes fully discriminated these DNA hairpins, while resulting in complete recognition of the complementary targets (Figure 5). Remarkably, this demonstrates that high-affinity Invader probes can distinguish targets with ~94% sequence homology (i.e., fifteen of the sixteen bps are identical between **DH2** and **DH2m**). This finding hints at interesting single nucleotide polymorphism (SNP) applications for Invader probes.

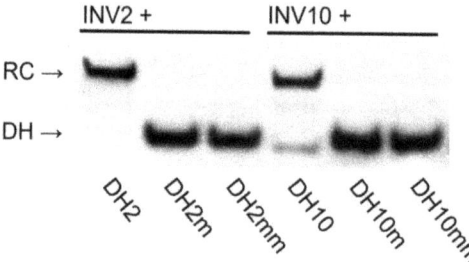

Figure 5. Binding specificity of Invader probes. A 100-fold molar probe excess was incubated with corresponding DNA hairpins featuring stems of identical sequence or differing in sequence at one ("m") or two positions ("mm") relative to the probes (37 °C, 15 h). For sequences of DNA hairpins, see Table S6. Conditions are as described in Figure 3. Data previously shown in [23]—reproduced with permission from the Royal Society of Chemistry.

2.5. Targeting Chromosomal DNA—Fluorescence In Situ Hybridization (FISH) Assays

Next, the ten Cy3-labeled Invader probes were evaluated for their ability to recognize corresponding DNA target regions within the *DYZ-1* gene of the bovine Y chromosome in the context of FISH assays. Thus, INV1–INV10 were incubated with fixed interphase nuclei from a male bovine kidney cell line under denaturing (d) or non-denaturing (nd) FISH conditions. The d-FISH assay was expected to yield information about the maximal recognition capacity of each probe, since access to the chromosomal DNA target regions is facilitated by high incubation temperatures. The nd-FISH experiments, on the other hand, were expected to reveal if a probe can recognize the corresponding target at more physiologically relevant conditions. Successful target recognition was expected to manifest itself in the form of a single, punctate fluorescent signal.

The two high-affinity probes, **INV2** and **INV10**, were found to recognize the DNA targets with excellent efficiency in d-FISH assays (i.e., ~90% of the analyzed nuclei displayed a single, intense, punctate signal against a low level of background; Figure 6 left column and Table 5). As previously reported [24], excellent target recognition was also observed with **INV4**. This was surprising considering the low driving force for dsDNA-recognition ($TA = 1.5\ °C$ and $\Delta G^{310}_{rec} = -7\ kJ/mol$, Table 1) and the lack of DNA hairpin recognition (Figure 3). While the reasons for the different performance in the DNA hairpin and d-FISH experiments observed with **INV4** are not fully understood, it should be noted that the experimental conditions (e.g., buffers, probe concentrations) are quite different, which may impact probe binding. Five probes (**INV3** and **INV6-INV9**) displayed single punctate signals in 40–60% of the analyzed nuclei under d-FISH conditions (Table 5 and Figures S15–S17 left column). Two probes failed to yield acceptable signal profiles, i.e., **INV1**, resulting in the formation of multiple signal blotches indicative of non-specific binding, and **INV5**, which did not produce signals of any kind (Table 5 and Figures S14 and S15, respectively).

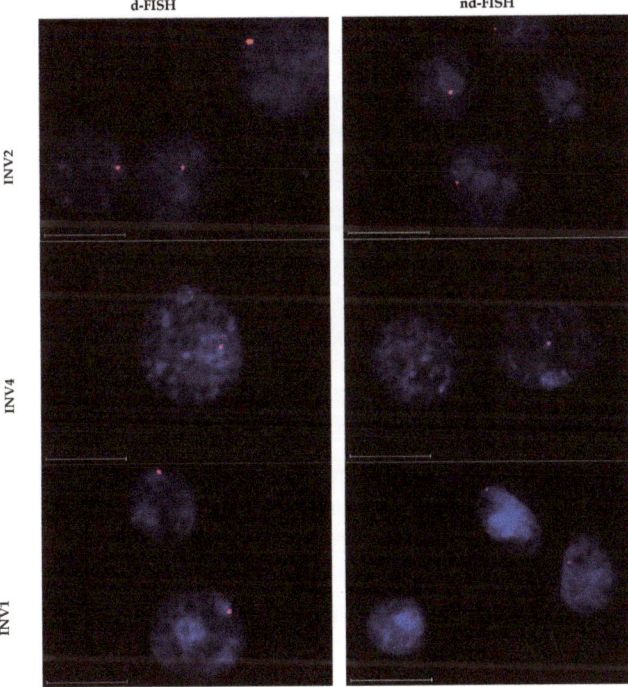

Figure 6. Representative images from FISH experiments in which *DYZ1*-targeting Invader probes

INV2, INV4, and INV10 (upper, middle and lower panels, respectively) were incubated with isolated nuclei from a bovine kidney cell line under denaturing (5 min, 80 °C, left) or non-denaturing (3 h, 37.5 °C, right) conditions. Fixed isolated nuclei were incubated with probes in a Tris buffer (20 mM of Tris-Cl, 100 mM of KCl, pH 8.0) and counterstained with DAPI. The images were obtained by overlaying Cy3 (red) and DAPI (blue) filter settings and adjusting the exposure. Nuclei are viewed at 60× magnification using a Nikon Eclipse T*i*-S inverted microscope. The scale bar represents 16 μm. For corresponding images for other Invader probes, see Figures S14–S17.

Table 5. Percent of nuclei presenting a single, punctate signal in d-FISH and nd-FISH assays when incubated with different Invader probes [a].

Probe	d-FISH	nd-FISH
INV1	0%	0%
INV2	~90%	~85%
INV3	~40%	~30%
INV4	~90%	~90%
INV5	0%	0%
INV6	~60%	~20%
INV7	~60%	~25%
INV8	~60%	~25%
INV9	~60%	0%
INV10	~90%	~90%
OPT6	~90%	~85%
OPT8	~90%	~75%
OPT9	~75%	~25%

[a] Incubation conditions are as described in Figure 6.

The probes largely retained their signaling capacities under nd-FISH conditions. Thus, **INV2**, **INV4**, and **INV10** yielded single, intense, punctate signals against a low background in 85%–90% of the analyzed nuclei (Figure 6 right column and Table 5). Moderately intense signals were observed for four of the probes in 20–30% of the nuclei (i.e., **INV3** and **INV6**-**INV8**, Figures S15 and S16, Table 5), while three of the probes (i.e., **INV1**, **INV5**, and **INV9**) did not produce discernable signals (Figures S14, S15 and S17 and Table 5). The diverging results observed for **INV9** under d-FISH vis-à-vis nd-FISH conditions indicate that this target region is inaccessible under non-denaturing conditions.

Hence, most of the studied Invader probes resulted in adequate-to-excellent recognition of chromosomal DNA targets under d-FISH and nd-FISH conditions. The Spearman's rank-order correlation analysis revealed that the signaling performance in the d-FISH and nd-FISH assays significantly correlates with the modification level of the probes and the observed C_{50} values (entries 15–18, Table 4). Along similar lines, correlations approaching significance were observed between the signaling performance in d-FISH and nd-FISH assays and the number of modifications or longest unmodified stretch (entries 19–22, Table 4). Correlations approaching significance were observed between nd-FISH signaling performance and T_A and ΔG_{rec}^{310} values, indicating that these metrics have some predictive value for nd-FISH, but not d-FISH, performance (entries 23–26, Table 4). Interestingly, signaling performance did not correlate with the GC-content of the target region (entries 27 and 28, Table 4).

The observed correlation with modification density provides a rationale for the excellent signaling characteristics of **INV2** and **INV10** (25–29% modified) and the moderate-to-poor signaling characteristics of most of the remaining probes. The signaling properties of two probes, i.e., **INV4** and **INV7**, however, are not easily rationalized. Thus, excellent signaling properties were observed for the sparsely modified **INV4** that failed to recognize the corresponding DNA hairpin target (Figure 3) and was far less activated for dsDNA-recognition than **INV2** and **INV10** (compare T_A and ΔG_{rec}^{310} values, Table 1). A distinguishing feature of **INV4** and its corresponding target region is the presence of two GGG/CCC-tracts, which we speculate may render the target region uniquely accessible due to the formation of non-canonical secondary structures [42]. An alternative explanation for the surprising signaling characteristics of **INV4** is that the corresponding target region is present six times within a single $DYZ-1$ repeat (which, in turn, is repeated ~6 × 10^4 times, Figure S1) [24], whilst the other target regions studied herein are only present once per

DYZ−1 repeat. The greater number of target sites may account for cooperative hybridization effects, resulting in a greater proportion of nuclei that present a signal. The modest signaling properties of **INV7** are perplexing given its high level of modification (25%), prominent activation for dsDNA-recognition ($TA = 38\,°C$ and $\Delta G_{rec}^{310} = -46$ kJ/mol, Table 1), and efficient hairpin recognition ($C_{50} = 0.7\,\mu M$, Table 3). We speculate that the corresponding chromosomal DNA target region is only partially accessible to **INV7** under these experimental conditions.

Control nd-FISH experiments, in which fixed nuclei were pre-treated with DNase I, RNase A, or Proteinase K prior to incubation with **INV2** or **INV10**, confirmed that the Invader probes target chromosomal DNA, rather than RNA or proteins. Thus, nuclei that were pre-treated with DNase I did not produce any signals (Figure S18), whereas pre-treatment with RNase A or Proteinase K continued to yield single punctate signals, albeit with lower intensity (Figure S19) [43].

Incubation of the Y-chromosome-targeting probes **INV2** and **INV10** with a female bovine endothelial cell line failed to produce signals under denaturing conditions (Figure 7), suggesting that Invader probes bind their chromosomal DNA targets with excellent specificity [43].

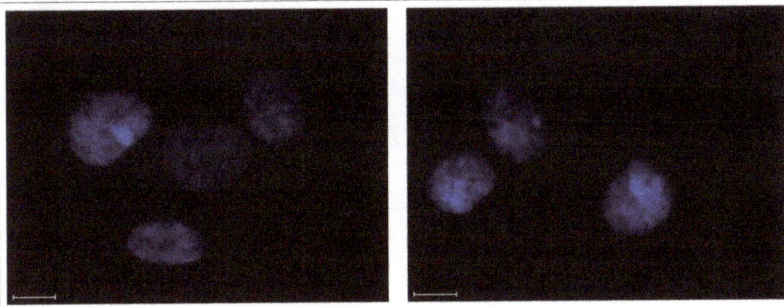

Figure 7. Images from d-FISH experiments in which **INV2** (left panel) and **INV10** (right panel) were incubated with fixed isolated female bovine endothelial nuclei. Note the absence of Cy3 signals. Incubation conditions and the image capture process were as described in Figure 6. Data previously shown in [23]—reproduced with permission from the Royal Society of Chemistry.

2.6. Design of Optimized Invader Probes

Having identified modification density as a key parameter for successful dsDNA-recognition, we set out to optimize three Invader probes that displayed poor-to-moderate signaling characteristics under nd-FISH conditions, i.e., **INV6**, **INV8**, and **INV9**. Thus, two or three additional hotspots were introduced to yield probes with modification densities of 27–33% (**OPT6**, **OPT8**, and **OPT9**; Table 1).

2.7. Thermal Denaturation and Thermodynamic Properties of Optimized Invader Probes

The more densely modified probes were found to be considerably less stable than the parent probes (T_ms ~20 °C lower and ΔG^{310} values ~12 kJ/mol higher on average; compare T_m and ΔG^{310} values for **INV6/8/9** and **OPT6/8/9**, Table 1, Tables S2 and S9, respectively). Moreover, the densely modified probe strands form more stable duplexes with cDNA than the parent counterparts (T_ms ~6 °C higher and ΔG^{310} values ~18 kJ/mol lower on average; compare T_ms and ΔG^{310} values, Table 1, Tables S2 and S9, respectively). Consequently, the driving forces for the recognition of isosequential dsDNA targets are substantially larger for the three redesigned probes compared to the parent counterparts (TA values between 43.0–55.5 °C vs. 13.5–26.5 °C and ΔG_{rec}^{310} values between −93 kJ/mol and −59 kJ/mol vs. between −52 kJ/mol and −19 kJ/mol, Table 1).

2.8. Recognition of Model DNA Hairpin Targets by Optimized Invader Probes

The dsDNA-recognition characteristics of the three optimized Invader probes were first evaluated using the aforementioned DNA hairpin assay (Figure 2). Thus, the probes were first screened at a 100-fold molar excess (Figure 8 and Figure S22) and then were more fully evaluated in dose–response experiments (Figures S23 and S24). Unlike the corresponding parent probes, **OPT8** and **OPT9** resulted in near-complete recognition of the hairpin targets when incubated at 100-fold molar excess (compare Rec$_{100x}$ values for **OPT8** and **OPT9** vs. **INV8** and **INV9**, Table 3). Surprisingly, **OPT6** resulted in similar levels of recognition of **DH6** as **INV6** (Rec$_{100x}$ ~40%, Table 3). The dose–response experiments verified these findings, as **OPT8** and **OPT9** displayed three- and five-fold reductions in their C$_{50}$ values relative to the parent probes, whilst **OPT6** displayed a C$_{50}$ value > 10 µM (Table 3).

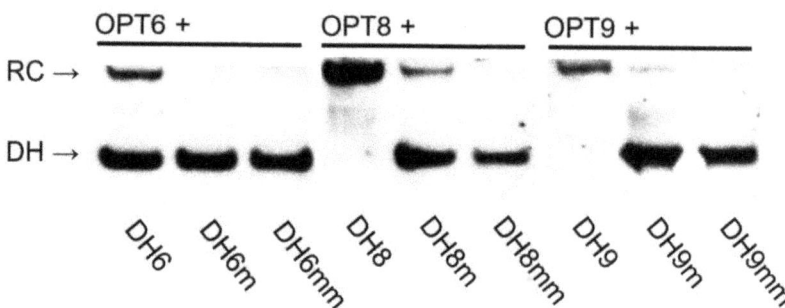

Figure 8. Representative electrophoretograms from recognition experiments in which a 100-fold molar excess of optimized Invader probes **OPT6**, **OPT8**, and **OPT9** was incubated with the corresponding DNA hairpins featuring stems of identical sequence or differing in sequence at one ("m") or two positions ("mm") relative to the probes (37 °C, 15 h). For sequences of DNA hairpins, see Table S6.

Importantly, complete discrimination of doubly mismatched DNA hairpins and merely trace recognition of the singly mismatched DNA hairpins was observed when the optimized high-affinity **OPT8** and **OPT9** probes were incubated at 100-fold molar excess (Figure 8).

2.9. Targeting Chromosomal DNA using Optimized Invader Probes

The optimized Invader probes were subsequently evaluated for their ability to recognize chromosomal DNA targets using the aforementioned d- and nd-FISH assays. Gratifyingly, improved signaling characteristics, relative to the parent probes, were observed for the optimized probes. Thus, 75%–90% of the nuclei display prominent, single, and punctate signals under d-FISH conditions (Figure 9 left column and Table 5). Along similar lines, ~85%, ~75% and ~25% of the nuclei displayed high-quality signals when **OPT6**, **OPT8** or **OPT9** were used under nd-FISH conditions, respectively, as compared to 0–25% with the parent probes (Figure 9 right column and Table 5). The higher signaling efficiency of **OPT6** vis-à-vis **OPT9** is surprising considering that the latter resulted in far more efficient recognition of the corresponding hairpin target. However, it should be noted that the experimental conditions (e.g., buffers, probe concentrations) are quite different between the two assays, which may impact the results. Nonetheless, the findings demonstrate that increasing the modification density of an Invader probe results in improved signaling characteristics, as per the conclusions of the Spearman's rank-order analysis. Thus, it is possible to design extensively modified Invader probes that enable sequence-unrestricted and highly specific recognition of chromosomal DNA targets. This important insight will facilitate future biotechnological applications utilizing Invader probes.

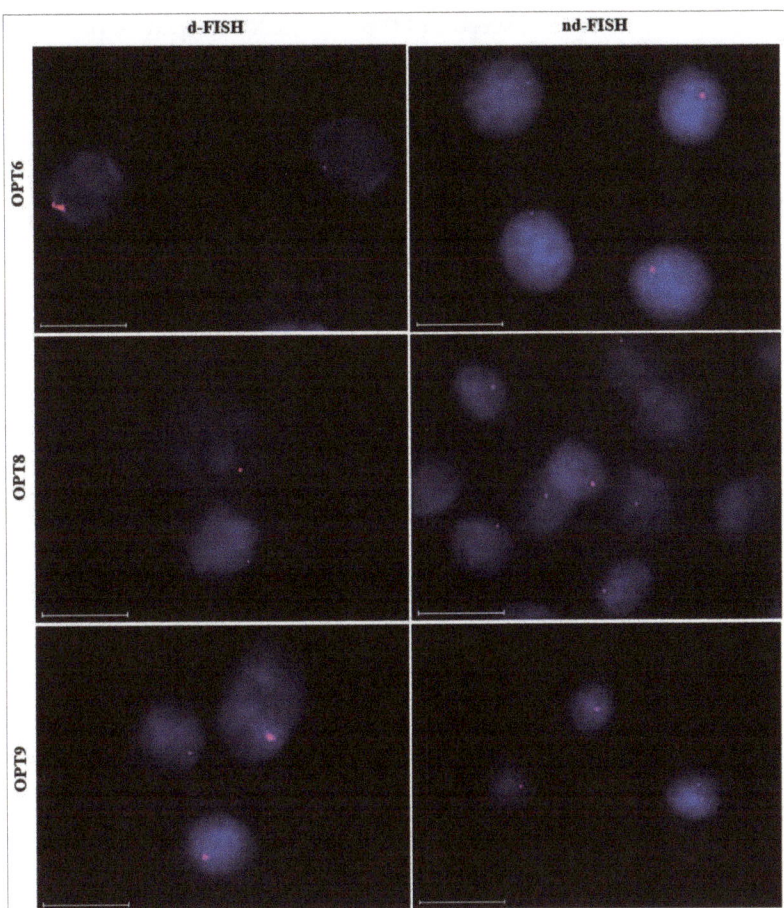

Figure 9. Representative images from FISH experiments in which optimized *DYZ1*-targeting Invader probes **OPT6**, **OPT8**, and **OPT9** (upper, middle, and lower panel, respectively) were incubated with isolated nuclei from a bovine kidney cell line under denaturing (5 min, 80 °C, left) or non-denaturing (3 h, 37.5 °C, right) conditions. Incubation conditions and the image capture process are as described in Figure 6.

3. Materials and Methods

3.1. Synthesis and Purification of Probe Strands

Individual Invader strands—i.e., oligodeoxyribonucleotides (ONs) modified with 2′-*O*-(pyren-1-yl)methyl-RNA monomers—were synthesized on an Expedite DNA synthesizer (0.2 μmol scale), using columns packed with long-chain alkylamine-controlled pore glass (LCAA-CPG, Glen Research, Sterling, VA, USA) solid support with a pore size of 500 Å. Standard protocols were used for the incorporation of DNA phosphoramidites. The 2′-*O*-(pyren-1-yl)methyl-RNA phosphoramidites were prepared as previously described for U monomer [44] and C/A monomers [37] and incorporated into ONs via extended hand-couplings (15 min, ~45-fold molar excess at a concentration of 0.02 M in anhydrous acetonitrile, using 0.01 M of 4,5-dicyanoimidazole as the activator) and oxidation (45 s), resulting in coupling yields of at least 85%. The Cy3-labeling of Invader strands was accomplished by incorporating a commercially available Cy3 phosphoramidite (Glen Research, Sterling, VA, USA) into ONs by hand-coupling (4,5-dicyanoimidazole, 3 min, anhydrous CH_3CN) (Glen Research, Sterling, VA, USA). Treatment with 32% aq. ammonia (55 °C, 17 h) ensured deprotection and cleavage from solid support. DMT-protected ONs were purified via ion-pair reverse-phase HPLC (Varian, Palo Alto, CA, USA) (Waters, XTerra MS C18 column: 0.05 M of triethyl ammonium acetate and acetonitrile gradient), followed by detritylation (80% acetic acid, 20 min) and precipitation (NaOAc, NaClO$_4$, acetone, −18 °C, 16 h). The purity (≥ 85%) and identity of the synthesized ONs was verified using analytical HPLC and MALDI-MS (Tables S1 and S7 and Figures S2–S4 and S16) recorded on a Quadrupole Time-of-Flight (Q-TOF) mass spectrometer (Waters Q-Tof Premier, Milford, MA, USA) using a 3-hydoxypicolinic acid matrix. Common reagents were obtained through VWR International (Radnor, PA) or Fisher Scientific (Hampton, NH).

3.2. Thermal Denaturation Experiments

ON concentrations were estimated using the following extinction coefficients (OD$_{260}$/μmol): G (12.01), A (15.20), T (8.40), C (7.05), pyrene (22.4) [45] and Cy3 (4.93) [46]. The thermal denaturation temperatures (T_ms) of the duplexes (1.0 μM final concentration of each strand) were determined on an UV/VIS spectrophotometer (Cary 100, Varian, Palo Alto, CA, USA) equipped with a 12-cell Peltier temperature controller and measured as the maximum of the first derivative of the thermal denaturation curves (A_{260} vs. T) recorded in medium salt buffer (T_m buffer: 100 mM of NaCl, 0.2 mM of EDTA, and pH 7.0 adjusted with 10 mM of Na$_2$HPO$_4$ and 5 mM of Na$_2$HPO$_4$). Strands were mixed in quartz optical cells with a path length of 1.0 cm and annealed by heating to 85 °C (2 min), followed by cooling to the starting temperature of the experiment. The temperature of the denaturation experiments ranged from at least 15 °C below the T_m to at least 15 °C above the T_m (although not above 95 °C). A temperature ramp of 1.0 °C/min was used in all experiments. The reported T_ms are the averages of at least two experiments within ± 1.0 °C.

3.3. Electrophoretic Mobility Shift Assays

The non-denaturing (nd)-PAGE assay was performed as previously described [24]. Thus, DNA hairpins (DH) (Integrated DNA Technologies, Coralville, IA, USA) were obtained from commercial sources and used without further purification. Hairpins were 3′-labeled with digoxigenin (DIG) using the 2nd generation DIG Gel Shift Kit (Roche Applied Bioscience, Penzberg, Germany), as recommended by the manufacturer. Briefly, 11-digoxigenin-ddUTP was incorporated at the 3′-end of the hairpin (100 pmol) using a recombinant terminal transferase. The reaction mixture was quenched through the addition of EDTA (0.05 M), diluted to 68.8 nM, and used without further processing. Solutions of Invader probes (concentrations as specified) were incubated with the corresponding DIG-labeled DNA hairpin (final concentration 34.4 nM) in HEPES buffer (50 mM of HEPES, 100 mM of NaCl, 5 mM of MgCl$_2$, pH 7.2, 10% sucrose, 1.44 mM of spermine tetrahydrochloride) at 37 °C for the specified time. Following incubation, loading dye (6 ×) was added and the mixtures were loaded onto 12% non-denaturing TBE-PAGE slabs (45 mM of tris-borate, 1 mM of EDTA; acrylamide:bisacrylamide (19:1)). Electrophoresis was performed using constant voltage (~70 V) at ~4 °C for ~1.5 h. The bands were subsequently blotted onto positively charged nylon membranes (~100 V, 30 min, ~4 °C) and cross-linked through exposure to UV light (254 nm, 5 × 15 W bulbs, 5 min). The membranes were then incubated with anti-digoxigenin-alkaline phosphatase F$_{ab}$ fragments, as recommended by the manufacturer, and transferred to a hybridization jacket. They were then incubated with the chemiluminescence substrate (CSPD) for 10 min at 37 °C, and chemiluminescence of the formed product was captured on X-ray films. Digital images of the developed X-ray films were obtained using a BioRad ChemiDoc^{TM} MP Imaging system (BioRad, Hercules, CA, USA), which was also used for densitometric quantification of the bands. The percentage of dsDNA-recognition was calculated as the intensity ratio between the recognition complex band and the unrecognized hairpin. An average of three independent experiments is reported along with standard deviations (±). The presented electrophoretograms are, in some instances, composite images of lanes from different runs. Non-linear regression was used to fit data points from the dose–response experiments. A script written for the "Solver" module in Microsoft Office Excel was used to fit data points from the dose–response experiments to the following equation: y = C + A (1 − e^-kt) where C, A, and k are fitting constants. The resulting equation was used to calculate C_{50} values by setting y = 50 and solving for t [47].

3.4. Spearman Rank-Order Correlation Analysis

A Spearman's rank-order correlation analysis was performed to identify correlations between parameter pairs and, ultimately, identify parameters that impact the Invader-mediated recognition of dsDNA targets. A wide range of parameters were considered. Spearman rank-order correlation coefficients (r_s) were calculated using the XRealStat function add-on for Microsoft Excel [48]. The ten Invader probes were ranked 1 to 10 for each studied parameter, and these rankings were compared to determine correlations between parameter pairs. For example, the probe with the highest C_{50} and most negative ΔG_{rec}^{310} values would be ranked "1", while the lowest C_{50} and least negative (or more positive) ΔG_{rec}^{310} values would be ranked 10. Invader probes with identical parameter values received averaged rankings for those parameters. The strength and direction of correlation between two ranked parameters was measured by Spearman's rank-order correlation coefficient r_s and deemed statistically significant if the associated p values were less than the α value of 0.05.

3.5. Cell Culture and Nuclei Preparation

Male bovine kidney cells (MDBK, ATCC: CCL-22, Bethesda, MD, USA) were maintained in DMEM with GlutaMax (Gibco, 10569-010) and 10% fetal bovine serum (Invitrogen, Waltham, MA, USA). Female bovine endothelial cells (CPAE, ATCC: CCL-209) were maintained in Eagle's Minimum Essential Medium (ATTC, 30-2003) and 20% fetal bovine serum (Invitrogen). The cells were cultured in separate 25 mL or 75 mL flasks at 38.5 °C in a 5% CO_2 atmosphere for 72–96 h to achieve 70–80% confluency. At this point, colcemid (Gibco KaryoMax, 15210-040) (65 µL per 5 mL of growth media) was added, and the cells were incubated at 37 °C and 5% CO_2 for an additional 20 min. At this point, the medium was replaced with pre-warmed 0.05% Trypsin-EDTA in DMEM to detach adherent cells (37 °C, up to 8 min). The cell suspension was transferred to a tube and centrifuged (10 min, 1000 rpm). The supernatant was discarded and the dislodged cell pellet was incubated with a hypotonic 75 mM KCl solution (5–8 mL, 20 min), followed by the addition of fixative (10 drops, MeOH:AcOH, 3:1 v/v) and further incubation with gentle mixing (10 min, room temperature). The suspension was centrifuged (1000 rpm, 10 min), the supernatant discarded, and additional fixative solution (5–8 mL) added to the nuclei suspension. This was followed by gentle mixing and incubation (30 min, room temperature). The centrifugation/resuspension/incubation with fixative solution steps was repeated three additional times. The final pellet—containing somatic nuclei—was resuspended in the fixative solution and stored at −20 °C until use.

3.6. Preparation of Slides for FISH Assays

The nuclei suspension was warmed to room temperature and resuspended in fresh fixative solution. Glass microscope slides were dipped in distilled water to create a uniform water layer across the slide. An aliquot of the nuclei suspension (3–5 µL or enough to cover the slide) was dropped onto the slide, while holding the slide at a 45° angle, and allowed to run down the length of the slide. The slides were then allowed to dry at a ~20° angle in an environmental chamber at 28 °C and a relative humidity of 38%.

3.7. Fluorescence In Situ Hybridization Experiments and Image Analysis

An aliquot of labeling buffer (~200 µL) consisting of 30 ng of Cy3-labeled Invader probe per 200 µL of PCR buffer (20 mM of Tris, 100 mM of KCl, pH 8.0) placed on each slide. Preliminary assay optimization studies (results not shown) revealed that this "1 × solution" resulted in the best qualitative signal-to-background ratio for the Invader probes under denaturing and non-denaturing conditions. As an exception hereto, **INV4** was used at 0.25× concentration to reduce background fluorescence.

When used in d-FISH assays, slides with labeling buffer were placed on a heating block (5 min, 80 °C) and covered with a lid to prevent evaporation of the labeling buffer. When used in nd-denaturing FISH assays, slides with labeling buffer were placed in a glass culture disk, covered with a lid, and incubated in an oven (3 h, 37.5 °C). Slides for both d-FISH and nd-FISH experiments were subsequently washed (3 min, 37.5 °C) in a chamber with TE Buffer (10 mM of Tris, 1 mM of EDTA, pH 8.0) and allowed to dry at room temperature. Once dried, Gold SlowFade plus DAPI (3 µL, Invitrogen, Waltham, MA, USA) was placed directly on each slide, and a round glass coverslip was mounted for fluorescence imaging. A Nikon Eclipse Ti-S Inverted Microscope (Nikon Instruments, Melville, NY, USA), equipped with a SOLA SMII LED light source system and Cy3 and DAPI filter sets, was used to visualize nuclei at 60× magnification. Images of the fluorophore labeled nuclei were captured using a 14-bit CoolSNAP HQ2 cooled CCD camera and processed with NIS-Elements BR 4.20 software.

Control experiments, in which fixed nuclei from the MDBK cells were pre-treated with DNase, RNase, or proteinase prior to incubation with Invader probes, were carried out as follows. DNase pre-treatment: 3 µL of cloned RNase-free DNase I (Takara N101 JF) was mixed with 50 µL 1 × Reaction Buffer (diluted 10 × Cloned DNase I Buffer II, Takara A301) per the manufacturer's recommendation. The solution was pipetted onto slides with fixed nuclei in 50 µL amounts. The slides were incubated with the DNase I solution for 20 min at 37.5 °C and then rinsed with TE buffer. RNase pre-treatment: 1 µL of RNase A (5 mg/mL, Fisher reagents BP2539-100) in 100 µL of buffer (10 mM of Tris-HCl, pH 6.5) was placed in 50 µL amounts on slides and incubated for 15 min at 37.5 °C and then rinsed with TE buffer. Proteinase pre-treatment: 1 µL of Proteinase K (6.25 µg/mL, Fisher BioReagents, BP1700-100) was added to 200 µL of buffer (10 mM of Tris-HCl, pH 7.5). The fixed nuclei were incubated with 50 µL of this solution for 10 min at 37.5 °C and then rinsed with TE buffer.

The assessment of signal coverage, i.e., the percentage of nuclei displaying representative signals, was based on an evaluation of >100 nuclei per Invader probe at d-FISH and nd-FISH assay conditions (Table 5).

4. Conclusions

Invader probes, i.e., DNA duplexes featuring +1 interstrand zipper arrangements of intercalator-functionalized nucleotides such as 2′-O-(pyren-1-yl)methyl-RNA, allow for the robust and highly specific, mixed-sequence recognition of complementary double-stranded DNA target regions. Thus, the successful recognition of a series of model DNA hairpins and chromosomal DNA regions is demonstrated. The modification density is the single-most important design parameter impacting the thermal denaturation and dsDNA-recognition properties of Invader probes. Thus, four of six densely modified Invader probes (modification densities $\geq 25\%$) displayed particularly promising signaling characteristics in FISH assays under non-denaturing conditions, i.e., the formation of intense, single, punctate signals against a low fluorescence background in $\geq 75\%$ of isolated interphase nuclei. The signaling performance is not limited by the GC-content of the target regions, as successful recognition was demonstrated for target regions with GC-contents between 36% and 71%. The modification density also impacts signaling performance in denaturing FISH assays, the efficiency of DNA hairpin recognition, as well as metrics quantifying the driving force for dsDNA-recognition (i.e., TA and ΔG_{rec}^{310} values) and the stability of probe-target duplexes (i.e., ΔT_m or $\Delta\Delta G^{310}$ values for probe-target duplexes). In contrast, the modification density has a limited impact on the stability of the probe (i.e., ΔT_m or $\Delta\Delta G^{310}$ values for probe duplexes). We speculate that a high modification density results in a perturbed probe that exposes the pyrene moieties, allowing them to contact the target dsDNA and initiate the unwinding process. Identification of the modification density as a key design parameter enabled improvement of three Invader probes with mediocre signaling characteristics in nd-FISH assays into probes displaying improved signaling performance.

Based on the findings from the present and prior studies, we offer the following recommendations for the design of nd-FISH Invader probes:

(i) Invader probes should be densely modified ($\geq 25\%$) and only feature short unmodified segments. Given the nature of the energetic hotspots (i.e., +1 interstrand zipper arrangements of 2′-O-(pyren-1-yl)methyl RNA monomers), an Invader probe can, at most, be 50% modified. Invader probes display exceptional binding specificity, though binding to singly mismatched dsDNA targets may be observed for very densely modified probes. If non-specific binding is observed, the modification density should be reduced.

(ii) The energetic hotspots of Invader probes should be constructed using 2′-O-(pyren-1-yl)methyl RNA pyrimidine monomers, whilst the corresponding guanine monomers are to be avoided; the adenine monomers are acceptable [37]. This maximizes the driving force for dsDNA-recognition as particularly stable probe-target duplexes are formed, since the intercalating pyrene moiety stacks strongly with 3′-flanking purines [37]. Thus, it is recommended that 5′-BC-3′ steps (and B = G in particular) are omitted for the introduction of energetic hotspots. This sets the practical upper limit of the probe's modification density [49].

These design recommendations, coupled with the straightforward synthesis of the requisite 2′-O-(pyren-1-yl)methyl RNA pyrimidine building blocks [37,44], is expected to facilitate the design and use of Invader probes for a broad range of applications in the life sciences.

Supplementary Materials: The following are available online at https://www.mdpi.com/article/10.3390/molecules28010127/s1, Figures S1–S24; Tables S1–S11; zipper nomenclature definition; additional discussion; supplementary references [23,24,38].

Author Contributions: Conceptualization, methodology, validation, formal analysis, and writing—review and editing, C.P.S., R.G.E. and P.J.H.; validation, investigation, and data curation, C.P.S. and R.G.E.; resources, S.K. and R.G.E.; writing—original draft preparation and visualization, C.P.S.; supervision, project administration, and funding acquisition, P.J.H. All authors have read and agreed to the published version of the manuscript.

Funding: Preliminary contributions by M.-G. Guthrie (Brigham Young Univ. Idaho)—supported by an INBRE Institutional Development Award (IDeA) from the National Institute of General Medicinal Science of the National Institute of Health under grant number P20 GM103408—are appreciated. This study was also supported by the Higher Education Research Council, Idaho State Board of Education [awards IF13-001, IF14-012, IGEM-004].

Institutional Review Board Statement: Not applicable.

Informed Consent Statement: Not applicable.

Data Availability Statement: Not applicable.

Conflicts of Interest: PJH is an inventor on patents pertaining to Invader probes, which have been issued to the University Idaho.

Sample Availability: Samples of Invader probes may be requested from the authors.

References and Notes

1. Muangkaew, P.; Vilaivan, T. Modulation of DNA and RNA by PNA. *Bioorg. Med. Chem. Lett.* **2020**, *30*, 127064. [CrossRef] [PubMed]
2. Brodyagin, N.; Katkevics, M.; Kotikam, V.; Ryan, C.A.; Rozners, E. Chemical approaches to discover the full potential of peptide nucleic acids in biomedical applications. *Beilstein J. Org. Chem.* **2021**, *17*, 1641–1688. [CrossRef] [PubMed]
3. Duca, M.; Vekhoff, P.; Oussedik, K.; Halby, L.; Arimondo, P.B. The triple helix: 50 years later, the outcome. *Nucleic Acids Res.* **2008**, *36*, 5123–5138. [CrossRef] [PubMed]
4. Hari, Y.; Obika, S.; Imanishi, T. Towards the sequence-selective recognition of double-stranded DNA containing pyrimidine-purine interruptions by triplex-forming oligonucleotides. *Eur. J. Org. Chem.* **2012**, 2875–2887. [CrossRef]

5. Dervan, P.B.; Edelson, B.S. Recognition of the DNA minor groove by pyrrole-imidazole polyamides. *Curr. Opin. Struct. Biol.* **2003**, *13*, 284–299. [CrossRef] [PubMed]
6. Kawamoto, Y.; Bando, T.; Sugiyama, H. Sequence-specific DNA binding pyrrole-imidazole polyamides and their applications. *Bioorg. Med. Chem.* **2018**, *26*, 1393–1411. [CrossRef]
7. Dragulescu-Andrasi, A.; Rapireddy, S.; Frezza, B.M.; Gayathri, C.; Gil, R.R.; Ly, D.H. A simple γ-backbone modification preorganizes peptide nucleic acid into a helical structure. *J. Am. Chem. Soc.* **2006**, *128*, 10258–10267. [CrossRef]
8. Bahal, R.; Sahu, B.; Rapireddy, S.; Lee, C.-M.; Ly, D.H. Sequence-unrestricted, Watson-Crick recognition of double helical B-DNA by (R)-MiniPEG-γPNAs. *ChemBioChem* **2012**, *131*, 56–60. [CrossRef]
9. Bohländer, P.R.; Vilaivan, T.; Wagenknecht, H.-A. Strand displacement and duplex invasion into double-stranded DNA by pyrrolidinyl peptide nucleic acids. *Org. Biomol. Chem.* **2015**, *13*, 9223–9230. [CrossRef]
10. Thadke, S.A.; Hridya, V.M.; Perera, J.D.R.; Gil, R.R.; Mukherjee, A.; Ly, D.H. Shape selective bifacial recognition of double helical DNA. *Commun. Chem.* **2018**, *1*, 79. [CrossRef]
11. Zheng, H.; Botos, I.; Clausse, V.; Nikolayevskiy, H.; Rastede, E.E.; Fouz, M.F.; Mazur, S.J.; Appella, D.H. Conformational constraints of cyclopentane peptide nucleic acids facilitate tunable binding to DNA. *Nucleic Acids Res.* **2021**, *49*, 713–725. [CrossRef] [PubMed]
12. Suparpprom, C.; Vilaivan, T. Perspectives on conformationally constrained peptide nucleic acid (PNA): Insights into the structural design, properties and applications. *RSC Chem. Biol.* **2022**, *3*, 648–697. [CrossRef] [PubMed]
13. Lohse, J.; Dahl, O.; Nielsen, P.E. Double duplex invasion by peptide nucleic acid: A general principle for sequence-specific targeting of double-stranded DNA. *Proc. Natl. Acad. Sci. USA* **1999**, *96*, 11804–11808. [CrossRef] [PubMed]
14. Aiba, Y.; Honda, Y.; Komiyama, M. Promotion of double-duplex invasion of peptide nucleic acids through conjugation with nuclear localization signal peptide. *Chem. Eur. J.* **2015**, *21*, 4021–4026. [CrossRef] [PubMed]
15. Aiba, Y.; Shibata, M.; Shoji, O. Sequence-specific recognition of double-stranded DNA by peptide nucleic acid forming double-duplex invasion complex. *Appl. Sci.* **2022**, *12*, 3677. [CrossRef]
16. Filichev, V.V.; Christensen, U.B.; Pedersen, E.B.; Babu, B.R.; Wengel, J. Locked nucleic acids and intercalating nucleic acids in the design of easily denaturing nucleic acids: Thermal stability studies. *ChemBioChem* **2004**, *5*, 1673–1679. [CrossRef] [PubMed]
17. Filichev, V.V.; Vester, B.; Hansen, L.H.; Pedersen, E.B. Easily denaturing nucleic acids derived from intercalating nucleic acids: Thermal stability studies, dual duplex invasion and inhibition of transcription start. *Nucleic Acids Res.* **2005**, *33*, 7129–7137. [CrossRef]
18. Yotapan, N.; Nim-anussornkul, D.; Vilaivan, T. Pyrrolidinyl peptide nucleic acid terminally labeled with fluorophore and end-stacking quencher as a probe for highly specific DNA sequence discrimination. *Tetrahedron* **2016**, *72*, 7992–7999. [CrossRef]
19. Asanuma, H.; Niwa, R.; Akahane, M.; Murayama, K.; Kashida, H.; Kamiya, Y. Strand-invading linear probe combined with unmodified PNA. *Bioorg. Med. Chem.* **2016**, *24*, 4129–4137. [CrossRef] [PubMed]
20. Nakamura, S.; Kawabata, H.; Fujimoto, K. Double duplex invasion of DNA induced by ultrafast photo-cross-linking using 3-cyanovinylcarbazole for antigene methods. *Chem. Commun.* **2017**, *53*, 7616–7619. [CrossRef]
21. Hibino, M.; Aiba, Y.; Shoji, O. Cationic guanine: Positively charged nucleobase with improved DNA affinity inhibits self-duplex formation. *Chem. Commun.* **2020**, *56*, 2546–2549. [CrossRef] [PubMed]
22. Emehiser, R.G.; Hrdlicka, P.J. Chimeric γPNA–Invader probes: Using intercalator-functionalized oligonucleotides to enhance the DNA-targeting properties of γPNA. *Org. Biomol. Chem.* **2020**, *18*, 1359–1368. [CrossRef] [PubMed]
23. Emehiser, R.G.; Dhuri, K.; Shepard, C.; Karmakar, S.; Bahal, R.; Hrdlicka, P.J. Serine-γPNA, Invader probes, and chimeras thereof: Three probe chemistries that enable sequence-unrestricted recognition of double-stranded DNA. *Org. Biomol. Chem.* **2022**, *20*, 8714–8724. [CrossRef] [PubMed]
24. Guenther, D.C.; Anderson, G.H.; Karmakar, S.; Anderson, B.A.; Didion, B.A.; Guo, W.; Verstegen, J.P.; Hrdlicka, P.J. Invader probes: Harnessing the energy of intercalation to facilitate recognition of chromosomal DNA for diagnostic applications. *Chem. Sci.* **2015**, *6*, 5006–5015. [CrossRef] [PubMed]
25. For a definition of the interstrand zipper nomenclature, see the Supporting Materials.
26. Crothers, D.M. Calculation of binding isotherms for heterogeneous polymers. *Biopolymers* **1968**, *6*, 575–584. [CrossRef]
27. Jain, S.C.; Tsai, C.; Sobell, H.M. Visualization of drug-nucleic acid interactions at atomic resolution. II. Structure of an ethidium-dinucleoside monophosphate crystalline complex, ethidium:5-iodocytidylyl (3'-5') guanosine. *J. Mol. Biol.* **1977**, *114*, 317–331. [CrossRef]
28. Williams, L.D.; Egli, M.; Gao, Q.; Rich, A. DNA intercalation: Helix unwinding and neighbor-exclusion. In *Structure and Function: Nucleic Acids*; Sarma, R.H., Sarma, M.H., Eds.; Adenine Press: Albany, NY, USA, 1992; Volume 1, pp. 107–125.
29. Ihmels, H.; Otto, D. Intercalation of organic dye molecules into double-stranded DNA-General principles and recent developments. *Top. Curr. Chem.* **2005**, *258*, 161–204.
30. Sau, S.P.; Madsen, A.S.; Podbevsek, P.; Andersen, N.K.; Kumar, T.S.; Andersen, S.; Rathje, R.L.; Anderson, B.A.; Guenther, D.C.; Karmakar, S.; et al. Identification and characterization of second-generation Invader locked nucleic acids (LNAs) for mixed-sequence recognition of double-stranded DNA. *J. Org. Chem.* **2013**, *78*, 9560–9570. [CrossRef]
31. Karmakar, S.; Madsen, A.S.; Guenther, D.C.; Gibbons, B.C.; Hrdlicka, P.J. Recognition of double-stranded DNA using energetically activated duplexes with interstrand zippers of 1-, 2- or 4-pyrenyl-functionalized O2'-alkylated RNA monomers. *Org. Biomol. Chem.* **2014**, *12*, 7758–7773. [CrossRef]

32. Denn, B.; Karmakar, S.; Guenther, D.C.; Hrdlicka, P.J. Sandwich assay for mixed-sequence recognition of double-stranded DNA: Invader-based detection of targets specific to foodborne pathogens. *Chem. Commun.* **2013**, *49*, 9851–9853. [CrossRef]
33. Emehiser, R.G.; Hall, E.; Guenther, D.C.; Karmakar, S.; Hrdlicka, P.J. Head-to-head comparison of LNA, $^{MP}\gamma$PNA, INA and Invader probes targeting mixed-sequence double-stranded DNA. *Org. Biomol. Chem.* **2020**, *18*, 56–65. [CrossRef] [PubMed]
34. Anderson, B.A.; Onley, J.J.; Hrdlicka, P.J. Recognition of double-stranded DNA using energetically activated duplexes modified with N2'-pyrene-, perylene-, or coronene-functionalized 2'-N-methyl-2'-amino-DNA monomers. *J. Org. Chem.* **2015**, *80*, 5395–5406. [CrossRef] [PubMed]
35. Guenther, D.C.; Karmakar, S.; Hrdlicka, P.J. Bulged Invader probes: Activated duplexes for mixed-sequence dsDNA recognition with improved thermodynamic and kinetic profiles. *Chem. Commun.* **2015**, *51*, 15051–15054. [CrossRef] [PubMed]
36. Adhikari, S.P.; Vukelich, P.; Guenther, D.C.; Karmakar, S.; Hrdlicka, P.J. Recognition of double-stranded DNA using LNA-modified toehold Invader probes. *Org. Biomol. Chem.* **2021**, *19*, 9276–9290. [CrossRef] [PubMed]
37. Karmakar, S.; Guenther, D.C.; Hrdlicka, P.J. Recognition of mixed-sequence DNA duplexes: Design guidelines for Invaders based on 2'-O-(pyren-1-yl)methyl-RNA monomers. *J. Org. Chem.* **2013**, *78*, 12040–12048. [CrossRef]
38. Perret, J.; Shia, Y.C.; Fries, R.; Vassart, G.; Georges, M. A polymorphic satellite sequence maps to the pericentric region of the bovine Y-chromosome. *Genomics* **1990**, *6*, 482–490. [CrossRef]
39. Mergny, J.L.; Lacroix, L. Analysis of thermal melting curves. *Oligonucleotides* **2003**, *13*, 515–537. [CrossRef]
40. The driving force for dsDNA-recognition with **INV4** and **INV9** is due to favorable changes in entropy.
41. Preliminary studies suggested that shorter incubation (i.e., 2.5 h) resulted in incomplete recognition for some probes. For more details, see the Supporting Materials.
42. Stadlbauer, S.; Kührová, P.; Wicherek, L.; Banáš, P.; Otyepla, M.; Trantírek, L.; Šponer, J. Parallel G-triplexes and G-hairpins as potential transitory ensembles in the folding of parallel-stranded DNA G-quadruplexes. *Nucleic Acids Res.* **2019**, *47*, 7276–7293. [CrossRef]
43. Similar results were previously reported for INV4 in [24].
44. Karmakar, S.; Anderson, B.A.; Rathje, R.L.; Andersen, S.; Jensen, T.B.; Nielsen, P.; Hrdlicka, P.J. High-affinity DNA targeting using readily accessible mimics of N2'-functionalized 2'-Amino-α-L-LNA. *J. Org. Chem.* **2011**, *76*, 7119–7131. [CrossRef] [PubMed]
45. Dioubankova, N.N.; Malakhov, A.D.; Stetsenko, D.A.; Gait, M.J.; Volynsky, P.E.; Efremov, R.G.; Korshun, V.A. Pyrenemethyl *ara*-uridine-2'-carbamate: A strong interstrand excimer in the major groove of a DNA duplex. *ChemBioChem* **2003**, *4*, 841–847. [CrossRef]
46. Morgan, M.A.; Okamoto, K.; Kahn, J.D.; English, D.S. Single-molecule spectroscopic determination of lac repressor-DNA loop conformation. *Biophys. J.* **2005**, *89*, 2588–2596. [CrossRef] [PubMed]
47. Brown, A.M. A step-by-step guide to non-linear regression analysis of experimental data using a Microsoft Excel spreadsheet. *Comput. Meth. Prog. Biomed.* **2001**, *65*, 191–200. [CrossRef] [PubMed]
48. Available online: https://www.real-statistics.com/free-download/real-statistics-resource-pack/ (accessed on 12 December 2022).
49. For example, it would only be possible to add one more hotspot to **INV4**, i.e., the 5'-TG-3' step at position 8-9 to give a hotspot content of ~29%.

Disclaimer/Publisher's Note: The statements, opinions and data contained in all publications are solely those of the individual author(s) and contributor(s) and not of MDPI and/or the editor(s). MDPI and/or the editor(s) disclaim responsibility for any injury to people or property resulting from any ideas, methods, instructions or products referred to in the content.

Article

Enzymatic Synthesis of Vancomycin-Modified DNA

Chiara Figazzolo [1,2], Frédéric Bonhomme [3], Saidbakhrom Saidjalolov [4], Mélanie Ethève-Quelquejeu [4] and Marcel Hollenstein [1,*]

[1] Institut Pasteur, Université de Paris Cité, CNRS UMR3523, Department of Structural Biology and Chemistry, Laboratory for Bioorganic Chemistry of Nucleic Acids, 28, rue du Docteur Roux, CEDEX 15, 75724 Paris, France
[2] Learning Planet Institute, 8, Rue Charles V, 75004 Paris, France
[3] Institut Pasteur, Université de Paris Cité, Department of Structural Biology and Chemistry, Unité de Chimie Biologique Epigénétique, UMR CNRS 3523, 28 rue du Docteur Roux, CEDEX 15, 75724 Paris, France
[4] Université Paris Cité, CNRS, Laboratoire de Chimie et Biochimie Pharmacologiques et Toxicologiques, 75006 Paris, France
* Correspondence: marcel.hollenstein@pasteur.fr

Abstract: Many potent antibiotics fail to treat bacterial infections due to emergence of drug-resistant strains. This surge of antimicrobial resistance (AMR) calls in for the development of alternative strategies and methods for the development of drugs with restored bactericidal activities. In this context, we surmised that identifying aptamers using nucleotides connected to antibiotics will lead to chemically modified aptameric species capable of restoring the original binding activity of the drugs and hence produce active antibiotic species that could be used to combat AMR. Here, we report the synthesis of a modified nucleoside triphosphate equipped with a vancomycin moiety on the nucleobase. We demonstrate that this nucleotide analogue is suitable for polymerase-mediated synthesis of modified DNA and, importantly, highlight its compatibility with the SELEX methodology. These results pave the way for bacterial-SELEX for the identification of vancomycin-modified aptamers.

Keywords: modified nucleic acids; DNA polymerases; nucleoside triphosphates; primer extension reactions; SELEX; antimicrobial resistance

1. Introduction

Bacterial antimicrobial resistance (AMR) is one of the biggest public health challenges of the 21st century and occurs when bacteria develop mechanisms that protect them against the effects of antibiotics [1]. AMR can be intrinsic to bacteria, or can be acquired through horizontal gene transfer or mutations in chromosomal genes [2]. This major threat to human health has caused 4.95 million deaths around the world in 2019, and these numbers are expected to rise steeply in the near future [3]. Consequently, an important first step in combatting AMR consists of refining our understanding of the mechanisms underlying drug failures [4,5]. This knowledge will, in turn, guide the development of alternative antimicrobial drugs as well as that of unconventional methods for drug design [6–10]. In this context, vancomycin is amongst the most clinically relevant and effective glycopeptide antibiotics used for the treatment of infections caused by Gram-positive pathogens [11,12]. This antibiotic is considered as a drug of last resort due to its high efficiency, particularly against methicillin-resistant *Staphylococcus aureus* (MRSA) and *Staphylococcus epidermidis* (MRSE) [13]. The mechanistic details of vancomycin-driven bactericidal activity have been resolved and involve tight binding to a terminal section of the precursor of the polymer peptidoglycan on the bacterial cell surface. This binding event further blocks transglycosylation and transpeptidation, which are essential for cell wall synthesis and maintenance (Figure 1) [11,12,14]. The efficiency and popularity of vancomycin has led to a clinical overuse [15] of this glycopeptide antibiotic which explains the emergence of resistance in bacterial pathogens. Vancomycin resistance is quite complex, but can essentially either be acquired by target modification or be intrinsic [12,16]. Acquired vancomycin resistance essentially occurs via a simple change in the amino acid composition of the peptidoglycan from D-Ala-D-Ala to D-Ala-D-Lac (X = NH and O in Figure 1, respectively). This minute alteration dramatically reduces the affinity of vancomycin for its peptidoglycan target which causes a significant loss in antimicrobial activity. Intrinsic resistance mechanisms include biofilm construction and formation of dormant stationary phase subpopulations [12,16–18]. In general, strategies implemented to combat vancomycin resistance and AMR strive to develop new, small molecule drugs [19–21], modify existing scaffolds to extend their lifetimes [13,22–24], or explore alternative routes such as the introduction of the multivalency concept [25–27] or prevention of biofilm formation [12,28]. Despite significant success of these approaches, universal and robust strategies to combat AMR are still in dire need. Herein, we present a first step towards the development of a novel strategy that combines aptamer recognition, vancomycin analogues tethered to DNA nucleobases, and bacterial-SELEX. Towards this aim, we have synthesized a nucleoside triphosphate equipped with a vancomycin moiety at the C5 site of the nucleobase. We then demonstrated the compatibility of this modified nucleotide with enzymatic DNA synthesis and with the SELEX methodology.

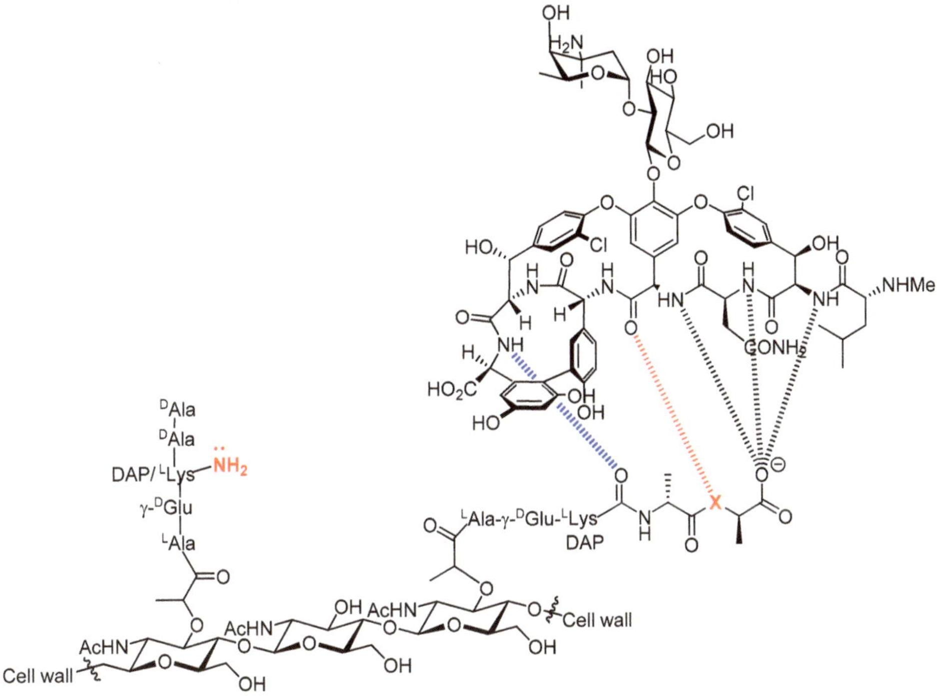

Figure 1. Chemical structure of vancomycin and description of the acquired resistance to vancomycin. The antibacterial activity of vancomycin stems from its high binding affinity(4.4×10^5 M^{-1}) for the terminal D-Ala-D-Ala sequence of the precursor peptidoglycan pentapeptide [29]. This binding creates a steric blockade which prevents the enzyme-mediated transpeptidation of the free amine of D-Lys or diamino pimelic acid (DAP) on the second last D-Ala residue from occurring. This in turn induces inhibition of bacterial cell wall biosynthesis. The origin of VanA and VanB bacterial resistance stems from a change in amino acid composition from D-Ala-D-Ala (X = NH) to D-Ala-D-Lac (X = O) [14]. The presence of an ester instead of an amide moiety suppresses a hydrogen bond and introduces a repulsive interaction between oxygen-centered lone pairs. This in turn is responsible for the 1000-fold reduction of binding affinity of vancomycin for the peptidoglycan [12,29,30].

2. Results and Discussion

2.1. Design and Chemical Synthesis of Vancomycin-Modified Nucleoside Triphosphate (dU^{Van}TP)

Aptamers are single-stranded nucleic acids capable of binding to targets with high affinity and specificity [31,32]. The first applications of aptamers in the context of microbial infections have previously been reported [33] including inhibition of biofilm formation, delivery of antibiotics conjugated to their scaffold, and participation in specific pathogen destruction [34–36]. In contrast, only a few reports have been dedicated to the application of aptamers to combat AMR, mainly as drug delivery systems [37]. For instance, aptamers conjugated to photosensitizers have been proposed as vectors for photodynamic therapy (PDT) to treat infections caused by antibiotic resistant pathogens [38–40]. In these examples, previously selected aptamers specific for particular bacterial targets are converted by post-SELEX modification to specific drug delivery agents. While this approach has met some success, post-SELEX modification protocols come with a lot of drawbacks [41,42]. First, the appendage of one or multiple drugs or reporter molecules on the termini of aptamers can lead to a loss in binding affinity (increase in K_D values) which is detrimental for such an approach. Secondly, most aptamers are often directed towards diagnostic detection purposes rather than as therapeutic agents and hence do not present a sufficiently developed modification pattern robust enough to resist nuclease-mediated degradation. To combat this, the implementation of a post-SELEX modification strategy to mitigate this limitation and improve their biostability will be required. Third, post-SELEX approaches only allow for the introduction of a limited number of antibiotics (mainly by connection via the 3'- and 5'-termini) which induces high-cost productions and relatively low topical drug concentration. Lastly, upon reaching its intended target, the drug will be released but unless this released compound is a new chemical analogue of the parent drug it is unlikely to elude the defense and resistance mechanisms of the pathogens. In light of these elements, we rationalized that the inclusion of antibiotics on the scaffold of nucleoside triphosphates will yield ways to generate modified aptamers specific for resistant bacterial pathogens and potentially restore the intended antimicrobial activity. By inspection of the vancomycin scaffold, we rationalized that the free carboxylic acid might be a favorable site for connection to a nucleotide. This moiety is not involved in target recognition and previously reported modification at this site usually yielded analogues that retain the original antimicrobial activity [13,27,30]. To form the connection between vancomycin and the aptamer we opted to use the copper(I)-catalyzed azide–alkyne cycloaddition (CuAAC) reaction, since this chemistry has previously been used to generate modified vancomycin derivatives [13]. In addition to this, nucleotides bearing a triazole-connecting linker arm are compatible with enzymatic DNA synthesis [43,44] and aptamer selection [45–47]. The synthesis of the vancomycin-bearing deoxyuridine triphosphate dU^{Van}TP (Figure 2) involved synthesis of the previously reported azide-modified vancomycin [13] and CuAAC reaction with 5-ethynyl-dUTP (EdUTP) [48] under standard conditions (see Materials and Methods (Section 4.1) and Supplementary Materials for full synthetic details). It is worth mentioning that the application of standard CuAAC reaction conditions yielded a mixture of species, including the expected nucleotide as well as a product corresponding to dU^{Van}TP coordinated to Cu^{2+} (Figures S1 and S2). Coordination of the divalent metal cation occurs preferentially through the N-terminal imino NH-CH$_3$ unit, two consecutive nitrogen atoms in the peptide chain, and one oxygen atom from the asparagine amide moiety [49,50]. The binding of vancomycin to Cu^{2+} is particularly efficient [50] and, hence, required the implementation of a thorough purification step to remove this undesired, cytotoxic transition metal cation from the modified nucleotide (see Materials and Methods (Section 4.1)). Alternatively, copper-free click chemistry may be considered in the future to avoid this rigorous purification step [51].

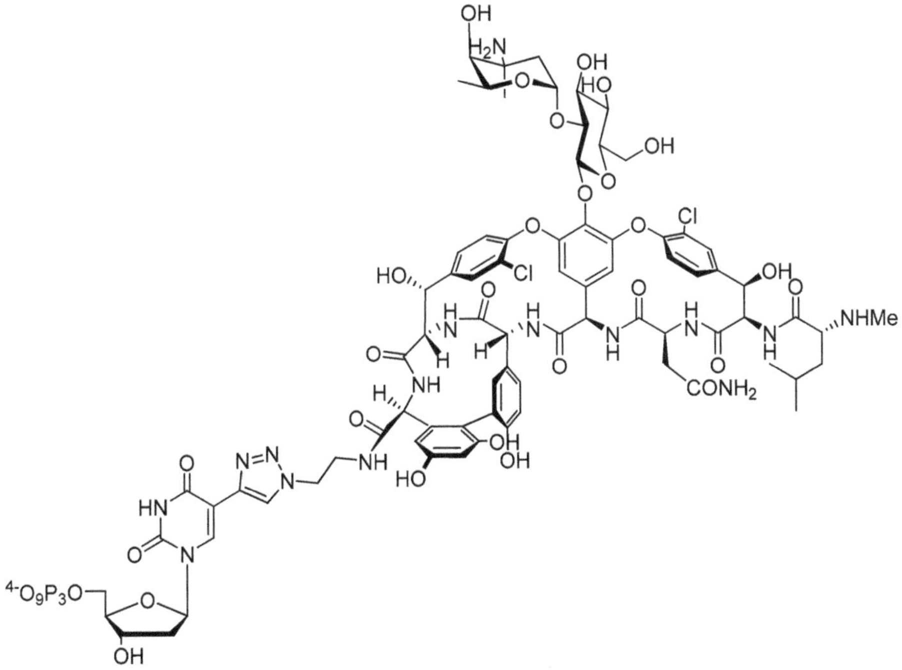

Figure 2. Chemical structure of modified nucleotide dU^{Van}TP.

2.2. Biochemical Characterization of the Modified Nucleotide dU^{Van}TP

With dU^{Van}TP in hand, we evaluated the possibility of using this analog to synthesize vancomycin-modified DNA via primer extension (PEX) reactions and PCR. To do so, we first carried PEX reactions using the 20-nucleotide long template **T1** and the 5′-FAM-labelled, 19-mer primer **P1** (Table 1). This rather simple system allows for the incorporation of a single, modified nucleotide and can be followed by gel electrophoresis, LC-MS analysis, and UV-melting experiments. PEX reactions were performed with Vent (*exo*$^-$) DNA polymerase (vide infra) and the products were analyzed by gel electrophoresis (Figure S9). This analysis revealed the formation of a product with a slower electrophoretic mobility compared to that of the product obtained with dTTP or the unmodified, single-stranded DNA template. In order to confirm the incorporation of a dU^{Van} nucleotide into DNA, we subjected the PEX reaction product to a digestion-LC-MS analysis protocol (see Materials and Methods (Section 4.2)) [52]. For the preparation of a standard for the modified nucleoside which is required for such an analysis, we dephosphorylated dU^{Van}TP with Shrimp Alkaline Phosphatase (rSAP) and analyzed the resulting nucleoside by LC-MS (Figures 3 and S6) and HRMS (Figure S7). PEX reaction products were nuclease-digested, dephosphorylated and the resulting deoxynucleosides were then analyzed by LC-MS (Figure 3).

The LC-MS chromatogram of the digested PEX reaction product using only dA/G/T/CTP displayed the expected four peaks corresponding to each of the four canonical nucleosides (Figure 3A). Similarly, the four canonical nucleosides were also detected in the LC-MS chromatogram of the digested reaction product obtained with dU^{Van}TP, however, an additional peak at longer retention time (Figure 3B) and with a mass corresponding to dU^{Van} was also detected (Figure S8 and Figure 3C). Therefore this analysis univocally confirmed the presence of the modified dU^{Van} nucleotide in the extended primer **P1** obtained after PEX reaction.

Table 1. DNA primer and templates used for primer extension reactions and PCR to evaluate the compatibility of dU^{Van}TP with enzymatic DNA synthesis [a].

P1	5′-TAC GAC TCA CTA TAG CCT C′
T1	5′-AGA GGC TAT AGT GAG TCG TA
T2	5′-CAC TCA CGT CAG TGA CAT GCA TGC CGA TGA CTA GTC GTC ACT AGT GCA CGT AAC GTG CTA GTC AGA AAT TTC GCA CCA C
T3	5′-CAC TCA CGT CAG TGA CAT GC N$_{40}$ GTC AGA AAT TTC GCA CCA C
T4	5′-Phos-AGA GGC TAT AGT GAG TCG TA
P2	5′-FAM-GTG GTG CGA AAT TTC TGA C
P3	5′-GTG GTG CGA AAT TTC TGA C
P4	5′-CAC TCA CGT CAG TGA CAT GC

[a] italicized letters represent primer binding sites and bold, red letters indicate sites where modified nucleotides can be incorporated. N40 represents the randomized region of the degenerate library.

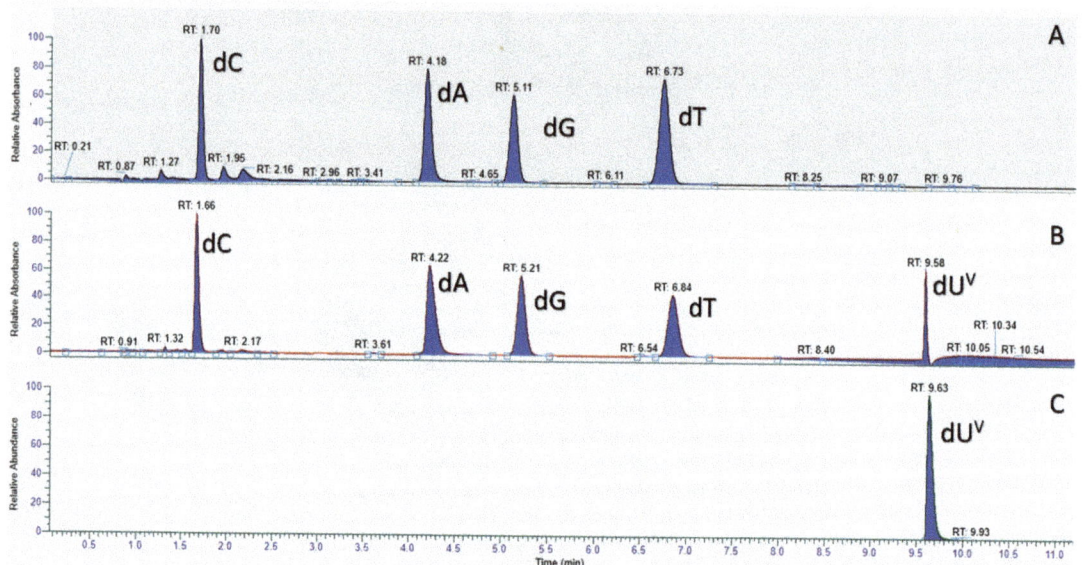

Figure 3. LC-MS chromatograms of deoxynucleosides from digested PEX reaction products obtained with dU^{Van}TP and dTTP. (**A**) LC-MS chromatogram of the digested product obtained with dTTP; (**B**) LC-MS chromatogram of the digested product obtained with dU^{Van}; (**C**) LC-MS chromatogram of the synthesized dU^{Van} standard.

Next, we sought to evaluate the effect of such a bulky modification on duplex stability. To do so, we performed large-scale PEX reactions on the aforementioned **P1/T1** primer/template duplex in the presence of dU^{Van}TP and Vent (*exo*$^-$) DNA polymerase [53,54]. The UV-melting curves of the resulting modified dsDNA were then measured under standard saline buffer conditions and analyzed using a newly written python script for T_m data analysis (see Figure 4 and Supplementary Materials) [55]. The presence of the vancomycin moiety led to a rather significant destabilization of the duplex ($\Delta T_m = -3.9\ °C$) compared to that of the unmodified dsDNA (62.7 ± 0.3 °C compared to 66.6 ± 0.1 °C). This destabilization might be caused either by the presence of the bulky vancomycin moiety since other large moieties such as polyethylene glycol [56] induced similar decreases in T_m values, or by the thermal penalty imparted by the introduction of a single triazole-containing nucleotide [57–59]. Nonetheless, this UV-melting analysis indicates that the modification did not significantly interfere with duplex formation.

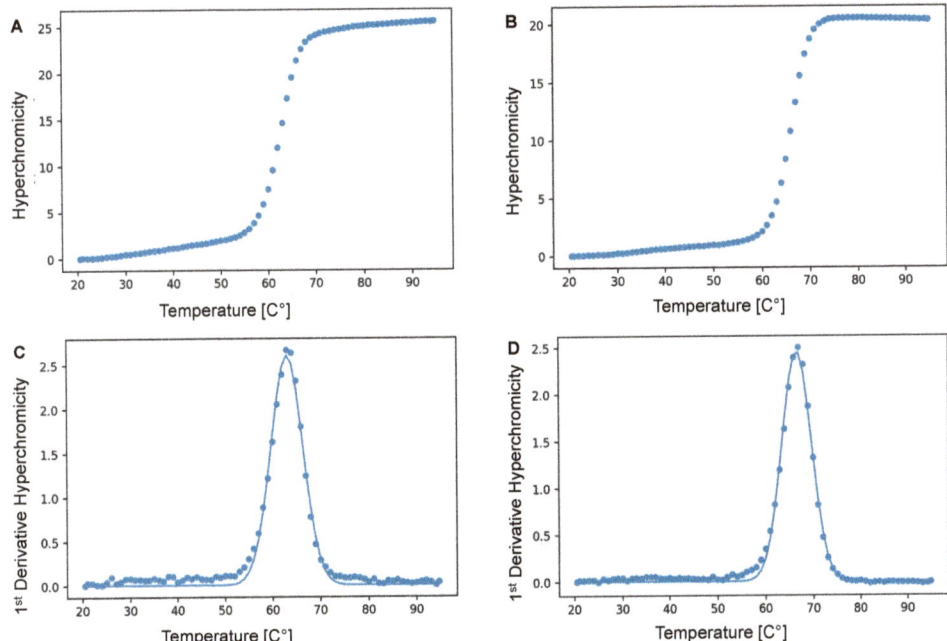

Figure 4. Illustrative UV melting curves (λ = 260 nm) of: (**A**) the vancomycin-modified duplex and (**B**) the corresponding natural DNA duplex. The corresponding first derivatives are shown in panels (**C**) and (**D**), respectively. The total duplex concentration was 2 µM in 137 mM NaCl, 2.7 mM KCl, 4.3 mM Na$_2$HPO$_4$, 1.47 mM KH$_2$PO$_4$, and at pH 7.4.

Next, we sought to evaluate the possibility of using dU^{Van}TP to synthesize longer vancomycin-containing DNA fragments using polymerase-mediated catalysis. To do so, we first carried out PEX reactions using the 19 nucleotide long 5′-FAM-labelled primer (**P2**) along with the corresponding 79-mer template (**T2**) that permits the incorporation of up to nine modified nucleotides (Table 1). A variety of polymerases were employed for PEX reactions, including the Klenow fragment of *E. coli* (Kf *exo*$^-$) DNA polymerase I, *Bst*, Hemo Klen Taq, and Taq (family A polymerases), the family B polymerases Phusion, Therminator, Vent (*exo*$^-$), phi29, and Deep Vent, and the Y-family polymerase *Sulfolobus* DNA Polymerase IV (Dpo4). Analysis by gel electrophoresis of the resulting PEX reaction products (Figure 5), revealed that a number of polymerases (Phusion, Q5, Therminator, Vent (*exo*$^-$) and Deep Vent) readily accepted dU^{Van}TP as a substrate. By contrast, polymerases such as Taq did not tolerate the modified nucleotide and no extended product was formed. The highest conversion yields were observed when Vent (*exo*$^-$) and Therminator were used as polymerases to catalyze the incorporation of the modified nucleotide into DNA. The bands corresponding to full length, modified products possessed a retarded gel electrophoretic mobility compared to that of control reactions performed with canonical nucleotides. This is expected when a bulky residue is attached to the nucleobase [56,60,61]. Despite the rather high substrate tolerance of Therminator for dU^{Van}TP, this polymerase has been shown to misincorporate natural and modified nucleotides [62–65], and to display an efficient pyrophosphorolysis capacity followed by incorporation of nucleotides [66]. Consequently, we believe that this catalyst might not be suitable for SELEX applications when used in conjunction with dU^{Van}TP and instead we used Vent (*exo*$^-$) for further experiments.

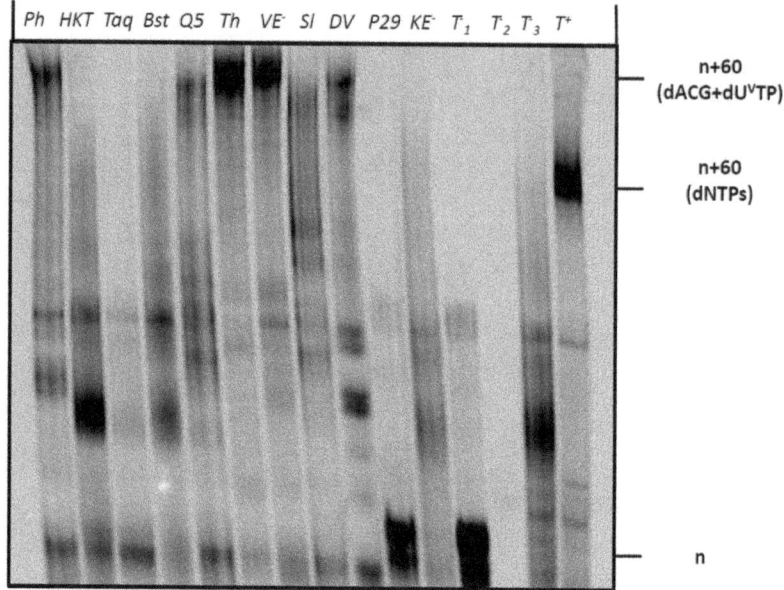

Figure 5. Gel analysis (PAGE 20%) of PEX reaction products carried out with the Phusion (Ph) (2U), Hemo Klem Taq (*HKT*) (8U), Taq (5U), Bst (8U), Q5 (2U), Therminator (Th) (2U), Vent (*exo*$^-$) (VE$^-$) (2U), Dpo4 (Sl) (2U), Deep Vent (DV) (2U), Phi29 (P29) (10U) and Kf (*exo*$^-$) (*KE*$^-$) (5U) DNA polymerases with primer **P2** and template **T2**. Each reaction mixture contained **T2** (15 pmol), **P2** (10 pmol), dATP, dCTP, dGTP (200 µM), dU^{Van}TP (200 µM), polymerase buffer (1 µL for 10X and 2 µL for 5X) and the respective polymerase and was incubated for 4 h at the appropriate temperature. The negative controls were performed without polymerase (T$^-$$_1$), primer **P2** (T$^-$$_2$) and dU^{Van}TP (T$^-$$_3$). The positive control (T$^+$) contained dTTP instead of dU^{Van}TP. n corresponds to unreacted primer **P2**, n+60 (dNTPs) corresponds to the expected full-length product in presence of dNTPs and n+60 (dATP, dCTP, dGTP + dU^{Van}TP) corresponds to the expected full-length product in presence of dU^{Van}TP instead of dTTP.

Having established conditions for the enzymatic preparation of long, vancomycin-containing DNA oligonucleotides via PEX reactions, we next set out to evaluate the compatibility of dU^{Van}TP with PCR. To do so, we performed PCR with template **T2** along with the forward and reverse primers **P3** and **P4**, respectively. We evaluated the capacity of three thermostable polymerases (i.e., Deep Vent, Vent (*exo*$^-$), and Phusion) to amplify template **T2** in the presence of the modified nucleotide, dU^{Van}TP. Surprisingly, the expected amplicons were produced in rather low yields, suggesting that the modified nucleotide is not very well tolerated by polymerases under PCR conditions (Figure S10). Despite the low synthetic yields, PCR represents an alternative to PEX reactions to produce vancomycin-containing dsDNA.

2.3. Compatibility of Vancomycin-Modified Nucleotide with SELEX

After establishing the conditions for the enzymatic synthesis of vancomycin-modified DNA, we next sought to evaluate the compatibility of dU^{Van}TP with the SELEX method for the identification of aptamers [45,67–70]. SELEX and combinatorial methods of in vitro evolution with modified nucleotides require (1) synthesis of modified, degenerate libraries using PEX reactions or PCR; (2) robust conversion from double-stranded to single-stranded modified libraries [71] and (3) an efficient and high-fidelity regeneration of unmodified DNA from modified templates [54,72–74]. In order to assess whether enzymatic synthesis with dU^{Van}TP fulfilled these criteria we first performed PEX reactions with a degenerate library (template **T3** in Table 1), a panel of polymerases and dTTP substituted by dU^{Van}TP. As with a defined sequence of the same length, the Therminator, Vent *(exo⁻)*, and Deep Vent DNA polymerases were capable of producing the expected full-length products in high yields (Figure 6). Thus these polymerases could produce fully modified dsDNA libraries and process these more demanding templates in the presence of the bulky modified nucleotide.

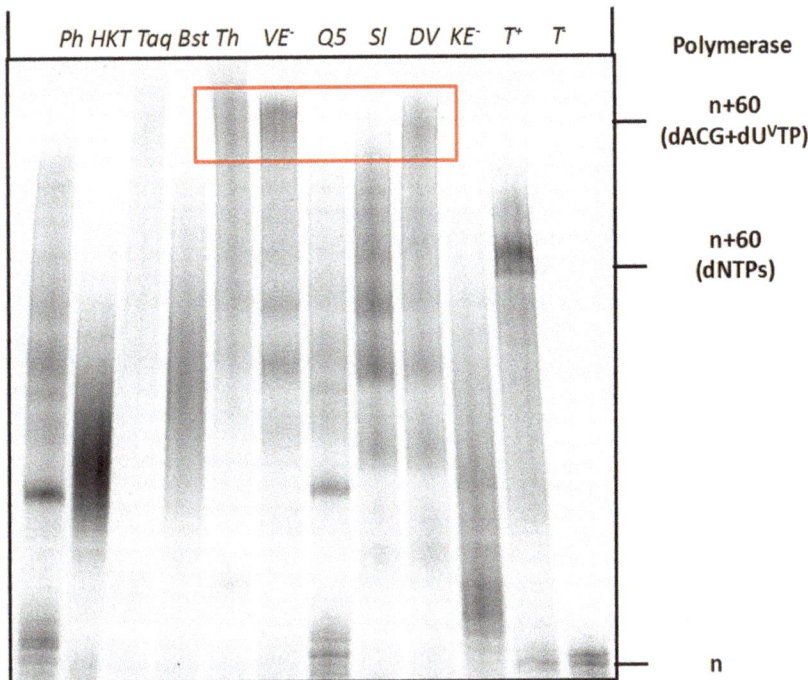

Figure 6. Gel analysis (PAGE 20%) of PEX reaction products carried out with the Phusion (**Ph**) (2U), Hemo Klem Taq (**HKT**) (8U), Taq (5U), Bst (8U), Q5 (2U), Therminator (**Th**) (2U), Vent *(exo⁻)* (**VE⁻**) (2U), Dpo4 (**Sl**) (2U), Deep Vent (**DV**) (2U), Phi29 (P29) (10U) and Kf *(exo⁻)* (**KE⁻**) (5U) DNA polymerases with primer **P2** and template **T3**. Each reaction mixture contained **T3** (15 pmol), **P2** (10 pmol), dATP, dCTP, dGTP (200 µM), dU^{Van}TP (200 µM), polymerase buffer (1 µL for 10X and 2 µL for 5X) and the respective polymerase and was incubated for 4 h at the appropriate temperature. The negative controls were performed without polymerase (T⁻$_1$), primer **P2** (T⁻$_2$) and dU^{Van}TP (T⁻$_3$). The positive control (T⁺) contained dTTP instead of dU^{Van}TP. n corresponds to unreacted primer **P2**, n+60 (dNTPs) corresponds to the expected full-length product in presence of dNTPs and n+60 (dATP, dCTP, dGTP + dU^{Van}TP) corresponds to the expected full-length product in presence of dU^{Van}TP instead of dTTP.

Next, we evaluated the possibility of reverse transcribing vancomycin-containing DNA oligonucleotides to unmodified sequences. To do so, we synthesized vancomycin-containing dsDNA by PCR as described above. After a thorough purification step, we subjected the modified product to PCR with canonical nucleotides and Taq (Figure 7A) or Vent *(exo⁻)* (data not shown) as DNA polymerases. Agarose gel electrophoretic analysis of the resulting reaction products clearly indicated the formation of the expected amplicon (Figure 7A). We next subjected the obtained PCR product with canonical nucleotides and the Taq polymerase to a TA-cloning step followed by Sanger sequencing of nine individual colonies. Sequencing of this small subset of colonies clearly indicated the absence of mutations compared to the expected parent sequence of template **T2** (Figure 7B). This analysis suggested a faithful reverse transcription of vancomycin-containing DNA to unmodified oligonucleotides. Lastly, we evaluated the possibility of converting double-stranded, modified sequences to single-stranded oligonucleotides, which is an important prerequisite for SELEX and a notoriously difficult step, especially for heavily modified sequences or bulky modifications such as vancomycin [71]. We first performed PEX reactions with dU^{Van}TP, Vent (exo^-), and a 5′-phosphorylated analogue of template T1 (template T4, Table 1). The resulting modified dsDNA was then converted to the corresponding ssDNA by λ-exonuclease digestion (Figure S11A). Gel analysis of the resulting products indicated formation of the expected single-stranded sequences (Figure S11B).

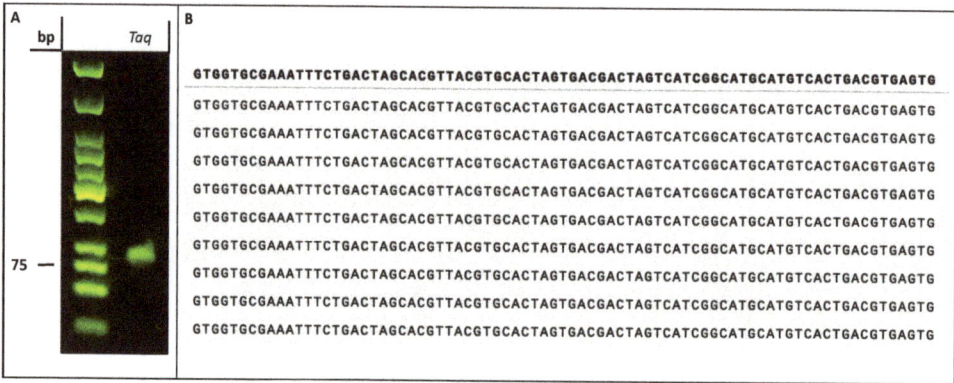

Figure 7. (**A**) Gel image (agarose 2%) analysis showing the product of the conversion of dU^{Van}-modified dsDNA into natural dsDNA by means of PCR in presence of Taq as a polymerase. The reaction was performed in 25 µL and contained purified dU^{Van}-modified dsDNA (10 nM), primers P3/P4 (500 nM), dNTPs (200 µM), Mg^{2+} (2 mM), Thermopol buffer (2.5 µL of 10X) and Taq (5U, 1 µL). The PCR program was 95 °C for 5 min, (95 °C 30″, 52 °C 30″, 75 °C 30″) X 10 cycles, and a final elongation at 75 °C for 5 min. (**B**) Sequence alignment of nine different plasmid colonies obtained by PCR conversion of dU^{Van}-modified dsDNA to natural DNA followed by cloning and sequencing. The first line in bold corresponds to the sequence of template T2, while the following nine lines correspond to the sequenced products.

3. Conclusions

AMR is a major threat to human health, and alternative treatment modalities and drugs are in dire need. Herein, we have presented the first step towards combining antibiotics and aptamers in order to potentially restore the effectiveness of drugs against resistant bacteria. Towards this aim, we have synthesized a deoxyuridine analogue bearing a vancomycin residue at position C5 of the nucleobase. The resulting nucleotide, dU^{Van}TP, was then shown to be compatible with enzymatic DNA synthesis under PEX reaction conditions and PCR (to a certain extent), despite the presence of such a bulky chemical modification. We have then demonstrated the compatibility of dU^{Van}TP with the SELEX methodology. In particular, we have shown that modified libraries could be generated by enzymatic synthesis and subsequently reverse transcribed into unmodified DNA without introducing mutations. Cell-SELEX experiments against vancomycin resistant strains and subsequent evaluation of the antibacterial activity of modified aptamers are currently underway. We expect the aptamer section of such a construct to bind to membrane and surface proteins, which are the typical targets of cell-SELEX experiments [75]. This binding event will then, in turn, restore binding of vancomycin to the peptidoglycan and, hence, restore its activity. A similar approach using a primer equipped with a hotspot peptide in SELEX was recently employed for the selection of aptamers against the human angiotensin-converting enzyme 2 (hACE2) [76] underscoring the usefulness of such a method. However, the advantage of our approach is that multiple vancomycin-bearing nucleotides will be integrated within the scaffold of the aptamer and will be involved in the binding interactions. Hence, uncertain, labor-intensive post-SELEX structure-activity relationship studies to optimize aptamers will not be required. Lastly, we believe that such an approach could easily be extended to other antibiotics such as β-lactamase inhibitors [7,8] or pentamidine-based scaffolds [77].

4. Materials and Methods

4.1. Chemical Synthesis of dU^{Van}TP

EdUTP [48] (2 mg, 0.004 mmol, 1.5 eq.) and azido-vancomycin [13] (4 mg, 0.0026 mmol, 1 eq.) were respectively dissolved in 100 µL H$_2$O each in two different Eppendorf tubes and degassed with argon. In a separated Eppendorf tube, CuI (3 mg, 0.02 mmol, 0.77 eq.) was suspended in DIPEA (10 µL, 0.06 mmol, 2.3 eq.), 50 µL H$_2$O and 50 µL CH$_3$CN and degassed with Argon. The three fractions were merged and 100 µL DMSO were added. The reaction mixture was degassed with Argon and left to shake at 1000 rpm, at r.t. for 3 h. The reaction mixture was added to 12 mL NaClO$_4$ 2% in acetone to precipitate the crude product, centrifuged at 4000 rpm for 15 min, and the surnatant was discarded. The crude was redissolved in 3 mL H$_2$O and purified by anion exchange analytical HPLC (100% B in 25 min; Buffer A, 10 mM TEAB in H$_2$O; Buffer B, 1 M TEAB in H$_2$O) with a DNA Pac PA 100 oligonucleotide column. Prior to each injection, EDTA (0.5 M, pH = 8) was added to the crude solution to coordinate the remaining Cu^{2+} present in the solution. The fractions corresponding to dU^{Van}TP were merged together and lyophilized overnight to give the product as 2.9 mg (55%) of a white powder. To remove the remaining coordinated Cu^{2+}, dU^{Van}TP was dissolved in 1 mL H$_2$O and stirred with 1 g of Chelex 100 resin for 2 h at rt.

^{31}P NMR (202.4 MHz, D$_2$O): −6.15 (d, J = 20.2 Hz, 1P), −11.17 (d, J = 18.2 Hz, 1P), −22.17 (t, J = 19.2 Hz, 1P); HR-ESI-MS (m/z): calcd for M = C$_{79}$H$_{94}$Cl$_2$N$_{15}$O$_{37}$CuP$_3$, [M] = 2071.00, found [M-2H]$^{2-}$ = 2068.3518. HR-ESI-MS (m/z): calcd for M = C$_{79}$H$_{94}$Cl$_2$N$_{15}$O$_{37}$P$_3$, [M] = 2009.48, found [M-2H]$^{2-}$ = 2007.4412.

4.2. Preparation of dU^{Van} Nucleoside Standard

dU^{Van}TP (10 nmol) was incubated with rSAP (10 µL) and rCutSmart buffer (2.2 µL of 10X) in a total volume of 22 µL for 1 h at 37 °C. Afterwards, rSAP was inactivated via incubation for 10 min at 65 °C. The product was purified and characterized via LCMS (Figure S6). HR-ESI-MS (m/z): calcd for M = C$_{79}$H$_{91}$Cl$_2$N$_{15}$O$_{28}$Cu, [M] = 1833.12, found [M]$^+$ = 1832.4842. HR-ESI-MS (m/z): calcd for M = C$_{79}$H$_{91}$Cl$_2$N$_{15}$O$_{28}$, [M] = 1769.57, found [M+2H]$^{2+}$ = 1770.5730.

4.3. Digestion and LC-MS Analysis of PEX Reaction Product

The primer **P1** (100 pmol) was annealed to the template **T1** (150 pmol) in H$_2$O by heating to 95 °C and then gradually cooling to room temperature (over 1 h). Vent (exo^-) polymerase (4U), Thermopol 10X buffer (provided by the supplier of the DNA polymerase) (2 µL) and dU^{Van}TP (or dTTP for the positive control) (200 µM) were added for a total volume of 10 µL. The reaction mixture was incubated for 4 h at 60 °C. The formation of the products was verified with an agarose E-GEL (4%) using the E-GEL sample loading buffer (1X). The products were purified via Monarch DNA Cleanup Columns (5 µg) (250 pmol product per column). The products (1 µL), the Nucleoside Digestion Mix buffer (2 µL of 10X) and the Nucleoside Digestion Mix (1 µL) were mixed in a total volume of 20 µL and incubated at 37 °C for 1 h. The digested DNA solution was then injected onto a ThermoFisher Hypersil Gold aQ chromatography column (100 × 2.1 mm, 1.9 µm particle size) heated at 30 °C without any further purification. The flow rate was set at 0.3 mL/min and run with an isocratic elution of 1% ACN in H$_2$O with 0.1% formic acid for 8 min then 100% ACN from 9 to 11 min. Parent ions were fragmented in positive ion mode with 10% normalized collision energy in parallel-reaction monitoring (PRM) mode. MS2 resolution was 17,500 with an AGC target of 2 × 10^5, a maximum injection time of 50 ms, and an isolation window of 1.0 m/z. The inclusion list contained the following masses: dC (228.1), dA (252.1), dG (268.1), dT (243.1) and 4 (884.8; z = 2). Extracted ion chromatograms of base fragments (±5 ppm) were used for detection (112.0506 Da for dC; 136.0616 for dA; 152.0565 Da for dG; 127.0501 Da for dT and 1625.4652 Da for 4). Synthetic standards were previously injected to confirm the assignment (fragment ion and retention time).

4.4. General Protocol for PEX Reactions

The 5′-FAM-labelled primer (10 pmol) was annealed to the template (15 pmol) in H_2O by heating to 95 °C and then gradually cooling to room temperature (1 h). The appropriate polymerase, 5X or 10X enzyme buffer (provided by the supplier of the DNA polymerase), the natural nucleotides (200 µM), and $dU^{Van}TP$ (or dTTP in the positive controls) (200 µM) were added for a total reaction volume of 10 µL. Following incubation for 4 h at the optimal temperature for the enzyme, the reactions were quenched by adding the stop solution (10 µL; formamide (70%), ethylenediaminetetraacetic acid (EDTA; 50 mM), bromophenol (0.1%), xylene cyanol (0.1%)). The reaction mixtures were subjected to gel electrophoresis in denaturing polyacrylamide gel (20%) containing trisborate- EDTA (TBE) 1× buffer (pH 8) and urea (7 M). Products were then analyzed on a phosphorimager.

4.5. Thermal Denaturation Experiments

Experiments were recorded in 1X phosphate-buffered saline (PBS) (137 mM NaCl, 2.7 mM KCl, 4.3 mM Na_2HPO_4, 1.47 mM KH_2PO_4, pH 7.4) in H_2O with 2 µM final duplex concentration. Paraffin oil (150 µL) was added to avoid evaporation. A control cell (1X PBS) was prepared into which the temperature probe was placed. Absorbance was monitored at 260 nm and the heating rate was set to 1 °C/min. A heating-cooling cycle in the temperature range of 20–95 °C was applied and repeated 3 times per sample. The data analysis was performed with a small Python (v3.8.5) script on the heating ramps (see Supplementary Materials). The absorbance melting curves were converted to hyperchromicities and the first derivative of each experimental hyperchromicity point was calculated. Afterwards, the Gaussian function that best fitted the first derivative curve was calculated and the melting temperature was obtained from it. The final T_m value was calculated as the mean of the three obtained values with the relative error.

Supplementary Materials: The following supporting information can be downloaded at: https://www.mdpi.com/article/10.3390/molecules27248927/s1, Figure S1: MALDI-TOF analysis of $dU^{Van}TP$ coordinated with Cu^{2+}; Figure S2: HR-ESI-MS analysis of $dU^{Van}TP$ coordinated with Cu^{2+}; Figure S3: MALDI-TOF analysis of $dU^{Van}TP$; Figure S4: HR-ESI-MS analysis of $dU^{Van}TP$; Figure S5: ^{31}P NMR spectrum of $dU^{Van}TP$; Figure S6: LC-MS analysis of dU^{Van}; Figure S7: HR-ESI-MS analysis of nucleoside dU^{Van}; Figure S8: Full LC-MS chromatogram of the nucleosides stemming from the digested and dephosphorylated PEX reaction product obtained with $dU^{Van}TP$; Figure S9: Gel image analysis of PEX reaction products obtained with $dU^{Van}TP$ and dTTP; Figure S10: Gel image analysis of PCR reaction products obtained with $dU^{Van}TP$ and dTTP; Figure S11: Schematic representation of the different steps involved in the generation of dU^{Van}-modified ssDNA from the corresponding dsDNA and gel image analysis of the products of the generation of the dU^{Van}-modified ssDNA; Figure S12: Gel image showing the results obtained by performing PCR on 19 different *E. Coli* colonies; Details of the python script for data analysis of UV-melting experiments.

Author Contributions: Conceptualization, M.H.; methodology, C.F., F.B., M.E.-Q., M.H.; software, C.F.; validation, C.F., F.B., S.S., M.E.-Q., M.H.; formal analysis, C.F., F.B., M.H.; investigation, C.F., F.B. and S.S.; resources, M.E.-Q. and M.H.; writing—original draft preparation, C.F. and M.H.; writing—review and editing, all authors.; visualization, M.H.; supervision, M.E.-Q. and M.H.; project administration, M.H.; funding acquisition, C.F. and M.H. All authors have read and agreed to the published version of the manuscript.

Funding: C.F. gratefully acknowledges the Pasteur-Paris University (PPU) International Ph.D. Program that has received funding from INCEPTION Project (PIA/ANR-16-CONV-0005), the "Ecole Doctorale Frontières de l'Innovation en Recherche et Education—Programme Bettencourt" for financial support. The authors gratefully acknowledge financial support from Institut Pasteur.

Data Availability Statement: The data presented in this study are available on request from the corresponding author.

Acknowledgments: Harley Betts is gratefully acknowledged for proof-reading our manuscript.

Conflicts of Interest: The authors declare no conflict of interest.

Sample Availability: Samples of the compounds are available from the authors.

References

1. Larsson, D.G.J.; Flach, C.-F. Antibiotic resistance in the environment. *Nat. Rev. Microbiol.* **2022**, *20*, 257–269. [CrossRef] [PubMed]
2. Blair, J.M.A.; Webber, M.A.; Baylay, A.J.; Ogbolu, D.O.; Piddock, L.J.V. Molecular mechanisms of antibiotic resistance. *Nat. Rev. Microbiol.* **2015**, *13*, 42–51. [CrossRef] [PubMed]
3. Murray, C.J.L.; Ikuta, K.S.; Sharara, F.; Swetschinski, L.; Robles Aguilar, G.; Gray, A.; Han, C.; Bisignano, C.; Rao, P.; Wool, E.; et al. Global burden of bacterial antimicrobial resistance in 2019: A systematic analysis. *Lancet* **2022**, *399*, 629–655. [CrossRef] [PubMed]
4. Saidjalolov, S.; Braud, E.; Edoo, Z.; Iannazzo, L.; Rusconi, F.; Riomet, M.; Sallustrau, A.; Taran, F.; Arthur, M.; Fonvielle, M.; et al. Click and Release Chemistry for Activity-Based Purification of β-Lactam Targets. *Chem. Eur. J.* **2021**, *27*, 7687–7695. [CrossRef] [PubMed]
5. Shinu, P.; Mouslem, A.K.A.; Nair, A.B.; Venugopala, K.N.; Attimarad, M.; Singh, V.A.; Nagaraja, S.; Alotaibi, G.; Deb, P.K. Progress Report: Antimicrobial Drug Discovery in the Resistance Era. *Pharmaceuticals* **2022**, *15*, 413. [CrossRef] [PubMed]
6. Namivandi-Zangeneh, R.; Sadrearhami, Z.; Dutta, D.; Willcox, M.; Wong, E.H.H.; Boyer, C. Synergy between Synthetic Antimicrobial Polymer and Antibiotics: A Promising Platform To Combat Multidrug-Resistant Bacteria. *ACS Infect. Dis.* **2019**, *5*, 1357–1365. [CrossRef]
7. Bouchet, F.; Atze, H.; Fonvielle, M.; Edoo, Z.; Arthur, M.; Ethève-Quelquejeu, M.; Iannazzo, L. Diazabicyclooctane Functionalization for Inhibition of β-Lactamases from Enterobacteria. *J. Med. Chem.* **2020**, *63*, 5257–5273. [CrossRef]
8. Iqbal, Z.; Sun, J.; Yang, H.; Ji, J.; He, L.; Zhai, L.; Ji, J.; Zhou, P.; Tang, D.; Mu, Y.; et al. Recent Developments to Cope the Antibacterial Resistance via β-Lactamase Inhibition. *Molecules* **2022**, *27*, 3832. [CrossRef]
9. Romero, E.; Oueslati, S.; Benchekroun, M.; D'Hollander, A.C.A.; Ventre, S.; Vijayakumar, K.; Minard, C.; Exilie, C.; Tlili, L.; Retailleau, P.; et al. Azetidinimines as a novel series of non-covalent broad-spectrum inhibitors of β-lactamases with submicromolar activities against carbapenemases KPC-2 (class A), NDM-1 (class B) and OXA-48 (class D). *Eur. J. Med. Chem.* **2021**, *219*, 113418. [CrossRef]
10. Fu, X.; Yu, J.; Dai, N.; Huang, Y.; Lv, F.; Liu, L.; Wang, S. Optical Tuning of Antibacterial Activity of Photoresponsive Antibiotics. *ACS Appl. Bio Mater.* **2020**, *3*, 4751–4755. [CrossRef]
11. Okano, A.; Isley, N.A.; Boger, D.L. Total Syntheses of Vancomycin-Related Glycopeptide Antibiotics and Key Analogues. *Chem. Rev.* **2017**, *117*, 11952–11993. [CrossRef]
12. Acharya, Y.; Dhanda, G.; Sarkar, P.; Haldar, J. Pursuit of next-generation glycopeptides: A journey with vancomycin. *Chem. Commun.* **2022**, *58*, 1881–1897. [CrossRef]
13. Silverman, S.M.; Moses, J.E.; Sharpless, K.B. Reengineering Antibiotics to Combat Bacterial Resistance: Click Chemistry [1,2,3]-Triazole Vancomycin Dimers with Potent Activity against MRSA and VRE. *Chem. Eur. J.* **2017**, *23*, 79–83. [CrossRef]
14. Bugg, T.D.H.; Wright, G.D.; Dutka-Malen, S.; Arthur, M.; Courvalin, P.; Walsh, C.T. Molecular basis for vancomycin resistance in Enterococcus faecium BM4147: Biosynthesis of a depsipeptide peptidoglycan precursor by vancomycin resistance proteins VanH and VanA. *Biochemistry* **1991**, *30*, 10408–10415. [CrossRef]
15. Walsh; Wright, Introduction: Antibiotic Resistance. *Chem. Rev.* **2005**, *105*, 391–394. [CrossRef]
16. Petchiappan, A.; Chatterji, D. Antibiotic Resistance: Current Perspectives. *ACS Omega* **2017**, *2*, 7400–7409. [CrossRef]
17. Müller, A.; Klöckner, A.; Schneider, T. Targeting a cell wall biosynthesis hot spot. *Nat. Prod. Rep.* **2017**, *34*, 909–932. [CrossRef]
18. Liu, Y.; Jia, Y.; Yang, K.; Li, R.; Xiao, X.; Wang, Z. Antagonizing Vancomycin Resistance in Enterococcus by Surface Localized Antimicrobial Display-Derived Peptides. *ACS Infect. Dis.* **2020**, *6*, 761–767. [CrossRef]
19. Melander, R.J.; Melander, C. The Challenge of Overcoming Antibiotic Resistance: An Adjuvant Approach? *ACS Infect. Dis.* **2017**, *3*, 559–563. [CrossRef]
20. Chang, M.; Mahasenan, K.V.; Hermoso, J.A.; Mobashery, S. Unconventional Antibacterials and Adjuvants. *Acc. Chem. Res.* **2021**, *54*, 917–929. [CrossRef]
21. Chiosis, G.; Boneca, I.G. Selective Cleavage of D-Ala-D-Lac by Small Molecules: Re-Sensitizing Resistant Bacteria to Vancomycin. *Science* **2001**, *293*, 1484–1487. [CrossRef] [PubMed]
22. Yarlagadda, V.; Sarkar, P.; Samaddar, S.; Haldar, J. A Vancomycin Derivative with a Pyrophosphate-Binding Group: A Strategy to Combat Vancomycin-Resistant Bacteria. *Angew. Chem. Int. Ed.* **2016**, *55*, 7836–7840. [CrossRef] [PubMed]
23. Ma, C.; He, N.; Ou, Y.; Feng, W. Design and Synthesis of New Vancomycin Derivatives. *ChemistrySelect* **2020**, *5*, 6670–6673. [CrossRef]
24. Okano, A.; Isley, N.A.; Boger, D.L. Peripheral modifications of [Ψ[CH2NH]Tpg4]vancomycin with added synergistic mechanisms of action provide durable and potent antibiotics. *Proc. Natl. Acad. Sci. USA* **2017**, *114*, E5052–E5061. [CrossRef] [PubMed]

25. Xing, B.; Ho, P.L.; Yu, C.-W.; Chow, K.-H.; Gu, H.; Xu, B. Self-assembled multivalent vancomycin on cell surfaces against vancomycin-resistant enterococci (VRE). *Chem. Commun.* **2003**, 2224–2225. [CrossRef]
26. Yang, C.; Ren, C.; Zhou, J.; Liu, J.; Zhang, Y.; Huang, F.; Ding, D.; Xu, B.; Liu, J. Dual Fluorescent- and Isotopic-Labelled Self-Assembling Vancomycin for in vivo Imaging of Bacterial Infections. *Angew. Chem. Int. Ed.* **2017**, *56*, 2356–2360. [CrossRef]
27. Griffin, J.H.; Linsell, M.S.; Nodwell, M.B.; Chen, Q.; Pace, J.L.; Quast, K.L.; Krause, K.M.; Farrington, L.; Wu, T.X.; Higgins, D.L.; et al. Multivalent Drug Design. Synthesis and In Vitro Analysis of an Array of Vancomycin Dimers. *J. Am. Chem. Soc.* **2003**, *125*, 6517–6531. [CrossRef]
28. Sarkar, P.; Basak, D.; Mukherjee, R.; Bandow, J.E.; Haldar, J. Alkyl-Aryl-Vancomycins: Multimodal Glycopeptides with Weak Dependence on the Bacterial Metabolic State. *J. Med. Chem.* **2021**, *64*, 10185–10202. [CrossRef]
29. McComas, C.C.; Crowley, B.M.; Boger, D.L. Partitioning the Loss in Vancomycin Binding Affinity for d-Ala-d-Lac into Lost H-Bond and Repulsive Lone Pair Contributions. *J. Am. Chem. Soc.* **2003**, *125*, 9314–9315. [CrossRef]
30. Flint, A.J.; Davis, A.P. Vancomycin mimicry: Towards new supramolecular antibiotics. *Org. Biomol. Chem.* **2022**, *20*, 7694–7712. [CrossRef]
31. Ellington, A.D.; Szostak, J.W. In vitro selection of RNA molecules that bind specific ligands. *Nature* **1990**, *346*, 818–822. [CrossRef]
32. Tuerk, C.; Gold, L. Systematic Evolution of Ligands by Exponential Enrichment: RNA Ligands to Bacteriophage T4 DNA Polymerase. *Science* **1990**, *249*, 505–510. [CrossRef]
33. Afrasiabi, S.; Pourhajibagher, M.; Raoofian, R.; Tabarzad, M.; Bahador, A. Therapeutic applications of nucleic acid aptamers in microbial infections. *J. Biomed. Sci.* **2020**, *27*, 6. [CrossRef]
34. Zhang, X.; Soontornworajit, B.; Zhang, Z.; Chen, N.; Wang, Y. Enhanced Loading and Controlled Release of Antibiotics Using Nucleic Acids As an Antibiotic-Binding Effector in Hydrogels. *Biomacromolecules* **2012**, *13*, 2202–2210. [CrossRef]
35. Catuogno, S.; Esposito, C.L.; De Franciscis, V. Aptamer-Mediated Targeted Delivery of Therapeutics: An Update. *Pharmaceuticals* **2016**, *9*, 69. [CrossRef]
36. Lijuan, C.; Xing, Y.; Minxi, W.; Wenkai, L.; Le, D. Development of an aptamer-ampicillin conjugate for treating biofilms. *Biochem. Biophys. Res. Commun.* **2017**, *483*, 847–854. [CrossRef]
37. Chen, X.-F.; Zhao, X.; Yang, Z. Aptamer-Based Antibacterial and Antiviral Therapy against Infectious Diseases. *J. Med. Chem.* **2021**, *64*, 17601–17626. [CrossRef]
38. Ocsoy, I.; Yusufbeyoglu, S.; Yılmaz, V.; McLamore, E.S.; Ildız, N.; Ülgen, A. DNA aptamer functionalized gold nanostructures for molecular recognition and photothermal inactivation of methicillin-Resistant Staphylococcus aureus. *Colloids Surf. B Biointerfaces* **2017**, *159*, 16–22. [CrossRef]
39. McKenzie, L.K.; Flamme, M.; Felder, P.S.; Karges, J.; Bonhomme, F.; Gandioso, A.; Malosse, C.; Gasser, G.; Hollenstein, M. A ruthenium–oligonucleotide bioconjugated photosensitizing aptamer for cancer cell specific photodynamic therapy. *RSC Chem. Biol.* **2022**, *3*, 85–95. [CrossRef]
40. Mallikaratchy, P.; Tang, Z.; Tan, W. Cell Specific Aptamer–Photosensitizer Conjugates as a Molecular Tool in Photodynamic Therapy. *ChemMedChem* **2008**, *3*, 425–428. [CrossRef]
41. Röthlisberger, P.; Hollenstein, M. Aptamer chemistry. *Adv. Drug Deliv. Rev.* **2018**, *134*, 3–21. [CrossRef] [PubMed]
42. McKenzie, L.K.; El-Khoury, R.; Thorpe, J.D.; Damha, M.J.; Hollenstein, M. Recent progress in non-native nucleic acid modifications. *Chem. Soc. Rev.* **2021**, *50*, 5126–5164. [CrossRef] [PubMed]
43. Figazzolo, C.; Ma, Y.; Tucker, J.H.R.; Hollenstein, M. Ferrocene as a potential electrochemical reporting surrogate of abasic sites in DNA. *Org. Biomol. Chem.* **2022**, *20*, 8125–8135. [CrossRef] [PubMed]
44. Leone, D.L.; Hubalek, M.; Pohl, R.; Sykorova, V.; Hocek, M. 1,3-Diketone-Modified Nucleotides and DNA for Cross-Linking with Arginine-Containing Peptides and Proteins. *Angew. Chem. Int. Ed.* **2021**, *60*, 17383–17387. [CrossRef] [PubMed]
45. Cheung, Y.W.; Röthlisberger, P.; Mechaly, A.E.; Weber, P.; Levi-Acobas, F.; Lo, Y.; Wong, A.W.C.; Kinghorn, A.B.; Haouz, A.; Savage, G.P.; et al. Evolution of abiotic cubane chemistries in a nucleic acid aptamer allows selective recognition of a malaria biomarker. *Proc. Natl. Acad. Sci. USA* **2020**, *117*, 16790–16798. [CrossRef]
46. Siegl, J.; Plückthun, O.; Mayer, G. Dependence of click-SELEX performance on the nature and average number of modified nucleotides. *RSC Chem. Biol.* **2022**, *3*, 288–294. [CrossRef]
47. Siegl, J.; Nikolin, C.; Phung, N.L.; Thoms, S.; Blume, C.; Mayer, G. Split–Combine Click-SELEX Reveals Ligands Recognizing the Transplant Rejection Biomarker CXCL9. *ACS Chem. Biol.* **2022**, *17*, 129–137. [CrossRef]
48. Röthlisberger, P.; Levi-Acobas, F.; Hollenstein, M. New synthetic route to ethynyl-dUTP: A means to avoid formation of acetyl and chloro vinyl base-modified triphosphates that could poison SELEX experiments. *Bioorg. Med. Chem. Lett.* **2017**, *27*, 897–900. [CrossRef]
49. Kucharczyk, M.; Brzezowska, M.; Maciąg, A.; Lis, T.; Jeżowska-Bojczuk, M. Structural features of the Cu^{2+}–vancomycin complex. *J. Inorg. Biochem.* **2008**, *102*, 936–942. [CrossRef]
50. Świątek, M.; Valensin, D.; Migliorini, C.; Gaggelli, E.; Valensin, G.; Jeżowska-Bojczuk, M. Unusual binding ability of vancomycin towards Cu^{2+} ions. *Dalton Trans.* **2005**, 3808–3813. [CrossRef]
51. Baskin, J.M.; Prescher, J.A.; Laughlin, S.T.; Agard, N.J.; Chang, P.V.; Miller, I.A.; Lo, A.; Codelli, J.A.; Bertozzi, C.R. Copper-free click chemistry for dynamic in vivo imaging. *Proc. Natl. Acad. Sci. USA* **2007**, *104*, 16793–16797. [CrossRef]
52. Nguyen, H.; Abramov, M.; Rozenski, J.; Eremeeva, E.; Herdewijn, P. In Vivo Assembly and Expression of DNA Containing Non-canonical Bases in the Yeast Saccharomyces cerevisiae. *ChemBioChem* **2022**, *23*, e202200060. [CrossRef]

53. Cahová, H.; Pohl, R.; Bednárová, L.; Nováková, K.; Cvačka, J.; Hocek, M. Synthesis of 8-bromo-, 8-methyl- and 8-phenyl-dATP and their polymerase incorporation into DNA. *Org. Biomol. Chem.* **2008**, *6*, 3657–3660. [CrossRef]
54. Hollenstein, M. Synthesis of Deoxynucleoside Triphosphates that Include Proline, Urea, or Sulfonamide Groups and Their Polymerase Incorporation into DNA. *Chem. Eur. J.* **2012**, *18*, 13320–13330. [CrossRef]
55. Zaboikin, M.; Freter, C.; Srinivasakumar, N. Gaussian decomposition of high-resolution melt curve derivatives for measuring genome-editing efficiency. *PLoS ONE* **2018**, *13*, e0190192.
56. Baccaro, A.; Marx, A. Enzymatic Synthesis of Organic-Polymer-Grafted DNA. *Chem. Eur. J.* **2010**, *16*, 218–226. [CrossRef]
57. Kočalka, P.; Andersen, N.K.; Jensen, F.; Nielsen, P. Synthesis of 5-(1,2,3-Triazol-4-yl)-2′-deoxyuridines by a Click Chemistry Approach: Stacking of Triazoles in the Major Groove Gives Increased Nucleic Acid Duplex Stability. *ChemBioChem* **2007**, *8*, 2106–2116. [CrossRef]
58. Gimenez Molina, A.; Raguraman, P.; Delcomyn, L.; Veedu, R.N.; Nielsen, P. Oligonucleotides containing 2′-O-methyl-5-(1-phenyl-1,2,3-triazol-4-yl)uridines demonstrate increased affinity for RNA and induce exon-skipping in vitro. *Bioorg. Med. Chem.* **2022**, *55*, 116559. [CrossRef]
59. Fan, H.; Sun, H.; Haque, M.M.; Peng, X. Effect of Triazole-Modified Thymidines on DNA and RNA Duplex Stability. *ACS Omega* **2019**, *4*, 5107–5116. [CrossRef]
60. Balintová, J.; Welter, M.; Marx, A. Antibody–nucleotide conjugate as a substrate for DNA polymerases. *Chem. Sci.* **2018**, *9*, 7122–7125. [CrossRef]
61. Hollenstein, M. Deoxynucleoside triphosphates bearing histamine, carboxylic acid, and hydroxyl residues—Synthesis and biochemical characterization. *Org. Biomol. Chem.* **2013**, *11*, 5162–5172. [CrossRef] [PubMed]
62. Gardner, A.F.; Jackson, K.M.; Boyle, M.M.; Buss, J.A.; Potapov, V.; Gehring, A.M.; Zatopek, K.M.; Corrêa, I.R., Jr.; Ong, J.L.; Jack, W.E. Therminator DNA Polymerase: Modified Nucleotides and Unnatural Substrates. *Front. Mol. Biosci.* **2019**, *6*, 28. [CrossRef]
63. Dunn, M.R.; Larsen, A.C.; Zahurancik, W.J.; Fahmi, N.E.; Meyers, M.; Suo, Z.; Chaput, J.C. DNA Polymerase-Mediated Synthesis of Unbiased Threose Nucleic Acid (TNA) Polymers Requires 7-Deazaguanine To Suppress G:G Mispairing during TNA Transcription. *J. Am. Chem. Soc.* **2015**, *137*, 4014–4017. [CrossRef] [PubMed]
64. Gardner, A.F.; Wang, J.; Wu, W.; Karouby, J.; Li, H.; Stupi, B.P.; Jack, W.E.; Hersh, M.N.; Metzker, M.L. Rapid incorporation kinetics and improved fidelity of a novel class of 3′-OH unblocked reversible terminators. *Nucleic Acids Res.* **2012**, *40*, 7404–7415. [CrossRef] [PubMed]
65. Diafa, S.; Evequoz, D.; Leumann, C.J.; Hollenstein, M. Enzymatic Synthesis of 7,5-Bicyclo-DNA Oligonucleotides. *Chem. Asian J.* **2017**, *12*, 1347–1352. [CrossRef]
66. Jang, M.-Y.; Song, X.-P.; Froeyen, M.; Marlière, P.; Lescrinier, E.; Rozenski, J.; Herdewijn, P. A Synthetic Substrate of DNA Polymerase Deviating from the Bases, Sugar, and Leaving Group of Canonical Deoxynucleoside Triphosphates. *Chem. Biol.* **2013**, *20*, 416–423. [CrossRef]
67. Renders, M.; Miller, E.; Lam, C.H.; Perrin, D.M. Whole cell-SELEX of aptamers with a tyrosine-like side chain against live bacteria. *Org. Biomol. Chem.* **2017**, *15*, 1980–1989. [CrossRef]
68. Chan, K.Y.; Kinghorn, A.B.; Hollenstein, M.; Tanner, J.A. Chemical Modifications for a Next Generation of Nucleic Acid Aptamers. *ChemBioChem* **2022**, *23*, e202200006. [CrossRef]
69. Gawande, B.N.; Rohloff, J.C.; Carter, J.D.; von Carlowitz, I.; Zhang, C.; Schneider, D.J.; Janjic, N. Selection of DNA aptamers with two modified bases. *Proc. Natl. Acad. Sci. USA* **2017**, *114*, 2898–2903. [CrossRef]
70. Minagawa, H.; Sawa, H.; Fujita, T.; Kato, S.; Inaguma, A.; Hirose, M.; Orba, Y.; Sasaki, M.; Tabata, K.; Nomura, N.; et al. A high-affinity aptamer with base-appended base-modified DNA bound to isolated authentic SARS-CoV-2 strains wild-type and B.1.617.2 (delta variant). *Biochem. Biophys. Res. Commun.* **2022**, *614*, 207–212. [CrossRef]
71. Ondruš, M.; Sýkorová, V.; Hocek, M. Traceless enzymatic synthesis of monodispersed hypermodified oligodeoxyribonucleotide polymers from RNA templates. *Chem. Commun.* **2022**, *58*, 11248–11251. [CrossRef]
72. Mei, H.; Liao, J.-Y.; Jimenez, R.M.; Wang, Y.; Bala, S.; McCloskey, C.; Switzer, C.; Chaput, J.C. Synthesis and Evolution of a Threose Nucleic Acid Aptamer Bearing 7-Deaza-7-Substituted Guanosine Residues. *J. Am. Chem. Soc.* **2018**, *140*, 5706–5713. [CrossRef]
73. Freund, N.; Taylor, A.I.; Arangundy-Franklin, S.; Subramanian, N.; Peak-Chew, S.-Y.; Whitaker, A.M.; Freudenthal, B.D.; Abramov, M.; Herdewijn, P.; Holliger, P. A two-residue nascent-strand steric gate controls synthesis of 2′-O-methyl- and 2′-O-(2-methoxyethyl)-RNA. *Nat. Chem.* **2022**. [CrossRef]
74. Hervey, J.R.D.; Freund, N.; Houlihan, G.; Dhaliwal, G.; Holliger, P.; Taylor, A.I. Efficient synthesis and replication of diverse sequence libraries composed of biostable nucleic acid analogues. *RSC Chem. Biol.* **2022**, *3*, 1209–1215. [CrossRef]
75. Sola, M.; Menon, A.P.; Moreno, B.; Meraviglia-Crivelli, D.; Soldevilla, M.M.; Cartón-García, F.; Pastor, F. Aptamers Against Live Targets: Is In Vivo SELEX Finally Coming to the Edge? *Mol. Ther. Nucleic Acids* **2020**, *21*, 192–204. [CrossRef]
76. Lee, M.; Kang, B.; Lee, J.; Lee, J.; Jung, S.T.; Son, C.Y.; Oh, S.S. De novo selected hACE2 mimics that integrate hotspot peptides with aptameric scaffolds for binding tolerance of SARS-CoV-2 variants. *Sci. Adv.* **2022**, *8*, eabq6207. [CrossRef]
77. Stokes, J.M.; MacNair, C.R.; Ilyas, B.; French, S.; Côté, J.-P.; Bouwman, J.; Farha, M.A.; Sieron, A.O.; Whitfield, C.; Coombes, B.K.; et al. Pentamidine sensitizes Gram-negative pathogens to antibiotics and overcomes acquired colistin resistance. *Nat. Microbiol.* **2017**, *2*, 17028. [CrossRef]

Article

Using the Intrinsic Fluorescence of DNA to Characterize Aptamer Binding

Chang Lu [1,2], Anand Lopez [2], Jinkai Zheng [1] and Juewen Liu [2,*]

[1] Institute of Food Science and Technology, Chinese Academy of Agricultural Sciences, Beijing 100193, China
[2] Department of Chemistry, Waterloo Institute for Nanotechnology, University of Waterloo, Waterloo, ON N2L 3G1, Canada
* Correspondence: liujw@uwaterloo.ca

Abstract: The reliable, readily accessible and label-free measurement of aptamer binding remains a challenge in the field. Recent reports have shown large changes in the intrinsic fluorescence of DNA upon the formation of G-quadruplex and i-motif structures. In this work, we examined whether DNA intrinsic fluorescence can be used for studying aptamer binding. First, DNA hybridization resulted in a drop in the fluorescence, which was observed for A30/T30 and a 24-mer random DNA sequence. Next, a series of DNA aptamers were studied. Cortisol and Hg^{2+} induced fluorescence increases for their respective aptamers. For the cortisol aptamer, the length of the terminal stem needs to be short to produce a fluorescence change. However, caffeine and adenosine failed to produce a fluorescence change, regardless of the stem length. Overall, using the intrinsic fluorescence of DNA may be a reliable and accessible method to study a limited number of aptamers that can produce fluorescence changes.

Keywords: intrinsic fluorescence; DNA; aptamer; binding

1. Introduction

DNA aptamers have many advantages compared to antibodies, such as much lower cost, higher stability and ease of modification [1–4]. The majority of research has been focused on a few model aptamers, although hundreds of other aptamers have been published [5–9]. A main issue in the field is a lack of reliable yet readily accessible methods to measure aptamer binding [10–12]. The versatility of DNA-based assays has sometimes worked against it due to a lack of quality control [13]. While the majority of immunoassays require immobilization of antibodies or antigens [14], homogeneous assays are preferred for its simplicity and avoiding nonspecific binding to surfaces [15–18].

Fluorescence spectroscopy is probably the most commonly used method to characterize aptamer binding using either covalently attached fluorophores or using DNA staining dyes [19–23] and cationic polymers [24]. For example, aptamers can be terminally labeled with a fluorophore/quencher pair or a FRET pair to form an aptamer beacon [25]. Another reliable method is to design structure-switching aptamers, where a quencher-labeled complementary DNA is hybridized with a fluorophore-labeled aptamer [5,26,27]. These methods require expensive covalent modifications, making it difficult to study different aptamer sequences. Using DNA staining dyes is cost-effective, but this method is less reliable and sometimes has a small signal change [28].

Recently, intrinsic fluorescence of DNA has been reported [29–31]. Although quite weak, with a few µM of DNA, a decent fluorescence can be achieved, and this concentration is comparable to that used for CD spectroscopy and is much less compared to ITC [32]. In addition, DNA hybridization, G-quadruplex formation and i-motif formation have all been shown to induce DNA fluorescence change [33–35]. These reactions are accompanied with a large conformational change of DNA. Aptamer binding, on the other hand, might have less conformational change. Thus far, whether such fluorescence can be used to study aptamer binding remains to be explored. In this work, we performed systematic studies of a series of DNA oligonucleotides and aptamers. We found that the majority of the tested aptamers failed to induce a change in the intrinsic fluorescence, and only two examples appeared to be successful.

2. Results

2.1. DNA Hybridization Drops Intrinsic Fluorescence

In this study, we chose a few target molecules in order to obtain a comprehensive understanding of DNA intrinsic fluorescence for aptamer binding. Before testing aptamers, we first examined the DNA hybridization reaction, since complementary DNA can also be considered as a special aptamer target. We first studied the hybridization of A30 with T30. By varying the excitation wavelength of A30, we observed two emission peaks at 387 nm and 432 nm, respectively (Figure 1A). For T30, the fluorescence was weaker, and the emission peaks continuously varied with the excitation peak (Figure 1B). When an equal concentration of A30 and T30 were mixed, the fluorescence increased slightly compared to that of A30 alone (Figure 1C). However, part of the fluorescence increase was due to the extra T30 added. To test the effect of hybridization, we then measured the fluorescence difference of 10 µM A30 and 5 µM A30 (equation 1), and compared it with the fluorescence difference of 5 µM A30/T30 hybrid with 5 µM T30 (equation 2). If these two differences were equal, then DNA hybridization had no effect on the intrinsic fluorescence of DNA, and we only observed a simple sum of the two strands (Figure 1D). The reason to compare the difference instead of directly comparing A30 plus T30 with A30/T30 hybrids is to avoid potential interference from background signals in the spectra. When the fluorescence intensity is low, contributions from the background cannot be neglected.

$$(F_{Background} + 2 \times F_{A30}) - (F_{background} + F_{A30}) = F_{A30} \quad (1)$$

$$(F_{Background} + F_{A30/T30}) - (F_{Background} + F_{T30}) = F_{A30/T30} - F_{T30} \quad (2)$$

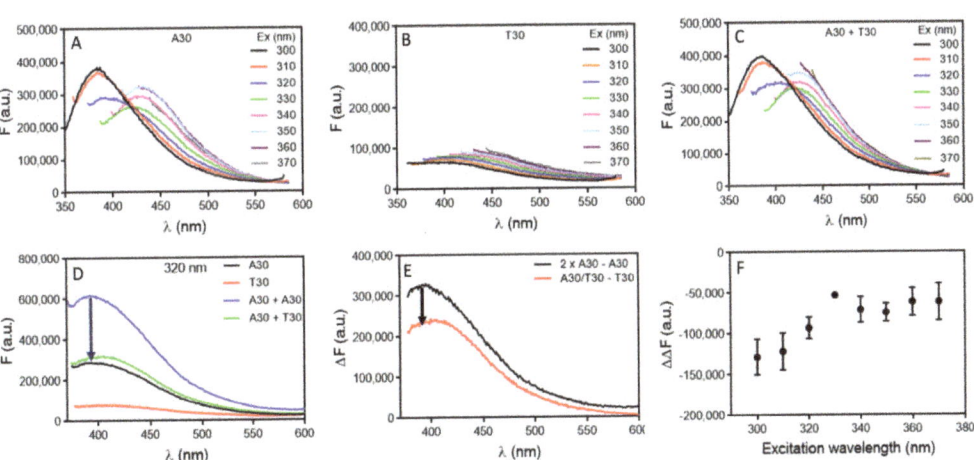

Figure 1. The fluorescence emission spectra of 5 µM (**A**) A30, (**B**) T30 and (**C**) annealed A30/T30 duplex excited at different wavelengths in buffer (10 mM PB, pH 7.0, 100 mM NaCl). (**D**) Comparison of fluorescence emission spectra of various DNA samples excited at 320 nm. (**E**) The fluorescence difference spectra: (2 × A30 − A30) and (A30/T30 − T30) with 320 nm excitation. (**F**) The difference of the difference spectra (the red line minus black line) in (**E**) at various excitation wavelengths.

Our results showed that the A30/T30 duplex DNA had a 27% lower fluorescence compared to the sum of the two strands (Figure 1E). We did the analysis for all the excitation wavelengths from 300 nm to 370 nm and the duplex fluorescence was lower at all the tested wavelengths (Figure 1F). Markovitsi and their coworkers studied the fluorescence of A20 and T20 DNA, and they found that with long wavelength UVA excitation (330 nm), the fluorescence of the A20/T20 duplex DNA was nearly 3-fold of their individual components [33]; however, we did not observe that. On the other hand, with UVC excitation (255 nm), the fluorescence yield dropped for the duplex, which was consistent with our observation, although we did not excite the sample at such a short wavelength to avoid the strong DNA absorbance at round 260 nm and hypochromicity associated with DNA hybridization. We did a control experiment with SYBR Green I (SGI) staining and confirmed formation of duplex DNA (Figure S1).

We then did the same experiment using a 24-mer random sequenced DNA named DNA1 and its complementary DNA (cDNA1), and dropped fluorescence was also observed upon hybridization (Figure S2). Therefore, we tend to believe that DNA hybridization would decrease the quantum yield of the intrinsic fluorescence of DNA when excited in the range of 300 nm to 370 nm.

2.2. Cortisol Binding Enhances Aptamer Fluorescence

After understanding DNA hybridization, we then focused on aptamers. We first tested the cortisol binding aptamer for its high binding affinity (K_d ~100 nM) [27,36]. The secondary structure of the aptamer and the structure of cortisol are shown in Figure 2A. The UV-vis spectrum of cortisol is shown in Figure 2B and a peak at 247 nm was observed. In addition, it does not have intrinsic fluorescence (Figure 2B, red spectrum).

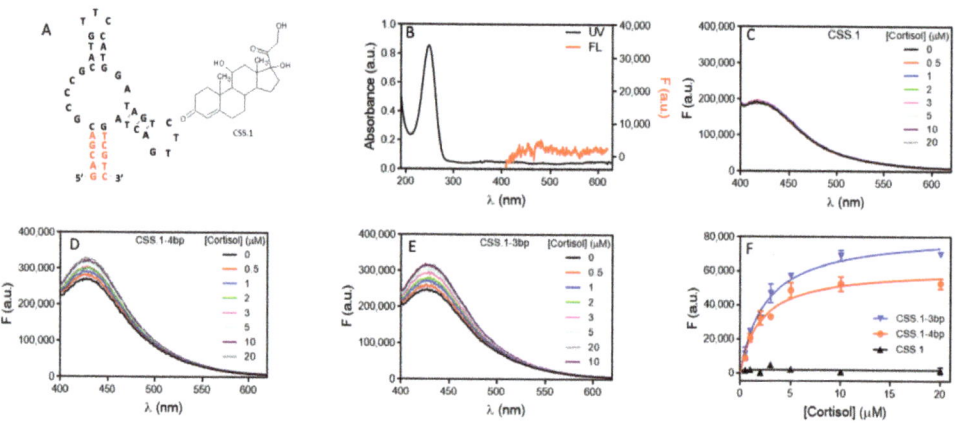

Figure 2. (**A**) The structure of the aptamer and cortisol. (**B**) UV-vis spectra of 50 uM cortisol and its fluorescence spectra when excited at 340 nm. The fluorescence of the (**C**) 5 bp (**D**) 4 bp and (**E**) 3 bp aptamers upon titration of cortisol. (**F**) The fitted binding curve of the three aptamers.

We first tested the aptamer in Figure 2A. When excited at 340 nm, a fluorescence peak from 5 µM of the aptamer solution was observed at 416 nm. However, when titrated with cortisol, no fluorescence change was observed (Figure 2C). This aptamer has a 5 bp stem. We suspected that the aptamer was already folded, and cortisol binding only induced some minor local conformational changes, which was too small to affect the aptamer fluorescence. To test this hypothesis, we then tested the same aptamer but with the stem shortened. Interestingly, we observed cortisol-dependent fluorescence enhancement in both the 4 bp (Figure 2D) and 3 bp aptamers (Figure 2E). We reason that the shorter aptamers were initially in an open conformation, which was closed upon cortisol binding. It is interesting that the fluorescence enhanced in this case. Since we expected DNA duplex formation to decrease fluorescence, the increase was likely from the formation of non-canonical base interactions. For example, we observed 9-fold fluorescence increased when K$^+$ was added to a G-quadruplex forming DNA [35].

The data were fitted and showed a similar K_d of around 2 µM (Figure 2F). Since the aptamer concentration was high (5 µM) compared to the expected K_d of 100 nM, the system is in the titration region and the K_d is roughly half of the aptamer concentration [37]. Therefore, for high affinity aptamers, it might not be possible to obtain an accurate K_d. Nevertheless, it can be used to confirm aptamer binding, and to study the effect of other conditions, such as mutation. We also tested a few other DNA sequences and when cortisol was added, no fluorescence change was observed (Figure S3).

2.3. Hg^{2+} Binding to Poly-T DNA Enhances Fluorescence

We then tested Hg^{2+} binding using a polythymine DNA, T30 [38]. Hg^{2+} binding to T30 was verified by the SGI staining (Figure S4) [39]. Since T30 is not a sequence derived from an aptamer selection, it is technically not an aptamer but can still serve as an interesting model system. Hg^{2+} can specifically bind between two thymine bases forming a T–Hg^{2+}–T base pair and fold the DNA into a hairpin structure (Figure 3A). This large conformational change of DNA may cause a fluorescence change. To test this hypothesis, we measure the fluorescence emission spectrum of T30 with 320 nm excitation, and the DNA showed an emission peak at 409 nm. The fluorescence increased when Hg^{2+} was titrated (Figure 3B), suggesting the T–Hg^{2+}–T binding-directed the fluorescence change. This is quite striking since Hg^{2+} is known for its fluorescence quenching property. As a control, we also tested a 24-mer random sequenced DNA, which was not expected to bind to Hg^{2+}. In this case, no fluorescence change was observed upon titrating Hg^{2+} (Figure 3C). Next, the relative fluorescence change of the two DNAs was calculated, and the intrinsic fluorescence of T30 increased up to 50% (Figure 3D). The fitted K_d for Hg^{2+} binding was 38 µM. Again, this is not the true K_d due to the high concentration of DNA used. Since each T30 DNA can bind around 13 Hg^{2+} ions (assuming a 4-nucleotide loop), half of that for 5 µM DNA is 32 µM, which is close to the observed K_d.

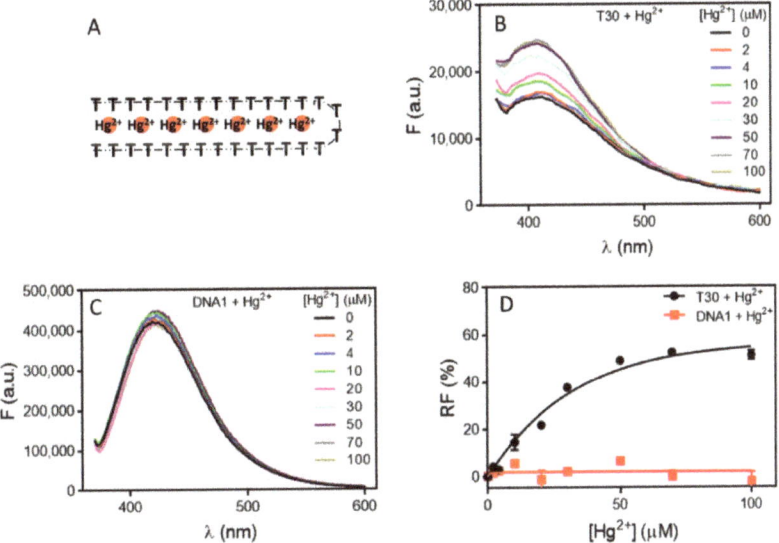

Figure 3. (**A**) A scheme showing Hg^{2+} binding to T30 DNA and each DNA was estimated to bind around 13 Hg^{2+} ions. The fluorescence spectra of (**B**) T30 and (**C**) DNA1 upon titration of Hg^{2+} with 320 nm excitation. (**D**) The fitted binding curve of the two DNAs.

2.4. Adenosine Binding Aptamer Fails to Show Aptamer Fluorescence Change

Next, the adenosine aptamer was studied, which is a model aptamer with a K_d around 7 µM [40]. The structures of adenosine and its aptamer are shown in Figure 4A. Adenosine has a UV absorption peak at 260 nm (Figure S5), and weak fluorescence when excited at various wavelength (Figure S6). We titrated adenosine to the aptamer at different excitation wavelengths (Figure S7), and similar trends were observed at all these wavelengths. To avoid the interference from adenosine absorption, we chose 350 nm as the excitation wavelength. Within 20 µM adenosine, no fluorescence change was observed for the aptamer (Figure 4B). We then shortened the aptamer to contain three or two base pairs in the stem. Surprisingly, we still did not observe any fluorescence change upon adding adenosine (Figure 4C,D). Next, we measured ThT fluorescence spectroscopy to verify the binding of adenosine and aptamer [41]. This aptamer is rich in guanine and can increase the fluorescence of associated ThT. Upon binding to adenosine, some ThT may be displaced from the aptamer/adenosine complex, leading to decreased fluorescence. Indeed, a large fluorescence drop was observed upon the addition of adenosine to the three aptamers (Figure 4E,F and Figure S8), confirming that the three aptamers can bind adenosine.

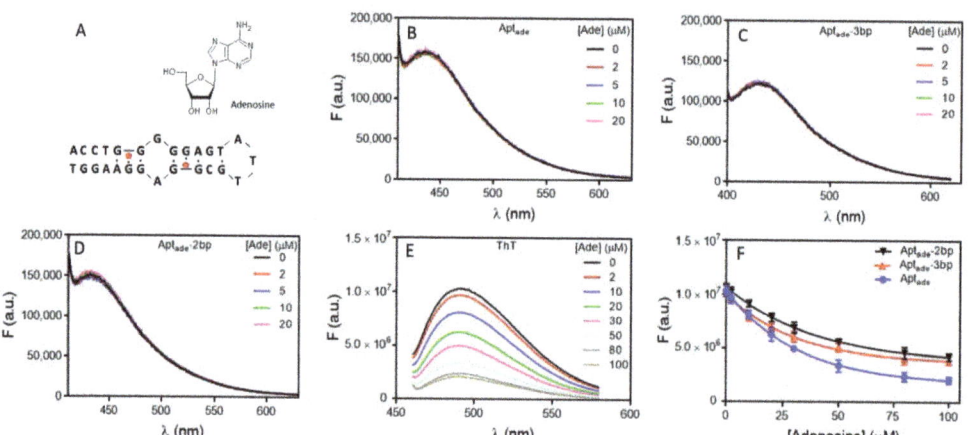

Figure 4. (A) The structures of the aptamer and adenosine. The fluorescence spectra of the (B) 4 bp (C) 3 bp and (D) 2bp adenosine binding aptamer upon titration of adenosine with 350 nm excitation. (E) The fluorescence emission spectra of ThT staining 100 nM of the adenosine aptamer after adding different concentration of adenosine in buffer (10 mM PB, 100 mM NaCl, 10 mM MgCl$_2$ pH 7) with 450 nm excitation. (F) The fitted binding curves based on ThT staining of the three aptamers.

Given the structure of this aptamer, one would expect a large conformational change upon target binding, especially for the shortened aptamers. The fact that no fluorescence change was observed could be related to the canceling of the fluorescence enhancement and dropping factors.

2.5. Caffeine Binding Aptamer Fails to Show Aptamer Fluorescence Change

We recently reported an aptamer for caffeine [6]. Its structure is shown in Figure 5A. Caffeine has strong absorption in the UV region with a peak at 273 nm (Figure 5B), and it also has fluorescence when excited at 300 nm or 310 nm (Figure S9). The interference from the intrinsic fluorescence of caffeine was minimal when excited at 340 nm or longer (Figure S10). Therefore, we chose to excite the aptamer at 340 nm. When we titrated caffeine to the aptamer; however, no change in fluorescence was observed (Figure 5C), and when we truncated the stem down from even to just one base pair, still no change was observed (Figure 5D, Figure S11).

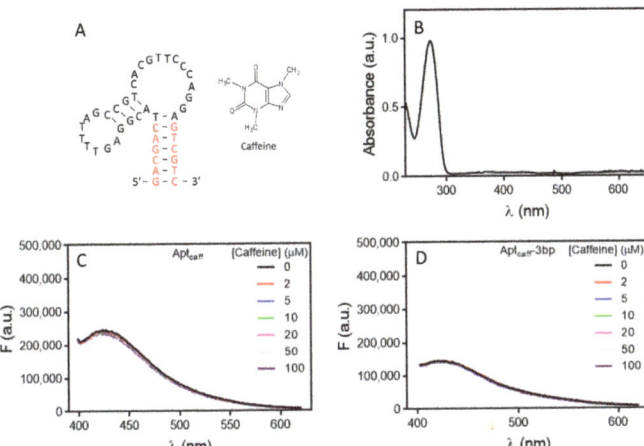

Figure 5. (**A**) The structures of the caffeine aptamer and caffeine. (**B**) UV-vis spectrum of 100 µM caffeine. The fluorescence spectra of the caffeine aptamer with (**C**) 6 bp and (**D**) 3 bp in the stem region upon titration of caffeine.

3. Discussion

In this work, we examined the change in the intrinsic fluorescence of DNA upon hybridization and aptamers upon target binding. In three cases, we observed fluorescence change, while in the other two, we did not. We summarized our results in Figure 6A. DNA hybridization decreased the fluorescence intensity (Figure 6A). The cortisol aptamer (Figure 6B) and Hg^{2+} aptamer (Figure 6C) showed binding induced fluorescence enhancement. However, adenosine and caffeine did not produce measurable fluorescence changes (Figure 6D). For typical small molecule binding aptamers, cortisol is the only example showing a fluorescence change, and the amount of fluorescence change was quite small. Even for the cortisol aptamer, when a stable aptamer with a long stem was used, no fluorescence was observed. Therefore, it is quite hard to predict the intrinsic fluorescence change of an aptamer upon binding to a small molecular target.

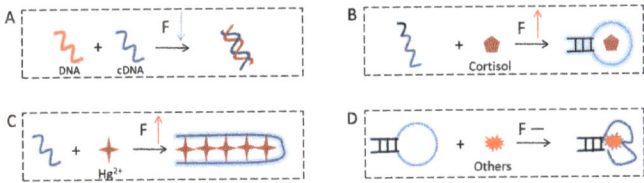

Figure 6. Schemes showing changes in DNA intrinsic fluorescence upon (**A**) DNA hybridization, and target binding to the (**B**) cortisol aptamer, (**C**) T30, and (**D**) the adenosine and caffeine aptamers that did not show a fluorescence change.

The question we intended to address in this work was whether the intrinsic fluorescence of aptamers could be used to study aptamer binding. Based on the above studies, the answer is likely to be no, especially for newly selected aptamers. A lack of target-dependent fluorescence change does not rule out aptamer binding. All the experiments here were performed using DNA and DNA aptamers. Since the fluorescence is related to DNA bases, we expect that RNA should have similar behavior.

For aptamers that show a fluorescence change, and if the fluorescence change is sufficiently large, this can be a cost-effective way to characterize aptamer binding and can provide useful information about binding kinetics, and buffer and salt requirement of binding. However, if the K_d is smaller than the DNA concentration, this method cannot be used to measure K_d.

If forming Watson-Crick base pair drops DNA's intrinsic fluorescence, then the increase we observed in the cortisol aptamer and T30/Hg^{2+} system was not due to the increase of the Watson-Crick base pair content. The reasons could be non-canonical base pairs and forming a more hydrophobic binding environment. This fluorescence increase in the cortisol case was quite small. For example, a large increase was seen when K^+ was added to G-rich DNAs (up to 9-fold), while a few fold increase was observed when pH was dropped for an i-motif forming DNA [29].

It is also important to pay attention to the fluorescence of target molecules. Although most of the molecules are not considered to be fluorescent (adenosine, caffeine), they have detectable emissions at around 10 μM or higher if excited at the right wavelength (Figures S6 and S9). Thus, it is important to choose an excitation wavelength to avoid such interference.

4. Materials and Methods

4.1. DNA Hybridization Drops Intrinsic Fluorescence

All of the DNA samples used in this work were purchased from Integrated DNA Technologies (Coralville, IA, USA) and their sequences are listed in Table S1. Mercury acetate ($Hg(Ac)_2$), magnesium chloride ($MgCl_2$), fluorescein sodium salt, cortisol, adenosine, caffeine, and thioflavin T (ThT) were from Sigma-Aldrich. Sodium chloride (NaCl), sodium nitrate ($NaNO_3$), sodium phosphate monobasic monohydrate, and sodium phosphate dibasic heptahydrate were obtained from Mandel Scientific (Guelph, ON, Canada). SYBR Green I (SGI) was purchased from Lonza (Rockland, ME, USA). Milli-Q water was used to prepare buffers and solutions.

4.2. Fluorescence Spectroscopy

All of the fluorescence spectra were recorded on a Horiba Fluoromax-4 spectrofluorometer (HORIBA Scientific, Edison, USA). The DNA was excited at 300–370 nm and its emission was recorded from 350 to 620 nm. 500 μL of DNA in buffer (100 mM NaCl, 10 mM $MgCl_2$, 1 mM PB pH 7) was put in a 1 cm × 1 cm quartz fluorescence cuvette, and then different concentrations of targets were added for fluorescence measurement. The measurements in this work were carried out in triplicate and the standard deviations were plotted as the error bars. All the experiments were performed at room temperature (~22 °C) unless otherwise indicated.

5. Conclusions

This work examined the change of the intrinsic fluorescence of DNA and aptamers upon hybridization and target binding. In contrast with the large fluorescence changes observed for the formation of G-quadruplex structures and i-motifs in previous work, we observed very small changes and even no change for aptamer binding. Our study indicated a slight fluorescence drop upon DNA hybridization. There is an increase in the fluorescence of the cortisol aptamer and T30/Hg^{2+} systems, which were attributable to the formation of non-canonical base pairs. However, we did not observe fluorescence change for caffeine or adenosine aptamers, even if we truncated the stems to afford larger conformational changes. Given the vast number of aptamers published, it is impossible to test them all in one paper. This paper has shown a few different types of behaviors and careful controls are needed to understand whether this method can be used for new aptamers. For systems that have a reliable fluorescence change, this method can be an effective way to study aptamer binding. Understanding the reason for the (lack of) fluorescence change upon target binding could be a topic for future research.

Supplementary Materials: The following supporting information can be downloaded at: https://www.mdpi.com/article/10.3390/molecules27227809/s1, Table S1: The DNA Sequences Used in This Work; Figure S1: The fluorescence emission spectra of ssDNA or dsDNA stained by SGI; Figure S2: The fluorescence emission spectrum of different DNA excited at different wavelength; Figure S3: Fluorescence of glucose and quinine aptamer upon titration of cortisol; Figure S4: The fluorescence emission spectrum of SYBR Green I after adding T30 and Hg^{2+}; Figure S5: UV-vis spectra of adenosine; Figure S6: The fluorescence emission spectrum of adenosine; Figure S7: The fluorescence emission spectrum of the adenosine aptamer upon titration of adenosine; Figure S8: The fluorescence emission spectrum of ThT after adding 3bp and 2bp adenosine aptamer and different concentration of adenosine; Figure S9: The fluorescence emission spectrum of caffeine; Figure S10: The fluorescence emission spectra of the caffeine aptamer upon titration of caffeine; Figure S11: Fluorescence of the 1bp, 2bp and 4bp aptamer upon titration of caffeine.

Author Contributions: Conceptualization, J.L.; methodology, J.L. and C.L.; investigation, C.L.; data curation, C.L.; writing—original draft preparation, C.L. and J.L.; writing—review and editing, C.L., A.L. and J.L.; supervision, J.L.; funding acquisition, J.L., C.L. and J.Z. All authors have read and agreed to the published version of the manuscript.

Funding: Funding for this work was from the Natural Sciences and Engineering Research Council of Canada (NSERC), the National Natural Science Foundation of China (31901776 and 32072181), and Agricultural Science and Technology Innovation Program (CAAS-ASTIP-2021-IFST-SN2021-05). C.L. was supported by a China Scholarship Council (CSC) scholarship to visit the University of Waterloo.

Institutional Review Board Statement: Not applicable.

Informed Consent Statement: Not applicable.

Data Availability Statement: The data presented in this study are available on request from the corresponding author.

Conflicts of Interest: The authors declare no conflict of interest.

Sample Availability: Samples of the compounds are available from the authors.

References

1. Yu, H.; Alkhamis, O.; Canoura, J.; Liu, Y.; Xiao, Y. Advances and Challenges in Small-Molecule DNA Aptamer Isolation, Characterization, and Sensor Development. *Angew. Chem. Int. Ed.* **2021**, *60*, 16800–16823. [CrossRef] [PubMed]
2. McConnell, E.M.; Nguyen, J.; Li, Y. Aptamer-Based Biosensors for Environmental Monitoring. *Front. Chem.* **2020**, *8*. [CrossRef] [PubMed]
3. Wu, L.; Wang, Y.; Xu, X.; Liu, Y.; Lin, B.; Zhang, M.; Zhang, J.; Wan, S.; Yang, C.; Tan, W. Aptamer-Based Detection of Circulating Targets for Precision Medicine. *Chem. Rev.* **2021**, *121*, 12035–12105. [CrossRef] [PubMed]
4. Dunn, M.R.; Jimenez, R.M.; Chaput, J.C. Analysis of aptamer discovery and technology. *Nat. Rev. Chem.* **2017**, *1*. [CrossRef]
5. Nakatsuka, N.; Yang, K.-A.; Abendroth, J.M.; Cheung, K.M.; Xu, X.; Yang, H.; Zhao, C.; Zhu, B.; Rim, Y.S.; Yang, Y.; et al. Aptamer–field-effect transistors overcome Debye length limitations for small-molecule sensing. *Science* **2018**, *362*, 319–324. [CrossRef]
6. Huang, P.-J.J.; Liu, J. Selection of Aptamers for Sensing Caffeine and Discrimination of Its Three Single Demethylated Analogues. *Anal. Chem.* **2022**, *94*, 3142–3149. [CrossRef]
7. Zhao, Y.; Yavari, K.; Liu, J. Critical evaluation of aptamer binding for biosensor designs. *TrAC Trends Anal. Chem.* **2021**, *146*, 116480. [CrossRef]
8. Mahmoudpour, M.; Ding, S.; Lyu, Z.; Ebrahimi, G.; Du, D.; Dolatabadi, J.E.N.; Torbati, M.; Lin, Y. Aptamer functionalized nanomaterials for biomedical applications: Recent advances and new horizons. *Nano Today* **2021**, *39*, 101177. [CrossRef]
9. Yu, H.; Luo, Y.; Alkhamis, O.; Canoura, J.; Yu, B.; Xiao, Y. Isolation of Natural DNA Aptamers for Challenging Small-Molecule Targets, Cannabinoids. *Anal. Chem.* **2021**, *93*, 3172–3180. [CrossRef]
10. Daems, E.; Moro, G.; Campos, R.; De Wael, K. Mapping the gaps in chemical analysis for the characterisation of aptamer-target interactions. *TrAC Trends Anal. Chem.* **2021**, *142*, 116311. [CrossRef]
11. McKeague, M.; De Girolamo, A.; Valenzano, S.; Pascale, M.; Ruscito, A.; Velu, R.; Frost, N.R.; Hill, K.; Smith, M.; McConnell, E.M.; et al. Comprehensive Analytical Comparison of Strategies Used for Small Molecule Aptamer Evaluation. *Anal. Chem.* **2015**, *87*, 8608–8612. [CrossRef]
12. Bottari, F.; Daems, E.; de Vries, A.-M.; Van Wielendaele, P.; Trashin, S.; Blust, R.; Sobott, F.; Madder, A.; Martins, J.C.; De Wael, K. Do Aptamers Always Bind? The Need for a Multifaceted Analytical Approach When Demonstrating Binding Affinity between Aptamer and Low Molecular Weight Compounds. *J. Am. Chem. Soc.* **2020**, *142*, 19622–19630. [CrossRef] [PubMed]
13. Zhao, Y.; Yavari, K.; Wang, Y.; Pi, K.; Van Cappellen, P.; Liu, J. Deployment of functional DNA-based biosensors for environmental water analysis. *TrAC Trends Anal. Chem.* **2022**, *153*. [CrossRef]

14. Cheng, Z.J.; Li, B.; Zhan, Z.; Zhao, Z.; Xue, M.; Zheng, P.; Lyu, J.; Hu, C.; He, J.; Chen, R. Clinical Application of Antibody Immunity against SARS-CoV-2: Comprehensive Review on Immunoassay and Immunotherapy. *Clin. Rev. Allergy Immunol.* **2022**. [CrossRef] [PubMed]
15. Ding, Y.; Liu, X.; Huang, P.-J.J.; Liu, J. Homogeneous assays for aptamer-based ethanolamine sensing: No indication of target binding. *Anal. Chim. Acta* **2022**, *1207*, 1348–1356. [CrossRef]
16. Nielsen, L.D.F.; Hansen-Bruhn, M.; Nijenhuis, M.A.D.; Gothelf, K.V. Protein-Induced Fluorescence Enhancement and Quenching in a Homogeneous DNA-Based Assay for Rapid Detection of Small-Molecule Drugs in Human Plasma. *ACS Sensors* **2022**, *7*, 856–865. [CrossRef]
17. Zhang, H.; Li, F.; Dever, B.; Li, X.-F.; Le, X.C. DNA-Mediated Homogeneous Binding Assays for Nucleic Acids and Proteins. *Chem. Rev.* **2012**, *113*, 2812–2841. [CrossRef]
18. He, L.; Huang, R.; Xiao, P.; Liu, Y.; Jin, L.; Liu, H.; Li, S.; Deng, Y.; Chen, Z.; Li, Z.; et al. Current signal amplification strategies in aptamer-based electrochemical biosensor: A review. *Chin. Chem. Lett.* **2021**, *32*, 1593–1602. [CrossRef]
19. Lu, C.; Huang, P.J.; Zheng, J.; Liu, J. 2-Aminopurine Fluorescence Spectroscopy for Probing a Glucose Binding Aptamer. *ChemBioChem* **2022**, *23*. [CrossRef]
20. Wang, X.X.; Zhu, L.J.; Li, S.T.; Zhang, Y.Z.; Liu, S.Y.; Huang, K.L.; Xu, W.T. Fluorescent Functional Nucleic Acid: Principles, Properties and Applications in Bioanalyzing. *TrAC Trends Anal. Chem.* **2021**, *141*, 116292. [CrossRef]
21. Zhang, J.; Wang, L.; Jäschke, A.; Sunbul, M. A Color-Shifting Near-Infrared Fluorescent Aptamer–Fluorophore Module for Live-Cell RNA Imaging. *Angew Chem. Int. Ed.* **2021**, *133*, 21611–21618. [CrossRef]
22. Billet, B.; Chovelon, B.; Fiore, E.; Oukacine, F.; Petrillo, M.; Faure, P.; Ravelet, C.; Peyrin, E. Aptamer Switches Regulated by Post-Transition/Transition Metal Ions. *Angew. Chem.* **2021**, *133*, 12454–12458. [CrossRef]
23. Han, A.; Hao, S.; Yang, Y.; Li, X.; Luo, X.; Fang, G.; Liu, J.; Wang, S. Perspective on Recent Developments of Nanomaterial Based Fluorescent Sensors: Applications in Safety and Quality Control of Food and Beverages. *J. Food Drug Anal.* **2020**, *28*, 486–507. [CrossRef] [PubMed]
24. Zhang, P.; Qin, K.; Lopez, A.; Li, Z.; Liu, J. General Label-Free Fluorescent Aptamer Binding Assay Using Cationic Conjugated Polymers. *Anal. Chem.* **2022**, *94*, 15456–15463. [CrossRef] [PubMed]
25. Li, J.J.; Fanga, X.; Tan, W. Molecular Aptamer Beacons for Real-Time Protein Recognition. *Biochem. Biophys. Res. Commun.* **2002**, *292*, 31–40. [CrossRef] [PubMed]
26. Nutiu, R.; Li, Y. Structure-Switching Signaling Aptamers. *J. Am. Chem. Soc.* **2003**, *125*, 4771–4778. [CrossRef]
27. Yang, K.-A.; Chun, H.; Zhang, Y.; Pecic, S.; Nakatsuka, N.; Andrews, A.M.; Worgall, T.S.; Stojanovic, M.N. High-Affinity Nucleic-Acid-Based Receptors for Steroids. *ACS Chem. Biol.* **2017**, *12*, 3103–3112. [CrossRef]
28. Khusbu, F.Y.; Zhou, X.; Chen, H.; Ma, C.; Wang, K. Thioflavin T as a fluorescence probe for biosensing applications. *TrAC Trends Anal. Chem.* **2018**, *109*, 1–18. [CrossRef]
29. Zuffo, M.; Gandolfini, A.; Heddi, B.; Granzhan, A. Harnessing intrinsic fluorescence for typing of secondary structures of DNA. *Nucleic Acids Res.* **2020**, *48*, e61. [CrossRef]
30. Chowdhury, M.H.; Ray, K.; Johnson, M.L.; Gray, S.K.; Pond, J.; Lakowicz, J.R. On the Feasibility of Using the Intrinsic Fluorescence of Nucleotides for DNA Sequencing. *J. Phys. Chem. C* **2010**, *114*, 7448–7461. [CrossRef]
31. Lakowicz, J.R.; Shen, B.; Gryczynski, Z.; D'Auria, S.; Gryczynski, I. Intrinsic Fluorescence from DNA Can Be Enhanced by Metallic Particles. *Biochem. Biophys. Res. Commun.* **2001**, *286*, 875–879. [CrossRef] [PubMed]
32. Gustavsson, T.; Markovitsi, D. Fundamentals of the Intrinsic DNA Fluorescence. *Accounts Chem. Res.* **2021**, *54*, 1226–1235. [CrossRef] [PubMed]
33. Banyasz, A.; Vayá, I.; Changenet-Barret, P.; Gustavsson, T.; Douki, T.; Markovitsi, D. Base Pairing Enhances Fluorescence and Favors Cyclobutane Dimer Formation Induced upon Absorption of UVA Radiation by DNA. *J. Am. Chem. Soc.* **2011**, *133*, 5163–5165. [CrossRef]
34. Xiang, X.; Li, Y.; Ling, L.; Bao, Y.; Su, Y.; Guo, X. Label-free and dye-free detection of target DNA based on intrinsic fluorescence of the (3+1) interlocked bimolecular G-quadruplexes. *Sensors Actuators B: Chem.* **2019**, *290*, 68–72. [CrossRef]
35. Lopez, A.; Liu, J. Probing metal-dependent G-quadruplexes using the intrinsic fluorescence of DNA. *Chem. Commun.* **2022**, *58*, 10225–10228. [CrossRef] [PubMed]
36. Niu, C.; Ding, Y.; Zhang, C.; Liu, J. Comparing two cortisol aptamers for label-free fluorescent and colorimetric biosensors. *Sensors Diagn.* **2022**, *1*, 541–549. [CrossRef]
37. Jarmoskaite, I.; Alsadhan, I.; Vaidyanathan, P.P.; Herschlag, D. How to measure and evaluate binding affinities. *eLife* **2020**, *9*. [CrossRef] [PubMed]
38. Ono, A.; Torigoe, H.; Tanaka, Y.; Okamoto, I. Binding of metal ions by pyrimidine base pairs in DNA duplexes. *Chem. Soc. Rev.* **2011**, *40*, 5855–5866. [CrossRef]
39. Wang, J.; Liu, B. Highly sensitive and selective detection of Hg^{2+} in aqueous solution with mercury-specific DNA and Sybr Green I. *Chem. Commun.* **2008**, 4759–4761. [CrossRef]
40. Huizenga, D.E.; Szostak, J.W. A DNA Aptamer That Binds Adenosine and ATP. *Biochemistry* **1995**, *34*, 656–665. [CrossRef]
41. Zhang, F.; Huang, P.-J.J.; Liu, J. Sensing Adenosine and ATP by Aptamers and Gold Nanoparticles: Opposite Trends of Color Change from Domination of Target Adsorption Instead of Aptamer Binding. *ACS Sensors* **2020**, *5*, 2885–2893. [CrossRef] [PubMed]

Review

Non-G Base Tetrads

Núria Escaja [1,2,*], Bartomeu Mir [1,2], Miguel Garavís [3] and Carlos González [3,*]

1 Organic Chemistry Section, Inorganic and Organic Chemistry Department, University of Barcelona, Martí i Franquès 1–11, 08028 Barcelona, Spain
2 Institute of Biomedicine, University of Barcelona, Av. Diagonal 645, 08028 Barcelona, Spain
3 Instituto de Química Física 'Rocasolano', CSIC, Serrano 119, 28006 Madrid, Spain
* Correspondence: nescaja@ub.edu (N.E.); cgonzalez@iqfr.csic.es (C.G.)

Abstract: Tetrads (or quartets) are arrangements of four nucleobases commonly involved in the stability of four-stranded nucleic acids structures. Four-stranded or quadruplex structures have attracted enormous attention in the last few years, being the most extensively studied guanine quadruplex (G-quadruplex). Consequently, the G-tetrad is the most common and well-known tetrad. However, this is not the only possible arrangement of four nucleobases. A number of tetrads formed by the different nucleobases have been observed in experimental structures. In most cases, these tetrads occur in the context of G-quadruplex structures, either inserted between G-quartets, or as capping elements at the sides of the G-quadruplex core. In other cases, however, non-G tetrads are found in more unusual four stranded structures, such as i-motifs, or different types of peculiar fold-back structures. In this report, we review the diversity of these non-canonical tetrads, and the structural context in which they have been found.

Keywords: DNA structure; RNA structure; quadruplex; tetrad

Citation: Escaja, N.; Mir, B.; Garavís, M.; González, C. Non-G Base Tetrads. *Molecules* **2022**, *27*, 5287. https://doi.org/10.3390/molecules27165287

Academic Editor: Eric Defrancq

Received: 11 July 2022
Accepted: 11 August 2022
Published: 19 August 2022

Publisher's Note: MDPI stays neutral with regard to jurisdictional claims in published maps and institutional affiliations.

Copyright: © 2022 by the authors. Licensee MDPI, Basel, Switzerland. This article is an open access article distributed under the terms and conditions of the Creative Commons Attribution (CC BY) license (https://creativecommons.org/licenses/by/4.0/).

1. Introduction

Interest in four-stranded nucleic acids structures has been constantly increasing in the last two decades. Although not always the case, four-stranded structures are usually stabilized by arrangements of four nucleobases commonly named as tetrads or quartets. The most extensively studied four-stranded structure is the G-quadruplex [1], the importance of which, in recent times, has been driven by its proven biological relevance and promising therapeutic applications [2–5].

G-quadruplexes exhibit wide structural diversity depending on strands orientation (parallel, antiparallel and hybrid structures) and connecting loop topologies (diagonal, lateral and propeller loops) [6–8]. Despite their structural variety, all G-quadruplexes share the same common element: the so-called G-tetrad. G-tetrads are platforms of four guanine nucleobases interacting through their Hoogsteen and Watson–Crick sides, following either a clockwise, or an anticlockwise arrangement (Figure 1A). These tetrads are stabilized by eight hydrogen bonds, and by electrostatic interactions with cations located between two consecutive tetrads. The cation, usually monovalent, compensates the negatively charged oxygens in the center of the tetrad. Among the different monovalent cations, K^+ is the most stabilizing, since its atomic radius fits very well in the central position of two consecutive tetrads. G-tetrads are planar, allowing an efficient stacking of multiple layers, and giving rise to extremely thermostable structures.

Although the G-quartet is the most common and well-studied tetrad, this is not the only possible arrangement of four nucleobases. In fact, a number of tetrads formed by the different nucleobases has been observed in experimental structures (Table 1). They can be classified in two main groups: (a) those formed by arrangements of the same nucleobase (homotetrads), and (b) those formed by the association of two base pairs. The latter can involve canonical G:C or A:T Watson-Crick base pairs, or a variety of mismatches.

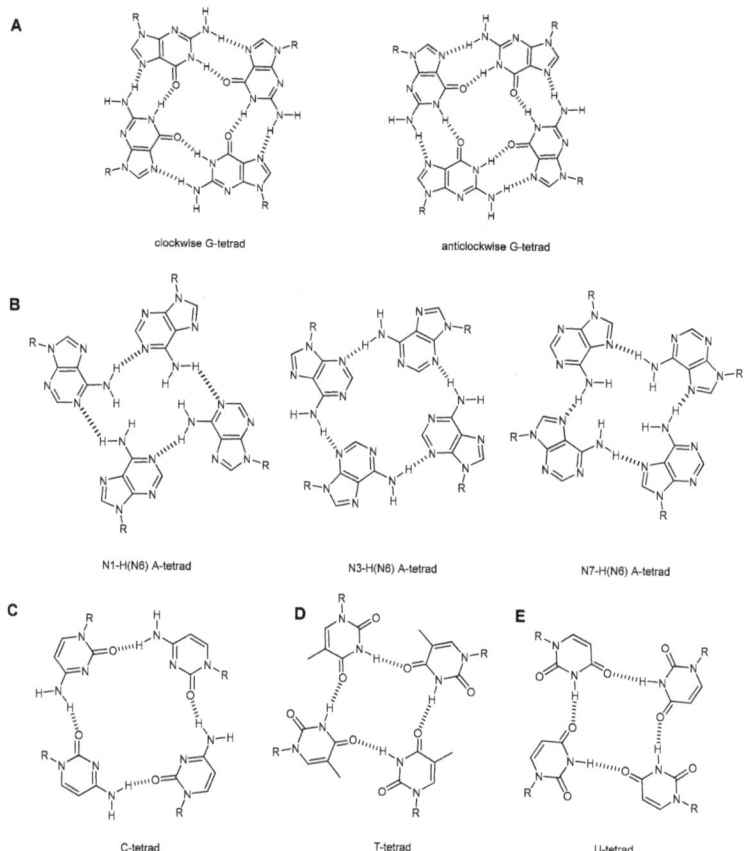

Figure 1. Schemes of the different homotetrads discussed in the text. (**A**) G-tetrads in their two possible orientations (clockwise and anticlokwise); (**B**) The three most common A-tetrads; (**C**–**E**) C-, T-, and U-homotetrads, respectively.

2. Results

2.1. Homotetrads

As in the case of G-tetrads, pure C-[9,10], T-[11,12] and A-tetrads [13] are formed by the arrangement of four identical nucleobases, as shown in Figure 1B–D. These homotetrads have always been observed in the context of parallel G-quadruplex structures, either in internal regions of the quadruplex or as capping elements interacting with terminal G-tetrads. Of particular interest is the case of the oligonucleotide of sequence d (AGAGA-GATGGGTGCGTT), which folds as a tetrameric parallel quadruplex. In this case, the four possible DNA homotetrads coexist in the same crystal structure [14]. Homotetrads have also been found in RNA G-quadruplex scaffolds. In this case, U-tetrads (Figure 1E) [15–17], usually located at the 3′-end of parallel quadruplexes [18], are the most common. However, RNA homotetrads formed by adenines [19,20] or inosines [21] have also been reported.

The conformation of the T-tetrad is shown in Figure 1D. T-tetrads were first observed in the solution structure of the tetrameric G-quadruplex formed by the sequence d(TGGTGGC) [11]. This tetrad is mainly stabilized by hydrogen bonds between imino protons H(N3) and O4 atoms of adjacent thymines. However, Liu et al. [14] suggested that weak hydrogen bond interactions between H(C5) and O2 atoms may also contribute to the stability. Similar T-tetrads have been observed in the crystal structure of d(TGGGGT)$_4$ [22] and in an all-LNA G-quadruplex from 5′-TGGGT-3′. In the latter case, a K$^+$ ion bridges the four O4 atoms [23]. Similar hydrogen bonds and K$^+$ interactions have been observed in RNA U-tetrads. Theoretical studies confirm the expected higher stability of U-tetrads due to the absence of methyl-O2 repulsions [24].

In the oligonucleotide d(TGGGCGGT), which also adopts a parallel tetrameric G-quadruplex structure, the central cytosines form the C-tetrad shown in Figure 1C [9]. In this structure, amino protons are at a hydrogen-bond distance from the O2 atoms. However, in the C-tetrad formed in the crystal structure of d(AGAGAGATGGGTGCGTT), the interactions between cytosines do not occur by direct hydrogen bonds, but through a network of highly conserved water molecules located in the middle of the tetrad [14]. In the crystal structure of the DNA decamer, d(CCA^{CNV}KGCGTGG) (^{CNV}K, 3-cyanovinylcarbazole), which forms a G-quadruplex structure in the presence of Ba^{2+}, a C-tetrad, is stabilized by water molecules-mediated contacts between the divalent cations and the cytosines, allowing Ba^{2+} ions to occupy the central ion channel [10]. Computational calculation suggests that C-tetrads in the contexts of G-quadruplex have a high propensity for binding alkaline earth cations [25].

A-tetrads are more common and more diverse (Figure 1B). They have been found in a number of DNA and RNA oligonucleotides derived from telomeric sequences, including the NMR structure of human telomere RNA ORN-1 [20], or the structure of (^{Br}dU)r(GAGGU) [19]. The A-tetrads observed in RNAs usually form hydrogen bonds between their amino H(N6) and N7 atoms (Figure 1B, right), although N3-H(N6) A-tetrads (Figure 1B, middle) have also been observed in the crystal structure of rU(^{Br}dG)r(AGGU) [26]. In the structure of d(AGAGAGATGGGTGCGTT) [14], the hydrogen bonds occur between amino H(N6) and the N3 atoms (Figure 1B, middle), resulting in a bigger central cavity. A third class of A-tetrad, in which hydrogen bonds are formed between H(N6) and N1 atoms (Figure 1B, left), has been reported in the solution structures of d(TGGAGGC)$_4$, a sequence that contains the GGAGG repeat present in the C-MYC oncogene [27], the telomeric repeat sequence [28], and the truncated telomeric sequence d(AG$_3$T) [13]. In this case, a dynamic exchange is observed between this tetrad and the N7-H(N6) A-tetrad [13]. Although in most A-tetrads adenines glycosidic angles are in anti, the syn conformation is also possible [13].

The overall size of T-, C- and A-tetrads, as estimated from diagonal C1'-C1' distance, is slightly smaller than in the case of the G-tetrad, being this distance around 1–2 Å shorter in T- and A-tetrads, and ~3 Å shorter in the case of the C-tetrad [14]. Due to their similar size, homotetrads provoke only minor distortions in the structure of parallel G-quadruplexes.

2.2. Base-Paired Tetrads

In addition to the homotetrads described above, tetrads can also be formed by the association of two base pairs. In most cases, the base pairs involved are Watson–Crick G:C or A:T base-pairs, although the isomorphous G:T base-pair and other mismatches have also been reported. Base-paired tetrads can be classified in two big groups depending on the relative orientation of the two base pairs: the so-called major groove and minor groove tetrads (Figure 2). Although most of the base-paired tetrads can be classified in these two categories, there are some exceptions of tetrads presenting very unusual nucleobase interactions [29] or water-mediated interactions in complex RNA structures [7].

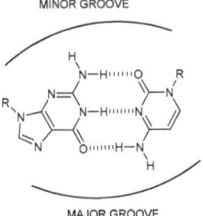

Figure 2. Major and minor groove sides of a G:C Watson-Crick base pair.

2.2.1. Major Groove Tetrads

In major groove tetrads, the interaction between base pairs occurs through their major groove side (the base-pair side that would be oriented towards the major groove in a canonical duplex, Figure 2). Among the first cases reported are the major groove G:C:G:C tetrads found in the structure of the fragile X syndrome $(CGG)_n$ triplet repeat [16], and in the d(GGGC) repeats of the adeno-associated viral DNA [30]. In both cases, the quadruplex arises from dimerization of two DNA hairpins, stabilized by several G-tetrads and G:C:G:C tetrads. The latter formed by two intra-molecular Watson–Crick base pairs directly facing each other and interacting through intermolecular bifurcated hydrogen bonds between the cytosine H(N4) and the guanine O6 and N7 atoms (Figure 3A). This arrangement gives rise to the so-called "direct" major groove tetrad. These tetrads have also been found in repetitive RNA sequences [31].

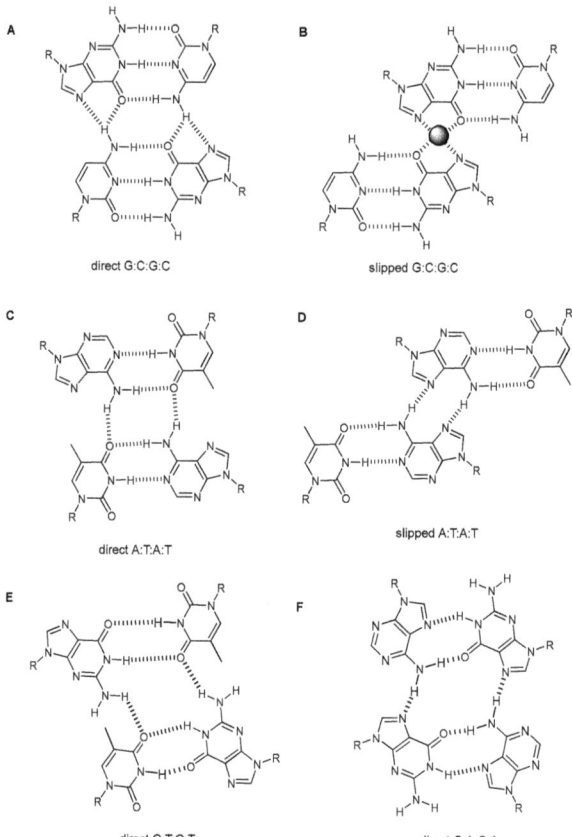

Figure 3. Schemes of the different major groove tetrads discussed in the text. (**A**) Direct major groove G:C:G:C tetrad; (**B**) Slipped major groove G:C:G:C tetrad; (**C**) Direct major groove A:T:A:T tetrad; (**D**) Slipped major groove G:C:G:C tetrad. (**E**) Direct major groove G:T:C:T. (**F**) Direct major groove G:A:G:A tetrad.

In other occasions, the relative position of the two G:C base pairs is shifted with respect to the direct tetrad (Figure 3B). The resulting tetrad has been named as "slipped" major groove tetrad [32]. In this case, the stabilization between base pairs does not arise from additional hydrogen bonds, but through cation coordination with O6(G) and N7(G) atoms from the two guanine residues. Interestingly, the conformations of the G:C:G:C tetrad may depend on the particular cation present in the sample. Thus, the antiparallel quadruplex structure of the d(GGGC)$_n$ repeats of the adeno-associated viral DNA in the sodium form contains two direct major groove G:C:G:C tetrads, whereas in the presence of potassium the alignment of the two G:C base pairs is shifted, with the two major groove edges of the guanines coordinating a K$^+$ cation [32]. Slipped major groove G:C:G:C tetrads formation has also been reported in a quadruplex-duplex hybrid structure, in which the slipped mixed tetrad is located between the G-quadruplex core and the duplex [33], and in an unusual quadruplex RNA structure formed by the major groove interaction of two antiparallel duplexes [34].

Direct and slipped major groove tetrads can be also formed by the association of A:T Watson-Crick base pairs (Figure 3C,D). The slipped alignment was found in the dimeric solution structure of the octamer d(GAGCAGGT) [35] and implies the formation of two N7(A)-H(N6)(A) additional hydrogen bonds. In the direct alignment, the inter-base pair hydrogen bonds are O4(T)-H(N6)(A). This tetrad has been observed as crystallographic contacts between unit cells in the dimeric propeller-like quadruplex structure of the human telomeric sequence, d(TAGGGTTAGGGT) [1]. Interestingly, these tetrads are not part of the G-quadruplex core but are formed between loop resides, connecting different quadruplexes and contributing to stabilize the crystal.

Major groove tetrads involving mismatch base pairs have also been found. G:T:G:T major groove tetrads (Figure 3E) have been proposed in the structures of some parallel G-quadruplexes [36] and in the interlocked G-quadruplex structure formed by d(GGGT) and derived sequences [37]. G:A:G:A tetrads involving G:A mismatches have been exhaustively explored by DFT methods [38]. These calculations indicate that squared planar G:A:G:A major groove tetrads are not favored, due to repulsion interactions between lone electron pairs of nitrogen atoms. However, an almost planar G:A:G:A tetrad with similar nucleobase orientation has been observed in the solution structure of d(GCGAGGGAGCGAGGG) [39]. This sequence is a variation in the d(GGGAGCG) repeat, found in the regulatory region of the PLEKHG3 human gene related to autism [40]. This structure is a dimer resulting from the self-association of two fold-back loops, and stabilized by a number of unusual elements, such as two G:A:G:A tetrads (Figure 3F) with all their glycosidic angles in anti, G:G base pairs, and direct minor groove G:C:G:C tetrads (Figure 4A).

2.2.2. Minor Groove Tetrads

The association between the two base pairs forming the tetrad can also occur through the minor groove side of the two base pairs (the side that would be oriented towards the minor groove in a canonical duplex, Figure 2). As in the previous case, the resulting minor groove tetrad can be direct or slipped, depending on the relative position of the two base pairs involved. G:C:G:C minor groove tetrads were found in the crystallographic structure of d(GCATGCT) [41,42], and in the solution structures of the cyclic oligonucleotides d<pTGCTCGCT> [43] and d<pCCGTCCGT> [44] (the notation d<p(sequence)> indicates cyclic deoxyoligonucleotide). All these structures are dimers stabilized by intermolecular Watson–Crick base-pairs. In the case of d(GCATGCT) and d<pTGCTCGCT>, the minor groove tetrads are direct, forming two bifurcated H(N2)(G)-O2(C) hydrogen bonds between the base pairs (Figure 4A). In contrast, in the structure of d<pCCGTCCGT> the minor groove tetrads are slipped, forming two H(N2)(G)-N3(G) hydrogen bonds (Figure 4B). No cation sensitivity between these two types of tetrads has been observed. However, the order of residues closing the loops seems to be crucial. Thus, in 5′-C-XX-G-3′ turns the tetrad is direct, and in 5′-G-XX-C-3′ slipped [45].

In most cases, the orientation of two Watson-Crick base pairs forming the tetrads is opposite to each other. However, in the dimeric structures of d(TGCTTCGT) and d(TCGTTGCT), and their cyclic analogue d<pCGCTCCGT> [45], the two G:C base pairs are in the same orientation, giving rise to a C:G:G:C tetrad. In this case, the hydrogen bonds between base pairs are H(N2)(G)-N3(G) and H(N2)(G)-O2(C) (Figure 4C).

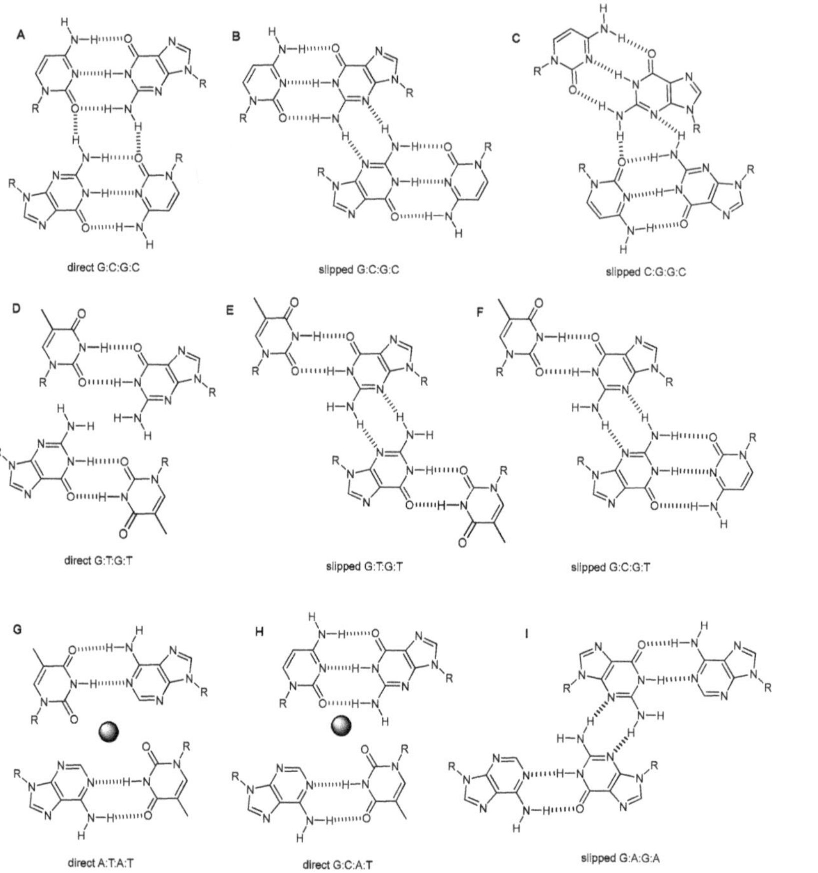

Figure 4. Schemes of the different minor groove tetrads discussed in the text. Direct (**A**) and slipped (**B**) minor groove G:C:G:C tetrads. (**C**) Minor groove C:G:G:C tetrad. Direct (**D**) and slipped (**E**) minor groove G:T:G:T tetrads. (**F**) Slipped minor groove G:C:G:T. Direct minor groove A:T:A:T (**G**) and G:C:A:T (**H**) tetrads stabilized by cation coordination (dark circle). (**I**) Slipped minor groove G:A:G:A tetrad.

Minor groove tetrads can also involve A:T base pairs, giving rise to A:T:A:T tetrads, such as in the dimeric structure of d<pCATTCATT> [43,46], or G:C:A:T mixed tetrads, observed in the dimeric structure of d<pCGCTCATT> [47]. In both cases, only the direct alignment has been observed (Figure 4G,H). A:T:A:T and G:C:A:T minor groove tetrads are stabilized by cation coordination, as shown by crystallographic and theoretical studies.

Likewise, minor groove tetrads can involve one or two G:T mismatches. As in the case of G:C:G:C, the two alignments (direct and slipped) have been observed in G:T:G:T tetrads [48] (Figure 4D,E). In mixed G:C:G:T tetrads, however, only the slipped alignment has been reported (Figure 4F) [48,49]. Minor groove tetrads with G:A mismatches (Figure 4I) have been observed in the dimeric fold-back structure of d(CGTAAGGCGTA) [50]. This structure is stabilized by minor groove G:C:G:C and G:A:G:A tetrads, both in the slipped configurations. The glycosidic angle of adenines in this tetrad is in syn.

In addition to the different hydrogen bonds pattern stabilizing the interaction between the two base pairs, there is a fundamental difference between major and minor groove tetrads, whereas in major groove tetrads, the two base-pairs are co-planar and in minor groove tetrads the two base pairs present a relative inclination that ranges from 20° to

40°, depending on the particular structure. The reason of this effect is not clear. It may be related to the close proximity between phosphate groups in structures stabilized by minor groove tetrads. In these cases, a relative inclination between base pairs may alleviate electrostatic repulsion between phosphates. However, it can also be an intrinsic property of purine-purine interactions through their minor groove side. The structures of parallel DNA duplexes stabilized by homopurine base pairs point towards this possibility, since G:G and A:A base pairs in these duplex structures exhibit a similar inclination [51]. Although a number of theoretical studies have shown that coplanarity between G:C base pairs is the most stable configuration in major groove tetrads [52,53], very little is known from a theoretical point of view about the geometry of minor groove G:C:G:C tetrads.

Table 1. Structures with the different non-G tetrads mentioned in the text with their corresponding references and their PDB codes, when available.

Type		Tetrad	Cation	PDB
Homo		N1-H(N6) A	-	1EVM [13], 1NP9 [28]
		N3-H(N6) A	Na$^+$	6A85 [14], 1MDG [26]
		N7-H(N6) A	-/Na$^+$	1EVN [13], 1J6S [19]
		C$_4$	-/Ba^{2+}/Na$^+$	1EVO [9], 4U92 [10], 6A85 [14]
		T$_4$	-/Na$^+$/K$^+$	1EMQ [11], 6A85 [14], 4L0A [23], 1S47 [24]
		U$_4$	-	1RAU [15], 6GE1 [18], 1J6S [19], 1MDG [26], 1J8G [54], 2AWE [55], 4RKV [56], 4RJ1 [56], 4RNE [56], 4XK0 [57]
		I$_4$	K$^+$	2GRB [21]
Base-paired	Major groove	Direct G:C:G:C	-/Na$^+$	1XCE [29], 1A8N [30], 1A6H [31], 3R1E [34], 1JVC [35], 1NYD [58], 6SYK [59], 6SX6 [59]
		Slipped G:C:G:C	K$^+$	1A8W [32], 7CV4 [33]
		Direct A:T:A:T	-/K$^+$	1K8P [1]
		Slipped A:T:A:T	-/Na$^+$	1XCE [29], 1JVC [35]
		Direct G:T:G:T	-	[36,37]
		Direct G:A:G:A	-	5M1L, 5M2L [39]
	Minor groove	Direct G:C:G:C	-	184D [41], 1MF5 [42], 1EU2 [43], 4ZKK [60], 5FHJ [61]
		Slipped G:C:G:C	-/divalent	6MC2, 6MC3, 6MC4, 6N4G [50], 2HK4 [45]
		Slipped C:G:G:C	-	2K8Z, 2K90, 2K97 [45]
		Direct G:T:G:T	-	-[48]
		Slipped G:T:G:T	-	1C11 [62], 2LSX [63], 7O5E [64]
		Slipped G:C:G:T	-	5OGA [65]
		Direct A:T:A:T	Na$^+$	1EU6 [43], 284D [46]
		Direct G:C:A:T	Na$^+$	1N96 [47]
		Slipped G:A:G:A	divalent	6MC2, 6MC3, 6MC4, 6N4G [50]

2.3. Structural Context

The different structural features between major and minor groove tetrads have important consequences. First, minor grove tetrads are not easily accommodated within G-tetrads scaffolds. In fact, base-paired tetrads found in the context of G-quadruplex structures are always of the major groove type. Secondly, minor groove tetrads cannot be piled up indefinitely due to the mutual inclination between the two base-pairs (between 20–40°, Figure 5A). No structure with more than two consecutive minor groove tetrads has been reported. Two is probably the maximum number of minor groove tetrads that can be assembled together.

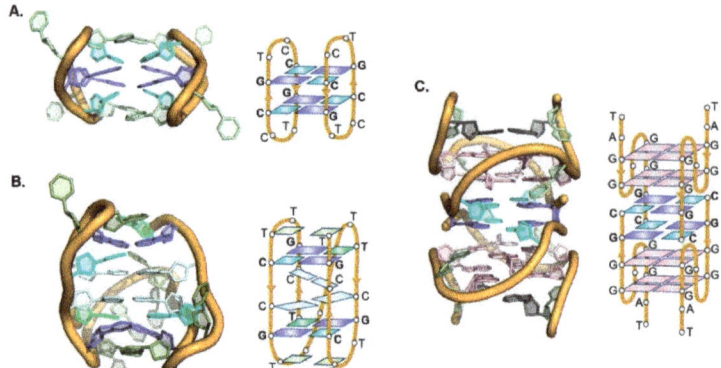

Figure 5. Examples of structures containing non-G tetrads. (**A**) Dimeric structures formed by two cyclic oligonucleotides <dCCGTCCGT> stabilized by two minor groove C:G:C:G tetrads (PDB 2HK4). (**B**) Minimal i-motif structure stabilized by two C:C$^+$ base pairs capped by two minor groove G:C:G:T tetrads (PDB 5OGA). (**C**) Association of two G-quadruplexes through formation of two major groove C:G:C:G tetrads (PDB 1NYD).

Major groove tetrads are planar and, in their direct configuration, are almost isomorphous with G-tetrads, with diagonal C1'C1' distance between purines almost identical to G-tetrads, and between pyrimidines around 1 Å shorter. This feature makes direct major groove tetrads perfectly suitable to be stacked between G-tetrads (Figure 5C) and, in fact, they are relatively common in G-quadruplex structures [1,16,30–37,39]. In contrast, slipped major groove tetrads are uncommon and only found in terminal positions [32]. The effect of the presence of G:C:G:C tetrads on G-quadruplexes recognition has been explored [66]. However, very few cases of ligands interacting with non-G tetrads has been reported to date [67,68].

On the other hand, minor groove tetrads have always been found in structures in which no G-tetrad is present. These tetrads appear to induce distinct DNA folding (the term "bi-loop" has been suggested for this motif) [46]. This motif has been usually found in homodimers formed by short linear or cyclic oligonucleotides (Figure 5A), in which dimerization occurs through the formation of two minor groove tetrads with intermolecular base pairs [41–49,69].

The tetrads are connected by short loops of one to three residues, with two-residue loops being the most stable [70]. The residues in the first position of the loops stack on top of the tetrads forming a cap at both ends of the structure. A common feature of all these structures is the great proximity of the backbones of the two strands in the same subunit. The very short distances between phosphates provokes unfavorable electrostatic interactions, which are partially alleviated by hydrophobic contacts between deoxyriboses.

This peculiar feature is common to other non-canonical DNA structure, the i-motif [71,72]. The i-motif is a four-stranded structure not stabilized by tetrads, but by intercalation of C:C$^+$ base pairs. These structures are formed by the association of two parallel-stranded duplexes through hemiprotonated C:C$^+$ base pairs. The two duplexes are intercalated in opposite orientations. Since the C:C$^+$ base pair requires partial cytosine protonation, i-motif structures are usually more stable at acidic pH. Interestingly, minor groove tetrads have been observed as capping elements in several i-motif structures. The first case reported was that minor grove G:T:G:T tetrad observed in dimeric structures of centromeric sequences [62]. Then, G:T:G:T [63] and G:C:G:T [65] tetrads were observed in other dimeric and monomeric i-motifs. In all cases, the tetrads are in the slipped configuration, and provoke a dramatic thermal and pH stabilization of the i-motif structure. This stabilization is most probably due to the favorable interaction of the charged C:C$^+$ base-pair with the guanines of the tetrad (Figure 5B). The capability of minor groove tetrads to stabilize i-motifs may have profound consequences in biology, since consensus sequences based on these interactions have been found to be prevalent in the human genome, occurring preferentially near regulatory regions [65].

Although several i-motifs have been found to be stable at a neutral pH, to date, the only i-motif structures determined at nearly physiological conditions are structures stabilized by minor groove tetrads. Of particular interest is the structure adopted by oligonucleotides containing several repeats of the sequence 5'-dCCGTTCCGT-3'. At a neutral pH, these sequences fold into i-motif structures stabilized by two C:C$^+$ base pairs and two C:G:C:G minor groove tetrads. This structure is peculiar since it contains neutral and hemiprotonated cytosines under the same experimental conditions. C:G:C:G tetrads are not stable at acidic pH, in which conditions cytosines are partially protonated. When lowering the pH from 7 to 5, these oligonucleotides undergo a large conformational transition towards a different i-motif structure, with four C:C$^+$ base pairs and capped by two G:T:G:T tetrads [73]. This conformational transition was further used to explore the use of a fluorescent cytosine analogue tricyclic 1,3-diaza-2-oxophenoxazine (tC^O) as an efficient internal probe. Interestingly, the increased stacking interactions between tC^O and the guanine residue of the tetrad, when located in contiguous positions, dramatically enhanced the thermal and pH stability of the i-motif structures [74].

Very recently, minor groove tetrads have been used to stabilize DNA constructs, in which i-motif and B-DNA coexist in the same structure, forming i-motif/duplex junctions [64]. These constructs consist of an i-motif moiety capped by a minor groove G:T:G:T tetrad at one of its ends, and an stem-loop hairpin at the other end of the i-motif. Stabilization conferred by the minor groove tetrad is essential to stabilize the structure at neutral pH, since analogous constructs with sequences unable to form the tetrad are not stable [64].

2.4. Base Paired Tetrads in Higher-Order Structures

In some occasions, major groove G:C:G:C tetrads are found in structures containing multiple interlocked G-quadruplexes in the regions connecting different units [58,75]. The ability to interlock multiple G-quadruplexes through these tetrads has been used to build nanowires [69,76,77]. In the case of sequences containing 5'-GC-3' in the terminal position, this assembly of multimeric G-quadruplexes is cation dependent [59]. Similar interlocked structures stabilized by major groove G:C:G:C tetrads have been also found in RNA G-quadruplexes [16]. Inclusion of G:C:G:C tetrads in G-quadruplex scaffolds can be also used for avoiding the inherent polymorphism of these structures when designing DNA nanoassemblies [78]. Furthermore, the formation of G:C:G:C tetrads has been proposed in higher order structures formed by multiple GGGGCC repeats found in the sequence of the C9ORF72 gene [79]. Expansion of these repeats is considered the most important genetic cause of the Amyotrophic Lateral Sclerosis (ALS).

On the other hand, G:C:G:C tetrads, usually in their minor groove conformation, are common between symmetry-related molecules in crystallographic structures. On many occasions, they are formed between terminal G:C base pairs in the crystal structures of DNA duplexes, contributing to stable crystal packing [80]. Slipped tetrads found in the context of a crystallographic network are more common between DNA duplexes that crystallize in B-form [81,82]. In contrast, direct minor groove tetrads are found in A-form DNA structures. The direct arrangement may be facilitated by the wider minor-groove characteristic of A-form double helices [83]. Although not so common, these interactions do not involve terminal residues. This is the case of the DNA-RNA chimeric duplex of sequence d(CCGGC)r(G)d(CCGG). Interestingly, direct minor grove G:C:G:C tetrads connecting different duplexes involve the single 2'-hydroxyl group per strand. This result indicates that minor groove tetrads can be formed with at least one of the four nucleotides being ribo-. However, studies on the association between cyclic and linear oligonucleotides suggest that minor groove tetrads containing ribonucleotides are much less stable than those formed by four dexoyribonucleotides [49].

Minor groove G:C:G:C tetrads have also been found involving terminal G and C residues that do not form base-pairs with their own duplex but with symmetry related ones, forming junction-like quadruplexes. Such structures have been found in pure DNA crystals and in several bisintercalative complexes involving acridine derivatives. In these complexes, the acridine derivatives do not intercalate in the usual way, but interact with the terminal nucleotides of four DNA duplexes forming a large intercalation platform between two G:C:G:C tetrads [84–88]. The structures of these complexes have attracted significant interest because some of the intercalating drugs involved in them are potent topoisomerase inhibitors. Interestingly, in most cases, these structures have been found in crystals containing Co^{2+} ions. Such counterions might be relevant to increase the stability of minor groove tetrads, since cobalt hexamine residues have also been found to stabilize the crystal lattice in one of the crystallographic structures of the dimeric linear heptamer d(GCATGCT) [41,42,60,61].

Base-paired tetrads have also been observed in ordered nanostructures based on G:C pairing. High resolution scanning tunnelling microscopy studies in adlayers formed by coadsorption of guanine and cytosine at Au(111) [89], and at graphite surfaces [90] have revealed the formation of well-ordered periodic structures based on G:C:G:C tetrads in the solid/liquid interface. Such structures are formed by alternate arrangements of G:C base pairs though their major and minor groove sides. Similar supramolecular nanopatterns based on A:T:A:T have also been reported [90]. More recently, base paired C:G:G:C, G:C:G:C and A:U:A:U tetrads have been used as building blocks for the formation of 2D-nanoporous networks [91]. Finally, G-quartets and other homotetrads have been also found in template-assembled synthetic supramolecular structures [92–94].

3. Conclusions

In summary, the number of arrangements of four nucleobases found in experimental structures amply exceeds the canonical G-tetrads. In this review we have classified these non-guanine tetrads in two main groups: homotetrads, formed by the arrangement of identical nucleobase, and base-paired tetrads, formed by the association of two base pairs (G:C, A:T or mismatches). Most of the tetrads in the second group can be classified in major groove or minor groove tetrads, depending on the side of the two base pairs interacting with each other. The structural contexts in which these tetrads occur is conditioned by their specific geometry. Thus, whereas homo- and major groove tetrads are mainly formed in the context of G-quadruplex structures, minor groove tetrads occur in i-motifs and, in peculiar, fold-back DNA structures. Interestingly, non-canonical tetrads are involved in different DNA-DNA recognition events, such as loop-loop association, formation of G-quadruplex assemblies, or DNA duplex packing. These tetrad-mediated interactions may be involved in biological processes and may have useful applications in DNA nanotechnology.

Author Contributions: Conceptualization, N.E. and C.G.; writing—review and editing, N.E., B.M., M.G. and C.G. All authors have read and agreed to the published version of the manuscript.

Funding: This research was funded by Ministerio de Ciencia e Innovación, grant number PID2020-116620GB-I00. B. M. acknowledges a "Margarita Salas" contract from the Spanish Ministerio de Ciencia e Innovación, and M. G. a European Union Marie Sklodowska Curie Action (799693).

Institutional Review Board Statement: Not applicable.

Informed Consent Statement: Not applicable.

Data Availability Statement: Not applicable.

Conflicts of Interest: The authors declare no conflict of interest. The funders had no role in the design of the study; in the collection, analyses, or interpretation of data; in the writing of the manuscript, or in the decision to publish the results.

References

1. Parkinson, G.N.; Lee, M.P.H.; Neidle, S. Crystal Structure of Parallel Quadruplexes from Human Telomeric DNA. *Nature* **2002**, *417*, 876–880. [CrossRef] [PubMed]
2. Spiegel, J.; Adhikari, S.; Balasubramanian, S. The Structure and Function of DNA G-Quadruplexes. *Trends Chem.* **2020**, *2*, 123–136. [CrossRef] [PubMed]
3. Varshney, D.; Spiegel, J.; Zyner, K.; Tannahill, D.; Balasubramanian, S. The Regulation and Functions of DNA and RNA G-Quadruplexes. *Nat. Rev. Mol. Cell Biol.* **2020**, *21*, 459–474. [CrossRef]
4. Frasson, I.; Pirota, V.; Richter, S.N.; Doria, F. Multimeric G-Quadruplexes: A Review on Their Biological Roles and Targeting. *Int. J. Biol. Macromol.* **2022**, *204*, 89–102. [CrossRef]
5. Kosiol, N.; Juranek, S.; Brossart, P.; Heine, A.; Paeschke, K. G-Quadruplexes: A Promising Target for Cancer Therapy. *Mol. Cancer* **2021**, *20*, 40. [CrossRef] [PubMed]
6. Lightfoot, H.L.; Hagen, T.; Tatum, N.J.; Hall, J. The Diverse Structural Landscape of Quadruplexes. *FEBS Lett.* **2019**, *593*, 2083–2102. [CrossRef] [PubMed]
7. Banco, M.T.; Ferré-D'Amaré, A.R. The Emerging Structural Complexity of G-Quadruplex RNAs. *RNA* **2021**, *27*, 390–402. [CrossRef]
8. Malgowska, M.; Czajczynska, K.; Gudanis, D.; Tworak, A.; Gdaniec, Z. Overview of the RNA G-Quadruplex Structures. *Acta Biochim. Pol.* **2016**, *63*, 609–621. [CrossRef]
9. Patel, P.K.; Bhavesh, N.S.; Hosur, R.V. NMR Observation of a Novel C-Tetrad in the Structure of the SV40 Repeat Sequence GGGCGG. *Biochem. Biophys. Res. Commun.* **2000**, *270*, 967–971. [CrossRef]
10. Zhang, D.; Huang, T.; Lukeman, P.S.; Paukstelis, P.J. Crystal Structure of a DNA/Ba2+ G-Quadruplex Containing a Water-Mediated C-Tetrad. *Nucleic Acids Res.* **2014**, *42*, 13422–13429. [CrossRef]
11. Patel, P.K.; Hosur, R. V NMR Observation of T-Tetrads in a Parallel Stranded DNA Quadruplex Formed by Saccharomyces Cerevisiae Telomere Repeats. *Nucleic Acids Res* **1999**, *27*, 2457–2464. [CrossRef] [PubMed]
12. Oliviero, G.; Amato, J.; Borbone, N.; Galeone, A.; Varra, M.; Piccialli, G.; Mayol, L. Synthesis and Characterization of DNA Quadruplexes Containing T-Tetrads Formed by Bunch-Oligonucleotides. *Biopolymers* **2006**, *81*, 194–201. [CrossRef] [PubMed]
13. Patel, P.K.; Koti, A.S.R.; Hosur, R.V. NMR Studies on Truncated Sequences of Human Telomeric DNA: Observation of a Novel A-Tetrad. *Nucleic Acids Res.* **1999**, *27*, 3836–3843. [CrossRef] [PubMed]
14. Liu, H.; Wang, R.; Yu, X.; Shen, F.; Lan, W.; Haruehanroengra, P.; Yao, Q.; Zhang, J.; Chen, Y.; Li, S.; et al. High-Resolution DNA Quadruplex Structure Containing All the A-, G-, C-, T-Tetrads. *Nucleic Acids Res.* **2018**, *46*, 11627–11638. [CrossRef] [PubMed]
15. Cheong, C.; Moore, P.B. Solution Structure of an Unusually Stable RNA Tetraplex Containing G-and U-Quartet Structures. *Biochemistry* **1992**, *31*, 8406–8414. [CrossRef]
16. Malgowska, M.; Gudanis, D.; Kierzek, R.; Wyszko, E.; Gabelica, V.; Gdaniec, Z. Distinctive Structural Motifs of RNA G-Quadruplexes Composed of AGG, CGG and UGG Trinucleotide Repeats. *Nucleic Acids Res.* **2014**, *42*, 10196–10207. [CrossRef]
17. Xu, Y.; Ishizuka, T.; Kimura, T.; Komiyama, M. A U-Tetrad Stabilizes Human Telomeric RNA G-Quadruplex Structure. *J. Am. Chem. Soc.* **2010**, *132*, 7231–7233. [CrossRef]
18. Andrałojć, W.; Małgowska, M.; Sarzyńska, J.; Pasternak, K.; Szpotkowski, K.; Kierzek, R.; Gdaniec, Z. Unraveling the Structural Basis for the Exceptional Stability of RNA G-Quadruplexes Capped by a Uridine Tetrad at the 3′ Terminus. *RNA* **2019**, *25*, 121–134. [CrossRef]
19. Pan, B.; Xiong, Y.; Shi, K.; Deng, J.; Sundaralingam, M. Crystal Structure of an RNA Purine-Rich Tetraplex Containing Adenine Tetrads: Implications for Specific Binding in RNA Tetraplexes. *Structure* **2003**, *11*, 815–823. [CrossRef]
20. Xiao, C.D.; Ishizuka, T.; Zhu, X.Q.; Li, Y.; Sugiyama, H.; Xu, Y. Unusual Topological RNA Architecture with an Eight-Stranded Helical Fragment Containing A-, G-, and U-Tetrads. *J. Am. Chem. Soc.* **2017**, *139*, 2565–2568. [CrossRef]
21. Pan, B.; Shi, K.; Sundaralingam, M. Crystal Structure of an RNA Quadruplex Containing Inosine Tetrad: Implications for the Roles of NH2 Group in Purine Tetrads. *J. Mol. Biol.* **2006**, *363*, 451–459. [CrossRef] [PubMed]
22. Cáceres, C.; Wright, G.; Gouyette, C.; Parkinson, G.; Subirana, J.A. A Thymine Tetrad in d(TGGGGT) Quadruplexes Stabilized with Tl+/Na+ Ions. *Nucleic Acids Res.* **2004**, *32*, 1097. [CrossRef] [PubMed]
23. Russo Krauss, I.; Parkinson, G.N.; Merlino, A.; Mattia, C.A.; Randazzo, A.; Novellino, E.; Mazzarella, L.; Sica, F. A Regular Thymine Tetrad and a Peculiar Supramolecular Assembly in the First Crystal Structure of an All-LNA G-Quadruplex. *Acta Crystallogr. Sect. D Biol. Crystallogr.* **2014**, *70*, 362–370. [CrossRef] [PubMed]
24. Gu, J.; Leszczynski, J. A Theoretical Study of Thymine and Uracil Tetrads: Structures, Properties, and Interactions with the Monovalent K+ Cation. *J. Phys. Chem. A* **2001**, *105*, 10366–10371. [CrossRef]
25. Milovanović, B.; Petković, M.; Popov, I.; Etinski, M. Water-Mediated Interactions Enhance Alkaline Earth Cation Chelation in Neighboring Cavities of a Cytosine Quartet in the DNA Quadruplex. *J. Phys. Chem. B* **2021**, *125*, 11996–12005. [CrossRef]
26. Pan, B.; Xiong, Y.; Shi, K.; Sundaralingam, M. An Eight-Stranded Helical Fragment in RNA Crystal Structure: Implications for Tetraplex Interaction. *Structure* **2003**, *11*, 825–831. [CrossRef]
27. Searle, M.S.; Williams, H.E.L.; Gallagher, C.T.; Grant, R.J.; Stevens, M.F.G. Structure and K Ion-Dependent Stability of a Parallel-Stranded DNA Quadruplex Containing a Core A-Tetrad. *Org. Biomol. Chem.* **2004**, *2*, 810–812. [CrossRef]

28. Gavathiotis, E.; Searle, M.S. Structure of the Parallel-Stranded DNA Quadruplex d(TTAGGGT) 4 Containing the Human Telomeric Repeat: Evidence for A-Tetrad Formation from NMR and Molecular Dynamics Simulations. *Org. Biomol. Chem.* **2003**, *1*, 1650–1656. [CrossRef]
29. Da Silva, M.W. Experimental Demonstration of T:(G:G:G:G):T Hexad and T:A:A:T Tetrad Alignments within a DNA Quadruplex Stem. *Biochemistry* **2005**, *44*, 3754–3764. [CrossRef]
30. Kettani, A.; Bouaziz, S.; Gorin, A.; Zhao, H.; Jones, R.A.; Patel, D.J. Solution Structure of a Na Cation Stabilized DNA Quadruplex Containing G·G·G·G· and G·C·G·C· Tetrads Formed by G-G-G-C Repeats Observed in Adeno-Associated Viral DNA. *J. Mol. Biol.* **1998**, *282*, 619–636. [CrossRef]
31. Kettani, A.; Kumar, R.A.; Patel, D.J. Solution Structure of a DNA Quadruplex Containing the Fragile X Syndrome Triplet Repeat. *J. Mol. Biol.* **1995**, *254*, 638–656. [CrossRef] [PubMed]
32. Bouaziz, S.; Kettani, A.; Patel, D.J. A K Cation-Induced Conformational Switch within a Loop Spanning Segment of a DNA Quadruplex Containing G-G-G-C Repeats. *J. Mol. Biol.* **1998**, *282*, 637–652. [CrossRef] [PubMed]
33. Tan, D.J.Y.; Winnerdy, F.R.; Lim, K.W.; Tû An Phan, A. Coexistence of Two Quadruplex-Duplex Hybrids in the PIM1 Gene. *Nucleic Acids Res.* **2020**, *48*, 11162–11171. [CrossRef] [PubMed]
34. Gudanis, D.; Popenda, L.; Szpotkowski, K.; Kierzek, R.; Gdaniec, Z. Structural Characterization of a Dimer of RNA Duplexes Composed of 8-Bromoguanosine Modified CGG Trinucleotide Repeats: A Novel Architecture of RNA Quadruplexes. *Nucleic Acids Res.* **2016**, *44*, 2409–2416. [CrossRef]
35. Zhang, N.; Gorin, A.; Majumdar, A.; Kettani, A.; Chernichenko, N.; Skripkin, E.; Patel, D.J. Dimeric DNA Quadruplex Containing Major Groove-Aligned A·T·A·T and G·C·G·C Tetrads Stabilized by Inter-Subunit Watson-Crick A·T and G·C Pairs. *J. Mol. Biol.* **2001**, *312*, 1073–1088. [CrossRef]
36. Jing, N.; Hogan, M.E. Structure-Activity of Tetrad-Forming Oligonucleotides as a Potent Anti- HIV Therapeutic Drug. *J. Biol. Chem.* **1998**, *273*, 34992–34999. [CrossRef]
37. Krishnan-Ghosh, Y.; Liu, D.; Balasubramanian, S. Formation of an Interlocked Quadruplex Dimer by d(GGGT). *J. Am. Chem. Soc.* **2004**, *126*, 11009–11016. [CrossRef]
38. Megger, D.A.; Lax, P.M.; Paauwe, J.; Fonseca Guerra, C.; Lippert, B. Mixed Guanine, Adenine Base Quartets: Possible Roles of Protons and Metal Ions in Their Stabilization. *J. Biol. Inorg. Chem.* **2018**, *23*, 41–49. [CrossRef]
39. Kocman, V.; Plavec, J. Tetrahelical Structural Family Adopted by AGCGA-Rich Regulatory DNA Regions. *Nat. Commun.* **2017**, *8*, 15355. [CrossRef]
40. Kocman, V.; Plavec, J. A Tetrahelical DNA Fold Adopted by Tandem Repeats of Alternating GGG and GCG Tracts. *Nat. Commun.* **2014**, *5*, 5831. [CrossRef]
41. Leonard, G.A.; Zhang, S.; Peterson, M.R.; Harrop, S.J.; Helliwell, J.R.; Cruse, W.B.; Langlois d'Estaintot, B.; Kennard, O.; Brown, T.; Hunter, W.N. Self-Association of a DNA Loop Creates a Quadruplex: Crystal Structure of d(GCATGCT) at 1.8 å Resolution. *Structure* **1995**, *3*, 335–340. [CrossRef]
42. Thorpe, J.H.; Teixeira, S.C.M.; Gale, B.C.; Cardin, C.J. Crystal Structure of the Complementary Quadruplex Formed by d(GCATGCT) at Atomic Resolution. *Nucleic Acids Res.* **2003**, *31*, 844–849. [CrossRef] [PubMed]
43. Escaja, N.; Pedroso, E.; Rico, M.; González, C. Dimeric Solution Structure of Two Cyclic Octamers: Four-Stranded DNA Structures Stabilized by A:T:A:T and G:C:G:C Tetrads. *J. Am. Chem. Soc.* **2000**, *122*, 12732–12742. [CrossRef]
44. Viladoms, J.; Escaja, N.; Frieden, M.; Gómez-Pinto, I.; Pedroso, E.; González, C. Self-Association of Short DNA Loops through Minor Groove C:G:G:C Tetrads. *Nucleic Acids Res.* **2009**, *37*, 3264–3275. [CrossRef] [PubMed]
45. Escaja, N.; Gómez-Pinto, I.; Pedroso, E.; González, C. Four-Stranded DNA Structures Can Be Stabilized by Two Different Types of Minor Groove G:C:G:C Tetrads. *J. Am. Chem. Soc.* **2007**, *129*, 2004–2014. [CrossRef]
46. Salisbury, S.A.; Wilson, S.E.; Powell, H.R.; Kennard, O.; Lubini, P.; Sheldrick, G.M.; Escaja, N.; Alazzouzi, E.; Grandas, A.; Pedroso, E. The Bi-Loop, a New General Four-Stranded DNA Motif. *Proc. Natl. Acad. Sci. USA* **1997**, *94*, 5515–5518. [CrossRef]
47. Escaja, N.; Gelpí, J.L.; Orozco, M.; Rico, M.; Pedroso, E.; González, C. Four-Stranded DNA Structure Stabilized by a Novel G:C:A:T Tetrad. *J. Am. Chem. Soc.* **2003**, *125*, 5654–5662. [CrossRef]
48. Viladoms, J.; Escaja, N.; Pedroso, E.; González, C. Self-Association of Cyclic Oligonucleotides through G:T:G:T Minor Groove Tetrads. *Bioorganic Med. Chem.* **2010**, *18*, 4067–4073. [CrossRef]
49. Escaja, N.; Gómez-Pinto, I.; Viladoms, J.; Rico, M.; Pedroso, E.; González, C. Induced-Fit Recognition of DNA by Small Circular Oligonucleotides. *Chemistry* **2006**, *12*, 4035–4042. [CrossRef]
50. Chu, B.; Zhang, D.; Hwang, W.; Paukstelis, P.J. Crystal Structure of a Tetrameric DNA Fold-Back Quadruplex. *J. Am. Chem. Soc.* **2018**, *140*, 16291–16298. [CrossRef]
51. Luteran, E.M.; Paukstelis, P.J. The Parallel-Stranded d(CGA) Duplex Is a Highly Predictable Structural Motif with Two Conformationally Distinct Strands. *Acta Cryst.* **2022**, *D78*, 299–309. [CrossRef]
52. Meyer, M.; Schneider, C.; Brandl, M.; Sühnel, J. Cyclic Adenine-, Cytosine-, Thymine-, and Mixed Guanine—Cytosine-Base Tetrads in Nucleic Acids Viewed from a Quantum-Chemical and Force Field Perspective. *J. Phys. Chem. A* **2001**, *105*, 11560–11573. [CrossRef]
53. Gu, J.; Leszczynski, J. Structures and Properties of the Planar G·C·G·C Tetrads: Ab Initio HF and DFT Studies. *J. Phys. Chem. A* **2000**, *104*, 7353–7358. [CrossRef]

54. Deng, J.; Xiong, Y.; Sundaralingam, M. X-Ray Analysis of an RNA Tetraplex (UGGGGU)4 with Divalent Sr2+ Ions at Subatomic Resolution (0.61 Å). *Proc. Natl. Acad. Sci. USA* **2001**, *98*, 13665. [CrossRef]
55. Pan, B.; Shi, K.; Sundaralingam, M. Base-Tetrad Swapping Results in Dimerization of RNA Quadruplexes: Implications for Formation of the i-Motif RNA Octaplex. *Proc. Natl. Acad. Sci. USA* **2006**, *103*, 3130–3134. [CrossRef] [PubMed]
56. Fyfe, A.C.; Dunten, P.W.; Martick, M.M.; Scott, W.G. Structural Variations and Solvent Structure of r(UGGGGU) Quadruplexes Stabilized by Sr2 + Ions. *J. Mol. Biol.* **2015**, *427*, 2205–2219. [CrossRef]
57. Chen, M.C.; Murat, P.; Abecassis, K.; Ferre-D'Amare, A.R.; Balasubramanian, S. Insights into the Mechanism of a G-Quadruplex-Unwinding DEAH-Box Helicase. *Nucleic Acids Res.* **2015**, *43*, 2223–2231. [CrossRef]
58. Da Silva, M.W. Association of DNA Quadruplexes through G:C:G:C Tetrads. Solution Structure of d(GCGGTGGAT). *Biochemistry* **2003**, *42*, 14356–14365. [CrossRef]
59. Pavc, D.; Wang, B.; Spindler, L.; Drevenšek-Olenik, I.; Plavec, J.; Šket, P. GC Ends Control Topology of DNA G-Quadruplexes and Their Cation-Dependent Assembly. *Nucleic Acids Res.* **2020**, *48*, 2749–2761. [CrossRef]
60. Thirugnanasambandam, A.; Karthik, S.; Mandal, P.K.; Gautham, N. The Novel Double-Folded Structure of d(GCATGCATGC): A Possible Model for Triplet-Repeat Sequences. *Acta Crystallogr. Sect. D Biol. Crystallogr.* **2015**, *71*, 2119–2126. [CrossRef]
61. Thirugnanasambandam, A.; Karthik, S.; Artheswari, G.; Gautham, N. DNA Polymorphism in Crystals: Three Stable Conformations for the Decadeoxynucleotide d(GCATGCATGC): Three. *Acta Crystallogr. Sect. D Struct. Biol.* **2016**, *72*, 780–788. [CrossRef] [PubMed]
62. Gallego, J.; Chou, S.H.; Reid, B.R. Centromeric Pyrimidine Strands Fold into an Intercalated Motif by Forming a Double Hairpin with a Novel T:G:G:T Tetrad: Solution Structure of the d(TCCCGTTTCCA) Dimer. *J. Mol. Biol.* **1997**, *273*, 840–856. [CrossRef] [PubMed]
63. Escaja, N.; Viladoms, J.; Garavís, M.; Villasante, A.; Pedroso, E.; González, C. A Minimal I-Motif Stabilized by Minor Groove G:T:G:T Tetrads. *Nucleic Acids Res.* **2012**, *40*, 11737–11747. [CrossRef] [PubMed]
64. Serrano-Chacón, I.; Mir, B.; Escaja, N.; González, C. Structure of I-Motif/Duplex Junctions at Neutral PH. *J. Am. Chem. Soc.* **2021**, *143*, 12919–12923. [CrossRef] [PubMed]
65. Mir, B.; Serrano, I.; Buitrago, D.; Orozco, M.; Escaja, N.; González, C. Prevalent Sequences in the Human Genome Can Form Mini I-Motif Structures at Physiological PH. *J. Am. Chem. Soc.* **2017**, *139*, 13985–13988. [CrossRef] [PubMed]
66. Cao, Y.; Yang, L.; Ding, P.; Li, W.; Pei, R. Ligand Selectivity by Inserting GCGC-Tetrads into G-Quadruplex Structures. *Chem. A Eur. J.* **2020**, *26*, 14730–14737. [CrossRef]
67. Gavathiotis, E.; Heald, R.A.; Stevens, M.F.G.; Searle, M.S. Drug Recognition and Stabilisation of the Parallel-Stranded DNA Quadruplex d(TTAGGGT)4 Containing the Human Telomeric Repeat. *J. Mol. Biol.* **2003**, *334*, 25–36. [CrossRef]
68. Kotar, A.; Kocman, V.; Plavec, J. Intercalation of a Heterocyclic Ligand between Quartets in a G-Rich Tetrahelical Structure. *Chem. A Eur. J.* **2020**, *26*, 814–817. [CrossRef]
69. Troha, T.; Drevenšek-Olenik, I.; Webba Da Silva, M.; Spindler, L. Surface-Adsorbed Long G-Quadruplex Nanowires Formed by G:C Linkages. *Langmuir* **2016**, *32*, 7056–7063. [CrossRef]
70. Escaja, N.; Gómez-Pinto, I.; Viladoms, J.; Pedroso, E.; González, C. The Effect of Loop Residues in Four-Stranded Dimeric Structures Stabilized by Minor Groove Tetrads. *Org. Biomol. Chem.* **2013**, *11*, 4804–4810. [CrossRef]
71. Assi, H.A.; Garavís, M.; González, C.; Damha, M.J. I-Motif DNA: Structural Features and Significance to Cell Biology. *Nucleic Acids Res.* **2018**, *46*, 8038–8056. [CrossRef] [PubMed]
72. Benabou, S.; Aviñó, A.; Eritja, R.; González, C.; Gargallo, R. Fundamental Aspects of the Nucleic Acid I-Motif Structures. *RSC Adv.* **2014**, *4*, 26956. [CrossRef]
73. Serrano-Chacón, I.; Mir, B.; Orozco, M.; Escaja, N.; González, C. PH-Induced Largescale Structural Transitions in i-Motifs. *J. Am. Chem. Soc.* **2022**. submitted.
74. Mir, B.; Serrano-Chacón, I.; Terrazas, M.; Garavís, M.; Orozco, M.; Escaja, N.; González, C. Site-Specific Incorporation of TC^o Results into Highly Stable i-Motifs Amenable to in Vivo Detection. *Nucleic Acids Res.* **2022**. submitted.
75. Mergny, J.L.; De Cian, A.; Amrane, S.; da Silva, M.W. Kinetics of Double-Chain Reversals Bridging Contiguous Quartets in Tetramolecular Quadruplexes. *Nucleic Acids Res.* **2006**, *34*, 2386–2397. [CrossRef]
76. Ma'Ani Hessari, N.; Spindler, L.; Troha, T.; Lam, W.C.; Drevenšek-Olenik, I.; Webba Da Silva, M. Programmed Self-Assembly of a Quadruplex DNA Nanowire. *Chem. A Eur. J.* **2014**, *20*, 3626–3630. [CrossRef]
77. Ilc, T.; Šket, P.; Plavec, J.; Webba Da Silva, M.; Drevenšek-Olenik, I.; Spindler, L. Formation of G-Wires: The Role of G:C-Base Pairing and G-Quartet Stacking. *J. Phys. Chem. C* **2013**, *117*, 23208–23215. [CrossRef]
78. Zheng, J.; Du, Y.; Wang, H.; Peng, Y.; Shi, L.; Li, T. Ultrastable Bimolecular G-Quadruplexes Programmed DNA Nanoassemblies for Reconfigurable Biomimetic DNAzymes. *ACS Nano* **2019**, *13*, 11947–11954. [CrossRef]
79. Zamiri, B.; Mirceta, M.; Bomsztyk, K.; Macgregor, R.B.; Pearson, C.E. Quadruplex Formation by Both G-Rich and C-Rich DNA Strands of the C9orf72 (GGGGCC)8•(GGCCCC)8 Repeat: Effect of CpG Methylation. *Nucleic Acids Res.* **2015**, *43*, 10055–10064. [CrossRef]
80. Subirana, J.A.; Abrescia, N.G.A. Extra-Helical Guanine Interactions in DNA. *Biophys. Chem.* **2000**, *86*, 179189. [CrossRef]
81. Tereshko, V.; Subirana, J.A. Influence of Packing Interactions on the Average Conformation of B-DNA in Crystalline Structures. *Acta Cryst.* **1999**, *55*, 810–819. [CrossRef] [PubMed]

82. Grzechowiak, M.; Ruszkowska, A.; Sliwiak, J.; Urbanowicz, A.; Jaskolski, M.; Ruszkowski, M. New Aspects of DNA Recognition by Group II WRKY Transcription Factor Revealed by Structural and Functional Study of AtWRKY18 DNA Binding Domain. *Int. J. Biol. Macromol.* **2022**, *213*, 589–601. [CrossRef] [PubMed]
83. Wahl, M.C.; Sundaralingam, M. A-DNA duplexes in the crystal. In *Oxford Handbook of Nucleic Acid Sructures*; Neidle, S., Ed.; Oxford University Press: New York, NY, USA, 1999; pp. 389–453. ISBN 9780198500384.
84. Adams, A.; Guss, J.M.; Denny, W.A.; Wakelin, L.P.G. Structure of 9-Amino-[N-(2-Dimethylamino)Propyl]-Acridine-4-Carboxamide Bound to d(CGTACG)2: A Comparison of Structures of d(CGTACG) 2 Complexed with Intercalators in the Presence of Cobalt. *Acta Crystallogr. Sect. D Biol. Crystallogr.* **2004**, *60*, 823–828. [CrossRef]
85. Adams, A.; Guss, J.M.; Collyer, C.A.; Denny, W.A.; Wakelin, L.P.G. A Novel Form of Intercalation Involving Four DNA Duplexes in an Acridine-4-Carboxamide Complex of d(CGTACG)2. *Nucleic Acids Res.* **2000**, *28*, 4244–4253. [CrossRef] [PubMed]
86. Thorpe, J.H.; Hobbs, J.R.; Todd, A.K.; Denny, W.A.; Charlton, P.; Cardin, C.J. Guanine Specific Binding at a DNA Junction Formed by d[CG(5-BrU)ACG]2 with a Topoisomerase Poison in the Presence of Co2+ Ions. *Biochemistry* **2000**, *39*, 15055–15061. [CrossRef] [PubMed]
87. Teixeira, S.C.M.; Thorpe, J.H.; Todd, A.K.; Powell, H.R.; Adams, A.; Wakelin, L.P.G.; Denny, W.A.; Cardin, C.J. Structural Characterisation of Bisintercalation in Higher-Order DNA at a Junction-like Quadruplex. *J. Mol. Biol.* **2002**, *323*, 167–171. [CrossRef]
88. Yang, X.L.; Robinson, H.; Gao, Y.G.; Wang, A.H.J. Binding of a Macrocyclic Bisacridine and Ametantrone to CGTACG Involves Similar Unusual Intercalation Platforms. *Biochemistry* **2000**, *39*, 10950–10957. [CrossRef]
89. Ding, Y.; Xie, L.; Zhang, C.; Xu, W. Real-Space Evidence of the Formation of the GCGC Tetrad and Its Competition with the G-Quartet on the Au(111) Surface. *Chem. Commun.* **2017**, *53*, 9846–9849. [CrossRef]
90. Xu, S.; Dong, M.; Rauls, E.; Otero, R.; Linderoth, T.R.; Besenbacher, F. Coadsorption of Guanine and Cytosine on Graphite: Ordered Structure Based on GC Pairing. *Nano Lett.* **2006**, *6*, 1434–1438. [CrossRef]
91. Bilbao, N.; Destoop, I.; De Feyter, S.; González-Rodríguez, D. Two-Dimensional Nanoporous Networks Formed by Liquid-to-Solid Transfer of Hydrogen-Bonded Macrocycles Built from DNA Bases. *Angew. Chem. Int. Ed.* **2016**, *55*, 659–663. [CrossRef]
92. Nikan, M.; Sherman, J. Template-Assembled Synthetic G-Quartets (TASQs). *Angew. Chem. Int. Ed.* **2008**, *47*, 4900–4902. [CrossRef] [PubMed]
93. Hui, B.W.; Sherman, J.C. Synthesis and characterization of a template-assembled synthetic U-quartet. *Chem. Commun.* **2012**, *48*, 109–111. [CrossRef] [PubMed]
94. Murat, P.; Bonnet, R.; Van der Heyden, A.; Spinelli, N.; Labbé, P.; Monchaud, D.; Teulade-Fichou, M.P.; Dumy, P.; Defrancq, E. Template-Assembled Synthetic G-Quadruplex (TASQ): A Useful System for Investigating the Interactions of Ligands with Constrained Quadruplex Topologies. *Chem. A Eur. J.* **2010**, *16*, 6106–6114. [CrossRef] [PubMed]

Article

Exploring the Interaction of G-quadruplex Binders with a (3 + 1) Hybrid G-quadruplex Forming Sequence within the PARP1 Gene Promoter Region

Stefania Mazzini [1,*], Salvatore Princiotto [1,2], Roberto Artali [3], Loana Musso [1], Anna Aviñó [4], Ramon Eritja [4], Raimundo Gargallo [5] and Sabrina Dallavalle [1,6]

1. Department of Food, Environmental and Nutritional Sciences (DEFENS), University of Milan, 20133 Milan, Italy; salvatore.princiotto@unimi.it (S.P.); loana.musso@unimi.it (L.M.); sabrina.dallavalle@unimi.it (S.D.)
2. Center for Research in Biosciences & Health Technologies (CBIOS), Universidade Lusófona de Humanidades e Tecnologias, Campo Grande 376, 1749-024 Lisbon, Portugal
3. Scientia Advice di Roberto Artali, 20832 Desio, Italy; roberto.artali@scientia-advice.com
4. Institute for Advanced Chemistry of Catalonia (IQAC), CSIC, Networking Center on Bioengineering, Biomaterials and Nanomedicine (CIBER-BBN), ISCIII, 08034 Barcelona, Spain; aaagma@cid.csic.es (A.A.); recgma@cid.csic.es (R.E.)
5. Department of Chemical Engineering and Analytical Chemistry, University of Barcelona, 08028 Barcelona, Spain; raimon_gargallo@ub.edu
6. National Institute of Fundamental Studies, Kandy 20000, Sri Lanka
* Correspondence: stefania.mazzini@unimi.it

Abstract: The enzyme PARP1 is an attractive target for cancer therapy, as it is involved in DNA repair processes. Several PARP1 inhibitors have been approved for clinical treatments. However, the rapid outbreak of resistance is seriously threatening the efficacy of these compounds, and alternative strategies are required to selectively regulate PARP1 activity. A noncanonical G-quadruplex-forming sequence within the PARP1 promoter was recently identified. In this study, we explore the interaction of known G-quadruplex binders with the G-quadruplex structure found in the PARP gene promoter region. The results obtained by NMR, CD, and fluorescence titration, also confirmed by molecular modeling studies, demonstrate a variety of different binding modes with small stabilization of the G-quadruplex sequence located at the PARP1 promoter. Surprisingly, only pyridostatin produces a strong stabilization of the G-quadruplex-forming sequence. This evidence makes the identification of a proper (3+1) stabilizing ligand a challenging goal for further investigation.

Keywords: G-quadruplex; PARP1 promoter; stabilizing ligand; NMR; circular dichroism; fluorescence; molecular modeling

1. Introduction

Poly (ADP-ribose) polymerase-1 (PARP1) is a nuclear enzyme involved in DNA repair processes [1,2]. The enzyme repairs breaks in single-strand DNA through a base excision repair pathway. Additionally, PARP1 is implicated in other cellular processes, such as transcriptional regulation, chromatin remodeling, cell signaling and cell death [3,4].

PARP1 inhibition causes the so-called "synthetic lethality" in tumor cells with defective homologous recombination pathways and sensitizes the tumor cells to DNA-damaging chemotherapies, including multiple chemotherapy or radiotherapy approaches, which remain the backbone of treatment for most cancer patients [5]. Consequently, PARP1 has emerged as an attractive target for cancer therapy. Several PARP1 inhibitors have been successfully developed, including Olaparib, Veliparib, Rucaparib, Talazoparib and Niraparib [6,7].

However, the emergence of resistance to the PARP1 inhibitors, mediated by multiple molecular mechanisms, has generated the need for alternative approaches to interfere with PARP1 activity.

In this context, an unexplored and promising approach could be related to a transcriptional repression of the enzyme. Recently, Chambers et al. investigated the promoter region of the PARP1 gene and identified noncanonical G-quadruplex-forming sequences by genome-mapping experiments [8].

G-quadruplexes are four-stranded structures formed by G-rich nucleic acids comprising a stack of multiple guanine(G)-tetrads [9]. A wide variety of G-quadruplexes were elucidated. These structures were extensively associated with cancer, playing an important role in telomere maintenance and in controlling the expression of several oncogenes and tumor suppressors [10,11]. For these reasons, the ligand-mediated structure modulation of G4-DNA in gene promoter regions may represent an approach to control deleterious gene expression. Both the DNA and RNA G-quadruplex structures are being extensively investigated as potential therapeutic targets for small-molecule G-quadruplexes binders [12].

Notably, the topology and loop conformation of the G4-DNA structures are known to be highly polymorphic, with the folding conditions and the nature of the loop sequences determining the overall topology. Small molecules end-stacking with G4-DNA include planar aromatic surfaces that mimic the large planar surface of the G-quadruplex. As end-stacking only requires a G-tetrad (i.e., a planar aromatic surface) to stack with, the binding of these molecules may not require a specific G4-DNA topology, and thus may lack discrimination among the various G4-DNA conformations. Alternatively, the groove and loop regions of G4-DNA differ from the canonical DNA duplex, and even among the G4-DNA structures; thus, may provide a better approach for structural selectivity in binding. Recent reviews have suggested that the most thermally stabilizing and selective G4-DNA ligands often have the combined features of end-stacking and groove/loop binding [12]. Specifically, these include fused aromatic polycyclic and macrocyclic ligands, with electron withdrawing atoms/groups that can be fixed in a planar conformation for effective G-tetrad stacking, and with basic side chains targeting loops and grooves for better selective G4-DNA binding and stabilization [13]. A significant example is pyridostatin (Figure 1), a G4-selective stabilizing ligand, designed as a flexible molecule, yet able to adopt a flat conformation such that it can participate in $\pi-\pi$ interactions with the G-tetrad. At. the same time, the molecule establishes electrostatic interactions through the amino groups that are present on the side chains. Pyridostatin alters the transcription and replication of specific human genomic loci containing high G-quadruplex clustering within the coding region, which encompasses telomeres and selected genes, such as the proto-oncogene *SRC* [14]. The molecule is an excellent G-quadruplex DNA binder with high thermal stabilization ($\Delta Tm > 30\ °C$). Its interaction with an exposed planar G-quartet rather than loop sequences in solutions suggests that it might recognize a very broad spectrum of G4 structures in a living cell and interact with them [15].

RHPS4 (Figure 1) is another promising and potent G-quadruplex stabilizing ligand. The compound is a water-soluble polycyclic fluorinated acridinium cation, which carries a net positive charge in its small acridinium ring [16]. These features enhance its affinity to G-quadruplex and its ability to penetrate heavy tumor masses [17,18]. RHPS4 forms stacking interactions by binding with high affinity to the G-quadruplex at the terminal G-tetrads, with selectivity over duplex DNA [19,20]. The treatment of various cancer cell lines and tumor xenografts revealed that RHPS4 is a potent telomerase inhibitor at sub-micromolar concentrations [21]. Additionally, it causes irreversible proliferation arrest after long-term culture at non-cytotoxic concentrations, and exhibits antitumoral activity in vivo [22].

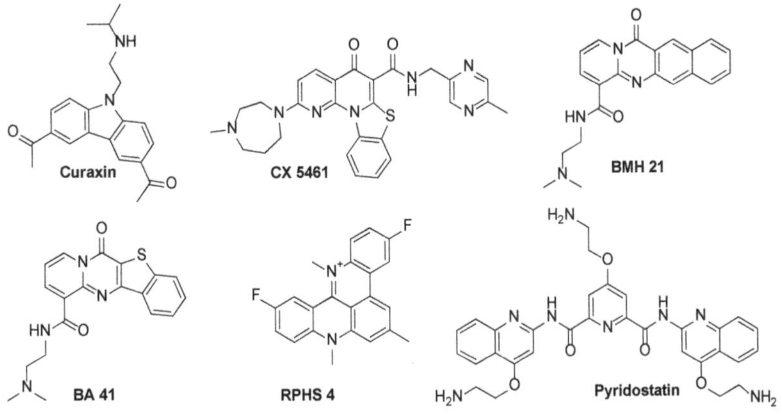

Figure 1. Structures of selected G-quadruplex binders.

Curaxin CBL0137 (Figure 1) belongs to a class of substituted carbazoles exerting anticancer activity by complex and diverse mechanisms. Recent findings have claimed the involvement of DNA binding in the curaxin antitumor activity [23,24]. The NMR studies by our group evidenced a significant binding of curaxin with the single repeat sequence of human telomeres (TTAGGGT)$_4$ and with Pu22T14T23, a model of the c-myc promoter Pu22 sequence [25]. Investigations of BMH21, BA41 and CX-5461 (Figure 1), [26] provided evidence that these compounds are also effective binders of the human telomeric and oncogene promoter G-quadruplexes, such as c-MYC and c-KYT.

Considering all of the above reported results, we were intrigued by the challenge to explore PARP1 modulation via G-quadruplex DNA targeting. In a recent paper, Sengar and coworkers [27] analyzed, by NMR spectroscopy, the G-quadruplex structure formed by a 23-nucleotide G-rich sequence (TGGGGGCCGAGGCGGGGCTTGGG (termed as TP3)), and located 125 nucleotides (nts) upstream of the transcription start site (TSS) in the PARP1 promoter. The study revealed a three-layered intramolecular (3 + 1) hybrid G-quadruplex-folding topology, in which three strands are oriented in one direction and the fourth in the opposite direction. It should be noted that this structure exhibits unique structural features, such as an adenine bulge and a GGT base triple capping structure taking place between the central edgewise loop, propeller loop, and a flanking 5′-end. This finding offered an attractive opportunity to target the PARP1 promoter with specificity. Thus, we investigated its interaction with the above reported G-quadruplex binders (curaxin, CX-5461, BMH21, BA41, RHPS4 and pyridostatin), characterized by a strong structural diversity, with the aim of defining the key molecular features for an effective binding.

2. Results and Discussion

To investigate the nature of the interaction of the curaxin, CX-5461, BMH21, BA41, RHPS4 and pyridostatin with the G-quadruplex (3+1) hybrid topology in the PARP1 promoter, a combination of CD, NMR and molecular modeling studies were used.

To this aim, the oligonucleotide (TP3-T6), which contains a G-to-T substitution at position 6, resulting in a high-quality NMR spectrum very similar to the spectrum of TP3, was employed (Figure 2). The modified oligomer shows the formation of the same G-quadruplex-folded structure; however, having an improved temporal stability upon exposure to room temperature [27]. As reported by Sengar et al. [27], the presence of 11 well-resolved distinct peaks in the characteristic region of ^{1}H NMR spectra between 10.5 and 12.5 ppm, derived from the Hoogsteen base pairs typical of G-quadruplex folding (corresponding to 12 guanines of three G-tetrad, NH of G21 and G22 being overlapped), indicate an intramolecular (3 + 1) hybrid G-quadruplex scaffold. The information concerning the mode of binding of the ligands can be obtained by analyzing the sharpening of the imino proton signals and their chemical shift variations.

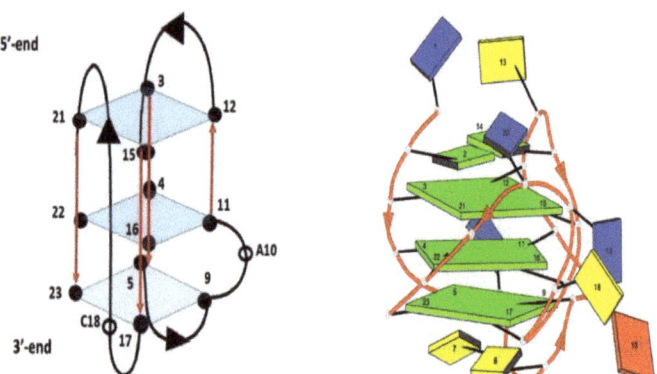

Figure 2. Schematic representation of TP3-T6 oligomer G-quadruplex. Adenine in red, Cytosine in yellow, Guanine in green and Thymine in blue.

The ^{1}H NMR titration experiments were performed tracking the guanine imino protons of TP3-T6, by adding increasing amounts of the ligand to the oligonucleotide solution. Differences in the ligand binding mode were observed for the considered compounds.

The titration of TP3-T6 with the curaxin displayed the broadening of the selected 1H imino proton signals, even at low ratio (R = (curaxin)/(TP3-T6) = 0.5). Specifically, the residues belonging to the 3'-end tetrad of the G-quadruplex G5, G9, G17 and G23 essentially disappeared at R = 1.0 (Figure 3). This behavior suggested that the curaxin mainly interacts with the 3'-end tetrad.

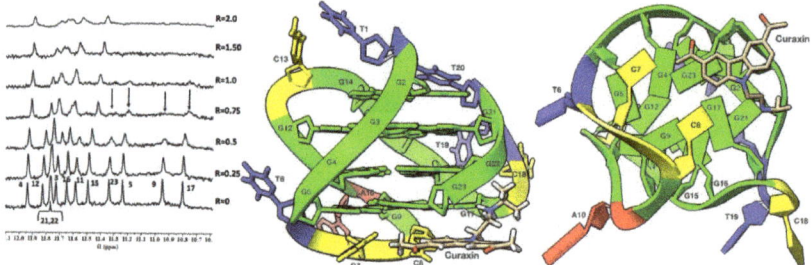

Figure 3. Imino proton region of the 1D NMR titration spectra of TP3-T6 with curaxin (Left), recorded at 25 °C and different R = (drug)/(DNA) ratios. The side (Center) and 3'-end (Right) views of the TP3-T6/curaxin complex obtained by Molecular Docking were created by using the Chimera-X software. In the side (Center) view, nucleotides are shown in sticks to reveal atoms and bonds, while in the 3'-end (Right) view the nucleotides are shown as slabs and filled sugars: Adenine in red, Cytosine in yellow, Guanine in green and Thymine in blue. The curaxin molecule is rendered in sticks and colored according to the atoms.

Adding the curaxin in excess (R > 1.50), a more significant broadening involved nearly all of the ^{1}H imino proton signals. This indicates an exchange process in the intermediate regime on the NMR time scale between free and bound state, suggesting the formation of different species in solution. The disappearance of some of the imino proton signals can be explained by the loss of the hydrogen bonding, resulting in a solvent exchange. The NMR behavior demonstrated that the curaxin binds at a ratio R = 1.0, to the most accessible 3′-end tetrad of TP3-T6. Conversely, at higher ratios (R > 1.0) non-specific external interactions occur between the ligand and the oligonucleotide.

The results of the molecular docking studies confirmed the preferential interaction of the curaxin with the 3′-end tetrad (see Figure 3). The side chain is oriented towards the groove, with which it interacts mainly through a strong H-bond between the nitrogen and G17N3. On the other hand, the curaxin polycyclic system interacts with the G23 ring through a π–π stacking interaction, and with G23OP$_2$ through an anion–π interaction. Despite this, the molecule fails to interact effectively with the G5-G9-G17-G23 tetrad, probably due to the steric hindrance exerted by C7 and C8.

The interaction of TP3-T6 with the curaxin was studied by molecular fluorescence spectroscopy to calculate the binding constant. The addition of TP3-T6 to a curaxin solution induced a decreased fluorescence signal intensity in the curaxin. The inverse titration of TP3-T6 with the curaxin was also performed (Supplementary Materials). In this case, an opposite effect was observed with an increased fluorescence signal with the addition of the ligand. An estimation of the stoichiometry and the binding constant (K_b) was calculated by the EQUISPEC program, starting from the multivariate analysis of the whole fluorescence spectra obtained during both of the titrations [28]. For the direct titration, the K_b value obtained by fitting with 1:1 model, was $6.72 \pm 0.20\ 10^5\ \text{M}^{-1}$ (Figure 4). For the inverse titration, a second binding site with K_b equal to $5.63 \pm 1.25\ 10^5\ \text{M}^{-1}$ was determined. All together, these values indicated a mild interaction of the curaxin with the TP3-T6.

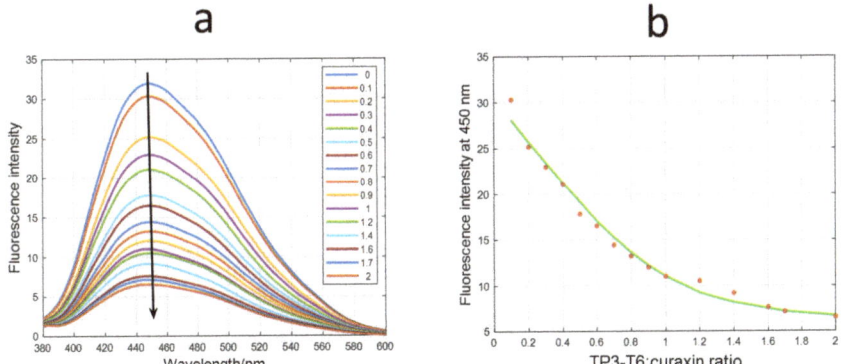

Figure 4. (a) Fluorescence spectra recorded along the titration of curaxin with TP3-T6; (b) Experimental (red symbols) and calculated (green line) fluorescence intensity at 450 nm for the titration of curaxin at different concentrations of TP3-T6. Conditions: 20 mM phosphate buffer (pH 7.1), 70 mM KCl of 3 μM curaxin and increasing amounts of 69 μM of TP3-T6. Excitation wavelength 334 nm.

CD-monitored melting experiments were used to investigate the potential stabilization of the G-quadruplex structure adopted by TP3-T6 treated with the curaxin (1:3 mixture) (Figure 5). The CD spectrum of TP3-T6 at 10 °C showed positive signals at 265 and 285 nm, which could be related to a hybrid parallel/antiparallel structure. Upon the addition of the curaxin, a small decrease in the intensity of the band at 285 nm was observed (Supplementary Materials). From the CD traces at 265 nm measured along the melting, the fraction of folded DNA was calculated, assuming a one-step process. This assumption was checked by means of multivariate analysis of the whole data set, as shown in the Supplementary Materials [29,30]. From the data in Figure 5b, it was deduced that the T_m values were 58.5 ± 0.8 and 56.7 ± 0.9 for TP3-T6, and for the 1:3 DNA/ligand mixture, respectively. Overall, these experiments showed small changes in the presence of the ligand, in agreement with the relatively low values of the binding constants previously determined.

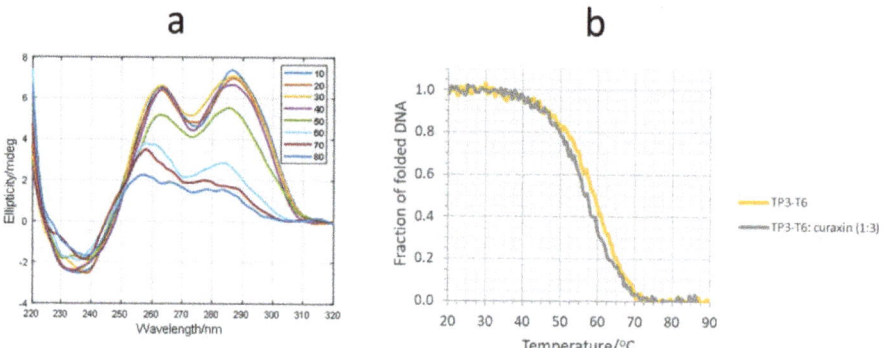

Figure 5. (a) Selected CD spectra recorded along the melting of the TP3-T6/curaxin 1:3 mixture. The whole data set is given in Supplementary Materials; (b) Fraction of folded DNA vs. temperature calculated from the melting of TP3-T6 and of the TP3-T6/curaxin 1:3 mixture at 265 nm. DNA and ligand concentration were 2 and 6 µM, respectively, 20 mM phosphate buffer (pH 7.1), 70 mM KCl.

All of the results did not evidence any significant stabilization of the G-quadruplex folding by the curaxin, in agreement with the external binding suggested by the NMR experiments and molecular modeling.

Similar results were found upon addition of 0.25 molar equivalent of CX-5461 to the G-quadruplex solution. A decrease in the intensity and/or a disappearance of the selected signals, such as G17, G9, G23, all belonging to the same tetrad, were observed (Figure 6). As in the previous case, at ratio R = [CX-5461]/[DNA] > 1.0, the profile of the G-quadruplex imino protons appeared altered, with a general signal broadening effect. The continuous titration to R = 2.0 stoichiometry accentuated the broadening of the imino proton resonances. The effects observed in this region suggested that the interaction of CX-5461 is with the 3'end G-tetrad (G17-G9-G23-G5).

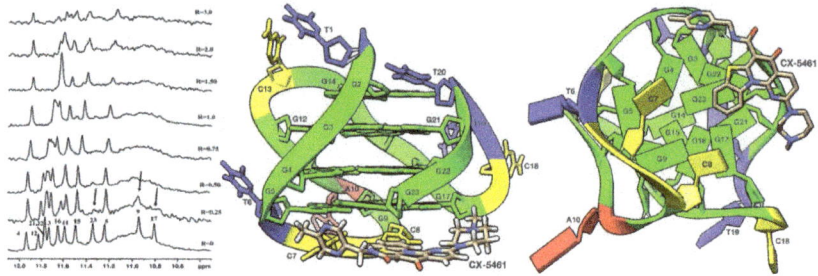

Figure 6. Imino proton region of 1D NMR titration spectra of TP3-T6 with CX-5461 at 25 °C at different R = (drug)/(DNA) ratios (Left). The complex between TP3-T6 and CX-5461 was obtained by molecular docking and the side (Center) and 3'-end (Right) views were created by using the ChimeraX software., nucleotides are shown in sticks to reveal atoms and bonds, the nucleotides in the side (Center) view are rendered in stick while in the 3'-end (Right) view they are shown as slabs and filled sugars. Adenine residues are colored in red, Cytosine in yellow, Guanine in green and Thymine in blue. In both views the ligand is represented as sticks and colored according to the atoms.

The complex obtained by the molecular docking between CX-5461 and the G-quadruplex confirmed the preferential interaction of the ligand with the 3'-end tetrad, as shown in Figure 6. The side chain containing the homo-piperazine ring is arranged along the groove, while the rest of the molecule lies below the G5-G9-G17-G23 tetrad. In particular, the [1,3] benzothiazolo [3,2-a]quinolin-5-one ring is quite coplanar with the 3'-end tetrad. In this orientation, the molecule establishes a network of π–π stackings with G23, as well as an anion–π interaction with the G23OP$_2$ atom.

BMH21 and BA41 displayed a very similar NMR behavior. Even at a low R ratio (0.5), all of the ^{1}H imino proton signals showed a significant generalized line broadening upon titration with both of the ligands, but the most affected were G5 and G23, which belong to the 3'-end tetrad (Figures 7 and 8). For R > 1.0 an increased broadening was observed, likely due to an intermediate exchange regime between the free and the bound states.

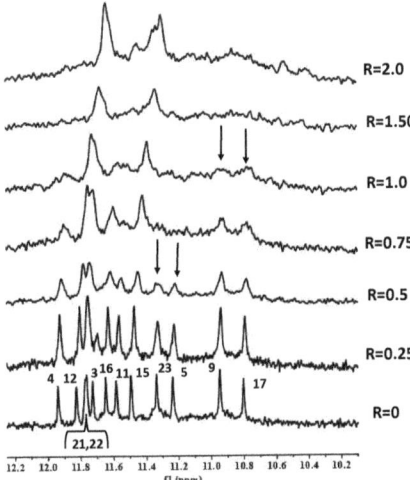

Figure 7. Imino proton region of 1D NMR titration spectra of TP3-T6 with BMH21 at 25 °C at different R = (drug)/(DNA) ratios.

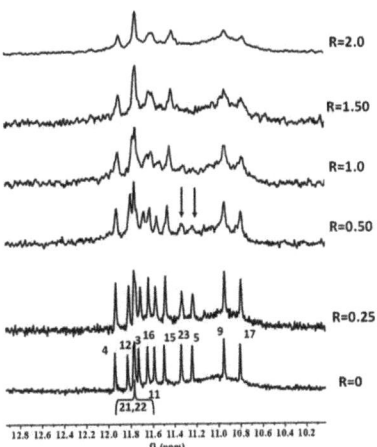

Figure 8. Imino proton region of 1D NMR titration spectra of TP3-T6 with BA41 at 25 °C at different R = (drug)/(DNA) ratios.

These findings suggested that BMH21 and BA41 at a low ratio mainly interact with the 3'-terminal end tetrad. As in the case of the curaxin, CD-monitored titrations and melting experiments did not show any stabilization of the DNA structure upon addition of BA41 (Figure S5, Supplementary Materials).

The complexes obtained by molecular docking of BMH21 and BA41 with TP3-T6 are depicted in Figure 9. The cyclic systems of BMH21 and BA41 are too bulky to allow the two molecules to fit under the 3′-end tetrad. As a result, although both of the molecules exhibit multiple interactions with the nucleotide atoms located on the groove surface, they are unable to properly stack with the planar portion of the G5-G9-G17-G23 tetrad.

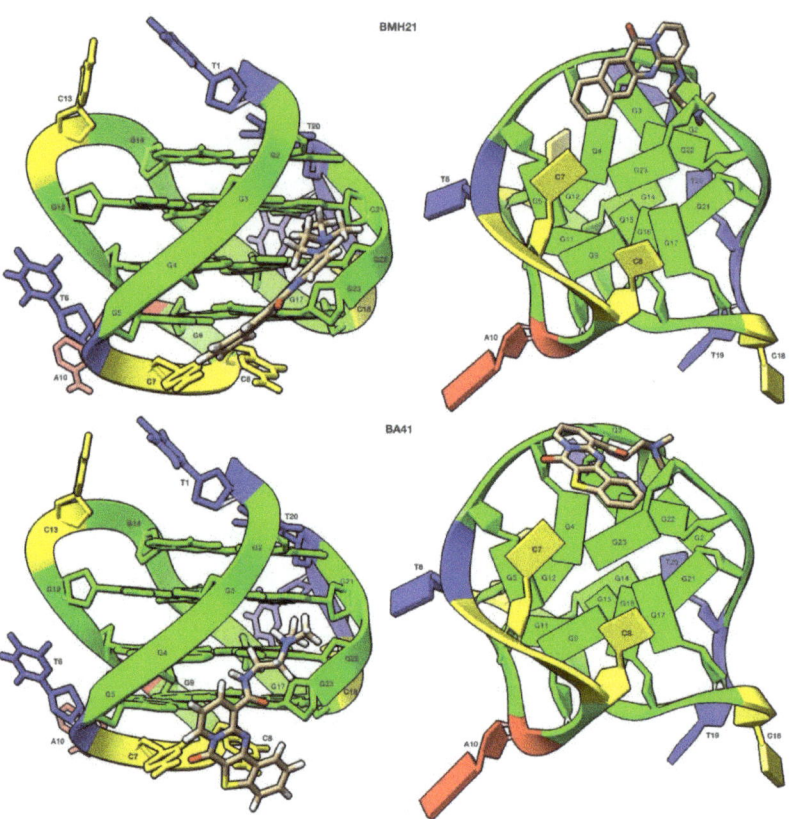

Figure 9. Side (Left) and 3′-end (Right) views of the complexes predicted by molecular docking of the TP3-T6 target with BMH21 and BA41. In the graphical representations, nucleotides in the side views are rendered in stick while those in the 3′-end views are rendered as slabs and filled sugars. Adenine residues are colored in red, Cytosine in yellow, Guanine in green and Thymine in blue. The ligands are always represented as sticks and colored according to the atoms. Both graphical representations were created with the ChimeraX software.

Concerning the titration experiments with RHPS4, we observed the broadening of selected ^{1}H imino proton signals at a very low ratio (R = 0.25). In particular G3, G12, G15 and G21, all belonging to the 5′-terminal end tetrad, became very broad, until they almost completely disappeared at R = 1.0. However, not very well-defined signals were detected again at R = 2.0 (Figure 10). A molecular docking investigation did not allow to obtain an unambiguous model of interaction of RHPS4 with the oligonucleotide.

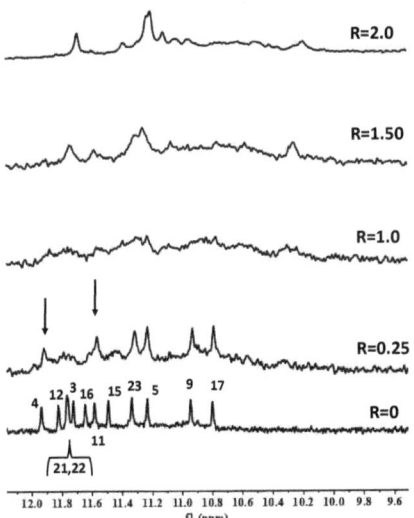

Figure 10. Imino proton region of 1D NMR titration spectra of TP3-T6 with RHPS4 at 25 °C at different R = (drug)/(DNA) ratios.

Upon titrating pyridostatin, minor changes were observed in the ^{1}H imino protons region of NMR spectra (Figure 11). In addition, a CD-monitored melting experiment was carried out (Figure S6, Supplementary Materials). In the presence of the pyridostatin, the intensity of the CD band at 290 nm was dramatically reduced, whereas the parallel contribution remained. Upon heating, a clear stabilization of the G-quadruplex structure was observed because of the presence of the ligand, being the ΔT_m > 30 °C. These observations indicate a strong stabilization of the parallel contribution, as opposed to the scarce interaction with the antiparallel contribution, as already described by other authors [31]. Overall, the results highlight the importance of integrating orthogonal experimental approaches (e.g., NMR studies to point out the ligand binding mode and CD spectroscopy to describe the thermal denaturation transition, which relates the different stabilization of the folded and unfolded forms of a sequence), to define the interaction of small molecules with G4-forming sequences [32].

The obtained findings demonstrated that the G-quadruplex structure located at the PARP1 promoter has a peculiar structure, which makes the identification of a proper 3 + 1 stabilizing ligand a challenging goal for further investigation. The results of the molecular modeling studies gave important information, clearly showing the impossibility for most of the ligands to give effective π–π stacking interactions with the tetrad at the 5′-end. The steric hindrance deriving from the presence of G2 and G14 (as also reported by Sengar [27]) prevented ligands from effectively interacting with the G3-G12-G15-G21 tetrad. On the other side, the bases C7 and C8 only partially prevented the π–π stacking interactions with tetrad at 3′-end. A summary of the interactions in the complexes between the ligands and TP3-T6 is shown in Table 1.

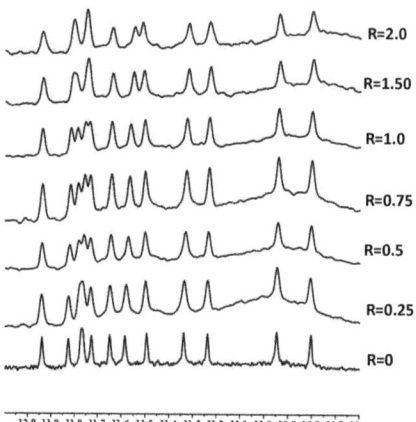

Figure 11. Imino proton region of 1D NMR titration spectra of TP3-T6 with pyridostatin at 25 °C at different R = (drug)/(DNA) ratios.

Table 1. Table summarizing the main interactions observed in the TP3-T6/Ligands complexes as obtained by molecular modeling. The table shows the nucleotides involved in the interactions together with the type of interactions.

Ligands	TP3-T6 3′-End Tetrad			TP3-T6 Groove						
	G4	G17	G23	G3	G4	G17	C18	G21	G22	G23
Curaxin	-	-	PP, AP	-	-	HB	-	-	-	-
CX-5461	-	-	PP, AP	-	-	-	-	-	-	-
BMX21	AP	-	HB	SB	-	-	-	-	-	HB
BA41	-	-	HB, PP	HB, AC	AC	-	-	-	-	-
RPHS4	-	-	-	-	-	-	-	-	AC, AP	-
Pyridostatin	-	HB	PP	-	-	-	HB, AP	SB	AP	-

Interaction Types: HB (Hydrogen Bond), AC (Attractive Charge), SB (Salt Bridge), AP (Anion-π), PP (π-π).

The NMR spectra of the curaxin complex suggested an interaction, for R = 1.0, with a specific G-tetrad residue belonging to the 3′-end. At a higher ratio, the NMR spectra provided the evidence for a groove mode of binding, which produced a degree of structural perturbation. This resulted in a destabilization of the G-quadruplex, as proved by the decease of T_m of the complex with curaxin. More interestingly, from our experiments it emerged that only the pyridostatin caused a stabilization of the PARP G-quadruplex structure. Pyridostatin is considered as an excellent G-quadruplex binder, mainly for its ability to adopt a flat but flexible conformation, prone to adapt to the dynamic and polymorphic nature of diverse G-quadruplex structures and for the presence of free nitrogen lone pairs [33] The evidence that only this compound was able to stabilize the PARP1 G-quadruplex confirms our previous findings about the peculiar characteristics of this structure [34], evidencing that the dimension and geometry of the ligand play a crucial role for an effective interaction. Significantly, BMH21 and BA41, which maintain a flat however small aromatic system, and an amino group in a proper position, were even able to disrupt the G-quadruplex tetrads, weakening the hydrogen bonds between the G-quartets.

3. Materials and Methods

Ligands. The compounds BMH21 and BA41 were synthesized as previously described [35]. Curaxin (CBL 0137) was purchased from Carbosynth Limited, Compton, UK. The corresponding hydrochloride was prepared by treatment with 4M HCl in dioxane.

Compound CX-5461 was purchased from ChemScene LLC, Monmouth Junction, NJ, USA. The corresponding hydrochloride was prepared by treatment with HCl in methanol. Pyridostatin and RPHS4 were kindly provided by Dr. Giovanni Beretta, Fondazione IRCCS Istituto Nazionale Tumori, Milan, Italy

Oligonucleotides. The synthesis of the oligonucleotides was performed by the ICTS NANBIOSIS Oligonucleotide Synthesis Platform. Unit 29. (CIBER-BBN).

NMR. The oligonucleotide was synthesized in 1 µmol scale on an Applied Biosystems DNA/RNA 3400 synthesizer by solid-phase 2-cyanoethylphosphoroamidite chemistry. The product obtained after the synthesis was passed through a cation exchange, Dowex 50WX2 resin, to exchange the counter ion to sodium form and then desalted with a Sephadex (NAP-10) G25 column.

The NMR spectra were recorded on a Bruker AV600 spectrometer operating at a frequency of 600.10 MHz for ^{1}H nucleus at 25 °C. The TP3-T6, 5′ –d(TGGGGTCCGAGGCGGGGCTTGGG) -3′ at 0.24–0.48 mM in G-quadruplex concentration range was prepared in 20 mM KH_2PO_4, 70 mM KCl, 90% H_2O, pH = 7.0. The samples were heated to 85 °C for 1 min and then cooled at room temperature overnight. The G-quadruplex TP3-T6, 5′-d(TGGGGTCCGAGGCGGGGCTTGGG)-3′ signals were previously assigned [27]. To the TP3-T6 solution, aliquots of ligands stock solutions in DMSO-d_6 (16–17 mM) were gradually added until R = (ligand)/(DNA) = 2.0. and ^{1}H NMR spectra were recorded.

CD and Fluorescence. The CD spectra were recorded on a Jasco J-810 spectropolarimeter equipped with a Peltier temperature control unit (Seelbach, Germany). The DNA solution of TP3-T6 was transferred to a covered cell and ellipticity was recorded with a heating rate of approximately 0.4 °C·min^{-1}. Simultaneously, the CD spectra were recorded every 5 °C from 220 to 310 nm. The spectrum of the buffer was subtracted. Each sample was allowed to equilibrate at the initial temperature for 30 min before the melting experiment began. In all of the experiments, the concentration of DNA was kept constant (2 µM) whereas the concentration of the considered ligands was increased. The medium consisted of 20 mM phosphate buffer (pH 7.1), 70 mM KCl [27].

The molecular fluorescence spectra were measured with a JASCO FP-6200 spectrofluorometer. The temperature was controlled at 20 °C, using a water bath. The fluorescence spectra were monitored using a quartz cuvette with a 10-mm path length, with the excitation and emission slits set at 10 nm, and the scan speed at 250 nm/min. The buffer consisted of 20 mM phosphate buffer (pH 7.1) and 70 mM KCl. In all of the experiments, the concentration of the ligands was 3 µM, whereas the concentration of TP3-T6 at 69 µM sequence was increased.

The determination of the ratio ligand: DNA and the calculation of the binding constants was completed from the fluorescence data recorded along the titrations of ligands with DNAs by using the EQUISPEC program [28]. This program, which is based on the multivariate analysis of the set of spectra measured along a titration, relies on the fulfillment of the law of mass action.

The CD spectra recorded alongside the melting experiments were analyzed by means of the Multivariate Curve Resolution, based on the Alternating Least Squares procedure (MCR-ALS). This is a soft-modeling multivariate data analysis method that is not based on the fulfilment of any physico-chemical model (studies on the interactions of Ag(I) with DNA and their implication on the DNA-templated synthesis of silver nanoclusters and on the interaction with complementary DNA and RNA sequences [30]).

Molecular modeling studies. The first model of the NMR ensemble deposited in the Protein Data Bank (PDB accession code: 6AC7 [27]) was used to obtain the starting TP3-T6 3D-structure.

The molecular docking calculations for each of the six ligands was performed by AutoDock 4.2, [36,37] using the Lamarckian Genetic Algorithm in combination with a grid-based energy evaluation method to calculate grid maps, in a 80 Å × 80 Å × 80 Å box with a spacing of 0.01 Å. The AutoDock Toolkit (ADT) [38] was used to add the Gasteiger–Marsili charges [39] to the ligands, and the phosphorus atoms in the DNA were parameterized, using the Cornell parameters. The Addsol utility of AutoDock was used to add the solvation parameters to the system. The initial population for each molecule consisted of 100 randomly placed individuals, a maximum number of 250 energy evaluations and an elitism value of 1, a mutation rate of 0.02 and a crossover rate of 0.80. The local search was conducted applying the so-called pseudo-Solis and Wets algorithm with a maximum of 250 iterations per local search and 250 independent docking runs. The docking results were scored by using an in-house version of the simpler intermolecular energy function, based on the Weiner force field, and the lowest energy conformations (differing by less than 1.0Å in positional root-mean-square deviation (rmsd)) were collected.

The resulting complexes were placed at the center of a box (boundaries at 2.0 nm apart from all atoms) and solvated with TIP3P water molecules. Amber ff99 force field [40] with bsc1 corrections [41] was used to describe the TP3-T6 G-Quadruples. To remove the bad contacts, 1000 minimization steps were performed on the initial systems, followed by a heating ramp of short (100 ps) consecutive simulations. The production simulations consisted of 5 ns of Langevin [42,43] Molecular Dynamics (MD) NPT equilibration at 298 K and 1 atm, as implemented in NAMD [44]. During this step, all of the bonds to hydrogen atoms were constrained using the SHAKE [45] algorithm. The water molecules were kept rigid with SETTLE [46], allowing an integration time step of 0.002 ps. The electrostatic interactions were calculated using the Particle Mesh Ewald (PME) [47,48] method (Coulomb cut-off radius of 1.2 nm). A Berendsen thermostat (coupling time of 0.1ps) was applied to the systems [49].

The molecular graphics and analyses performed with UCSF ChimeraX, developed by the Resource for Biocomputing, Visualization, and Informatics at the University of California, San Francisco, CA, USA, with support from the National Institutes of Health R01-GM129325 and the Office of Cyber Infrastructure and Computational Biology, National Institute of Allergy and Infectious Diseases [50].

4. Conclusions

The enzyme PARP1 emerged as an attractive target for cancer therapy, being involved in DNA repair processes. The use of PARP1 inhibitors, which have been recently developed and approved for clinical treatments, is threatened by the rapid insurgence of resistance. Recent experiments have identified a noncanonical G-quadruplex-forming sequence within the PARP1 promoter, opening a new avenue of investigation. We were particularly intrigued by the possibility of identifying compounds acting as PARP inhibitors by an alternative approach, e.g., via G-quadruplex DNA targeting.

One major task for the design of DNA ligands is not only the discrimination of the G-quadruplex vs. other DNA structures such as double or triple helix, but also the selectivity for a unique quadruplex topology. Considering that no PARP promoter modulator has been identified so far, we investigated the interaction of known G-quadruplex binders with the G-quadruplex-forming sequence within the PARP1 promoter.

The results obtained by NMR, CD and fluorescence titration emphasized that the structural requirements for the interaction are quite strict. In fact, compounds that strongly stabilize G-quadruplex structures had only a low affinity to the PARP1 promoter. The molecular modeling studies suggested the impossibility for most of the compounds to favorably interact with the 3'-end tetrad of this particular G-quadruplex structure. The position of the cytosine residues, C7 and C8, prevented an optimal interaction with G5 and G6, and partially also with G17. The ligands were therefore forced to interact mainly with the phosphate groups located on the surface of the groove. CX-5461 was capable of interacting promisingly with the G5-G9-G17-G23 tetrad. The shape of the [1,3]benzothiazolo

[3,2-a]quinolin-5-one ring permitted CX-5461 to fit into the pocket delimited by C7 and C8, allowing it to effectively overlap with the planar portion of G23 and originating a dense network of π–π stacking interactions. Pyridostatin, endowed with a flat but flexible conformation, gave a strong stabilization of the parallel contribution, as opposed to the scarce interaction with the antiparallel contribution.

The investigated molecules can serve as molecular tools toward the identification of reliable selective PARP G-quadruplex modulators, which could serve as an alternative therapeutic approach for the regulation of PARP1 activity.

Supplementary Materials: The following supporting information can be downloaded at: https://www.mdpi.com/article/10.3390/molecules27154792/s1, Figure S1: (a) Fluorescence spectra recorded along the inverse titration of TP3-T6 with curaxin. (b) Experimental (red symbols) and fitted (green line) fluorescence intensities at 450 nm for the titration of TP3-T6 at different concentration of curaxin; Figure S2: Complete set of CD spectra recorded along the melting of TP3-T6 (a) and of the TP3-T6:curaxin (1:3) mixture (b); Figure S3: Results of the analysis by means of MCR-ALS of the complete set of CD spectra recorded along the melting of TP3-T6; Figure S4: CD spectra measured along the titration of TP3-T6 with BA41. Figure S5: Melting experiments with BMH-21 and BA41; Figure S6: CD-monitored titration with PDS.

Author Contributions: Conceptualization, S.M., S.D., R.E. and R.G. methodology, S.M, S.P., R.A., L.M., A.A. and R.G.; investigation, S.M, S.P., R.A., L.M., A.A. and R.G resources, S.M., S.D., R.E., and R.G.; data curation, S.M., S.P., R.A., A.A. and R.G.; writing—original draft preparation, S.M., S.D., A.A, R.E. and R.G.; writing—review and editing, all authors. All authors have read and agreed to the published version of the manuscript.

Funding: This research was supported by the Italian MIUR Project PRIN 2017 2017SA5837 and by PIANO DI SOSTEGNO ALLA RICERCA 2020—Linea 2 azione B (DEFENS). This investigation was also supported by a research grant from the Spanish Ministerio de Ciencia e Innovación (PID2019-107158GB-I00).

Institutional Review Board Statement: Not applicable.

Informed Consent Statement: Not applicable.

Data Availability Statement: Not applicable.

Conflicts of Interest: The authors declare no conflict of interest.

Sample Availability: Samples of the compounds tested are available from the authors.

References

1. Morales, J.; Li, L.; Fattah, F.J.; Dong, Y.; Bey, E.A.; Patel, M.; Gao, J.; Boothman, D.A. Review of Poly (ADP-Ribose) Polymerase (PARP) Mechanisms of Action and Rationale for Targeting in Cancer and Other Diseases. *Crit. Rev. Eukaryot. Gene Expr.* **2014**, *24*, 15–28. [CrossRef] [PubMed]
2. Pazzaglia, S.; Pioli, C. Multifaceted Role of PARP-1 in DNA Repair and Inflammation: Pathological and Therapeutic Implications in Cancer and Non-Cancer Diseases. *Cells* **2020**, *9*, 41. [CrossRef] [PubMed]
3. Kraus, W.L. Transcriptional Control by PARP-1: Chromatin Modulation, Enhancer-Binding, Coregulation, and Insulation. *Curr. Opin. Cell Biol.* **2008**, *20*, 294–302. [CrossRef] [PubMed]
4. Ray Chaudhuri, A.; Nussenzweig, A. The Multifaceted Roles of PARP1 in DNA Repair and Chromatin Remodelling. *Nat. Rev. Mol. Cell Biol.* **2017**, *18*, 610–621. [CrossRef] [PubMed]
5. Ashworth, A. A Synthetic Lethal Therapeutic Approach: Poly(ADP) Ribose Polymerase Inhibitors for the Treatment of Cancers Deficient in DNA Double-Strand Break Repair. *J. Clin. Oncol.* **2008**, *26*, 3785–3790. [CrossRef] [PubMed]
6. Murai, J.; Huang, S.-Y.N.; Renaud, A.; Zhang, Y.; Ji, J.; Takeda, S.; Morris, J.; Teicher, B.; Doroshow, J.H.; Pommier, Y. Stereospecific PARP Trapping by BMN 673 and Comparison with Olaparib and Rucaparib. *Mol. Cancer Ther.* **2014**, *13*, 433–443. [CrossRef]
7. Murai, J.; Huang, S.N.; Das, B.B.; Renaud, A.; Zhang, Y.; Doroshow, J.H.; Ji, J.; Takeda, S.; Pommier, Y. Trapping of PARP1 and PARP2 by Clinical PARP Inhibitors. *Cancer Res.* **2012**, *72*, 5588–5599. [CrossRef]
8. Chambers, V.S.; Marsico, G.; Boutell, J.M.; di Antonio, M.; Smith, G.P.; Balasubramanian, S. High-Throughput Sequencing of DNA G-Quadruplex Structures in the Human Genome. *Nat. Biotechnol.* **2015**, *33*, 877–881. [CrossRef]
9. Lipps, H.J.; Rhodes, D. G-Quadruplex Structures: In Vivo Evidence and Function. *Trends Cell Biol.* **2009**, *19*, 414–422. [CrossRef] [PubMed]
10. Rhodes, D.; Lipps, H.J. G-Quadruplexes and Their Regulatory Roles in Biology. *Nucleic Acids Res.* **2015**, *43*, 8627–8637. [CrossRef]

11. Eddy, J.; Maizels, N. Gene Function Correlates with Potential for G4 DNA Formation in the Human Genome. *Nucleic Acids Res.* **2006**, *34*, 3887–3896. [CrossRef]
12. Del Mundo, I.M.A.; Vasquez, K.M.; Wang, G. Modulation of DNA Structure Formation Using Small Molecules. *Biochim. et Biophys. Acta* **2019**, *1866*, 118539. [CrossRef] [PubMed]
13. Duarte, A.R.; Cadoni, E.; Ressurreição, A.S.; Moreira, R.; Paulo, A. Design of Modular G-Quadruplex Ligands. *ChemMedChem* **2018**, *13*, 869–893. [CrossRef] [PubMed]
14. Müller, S.; Sanders, D.A.; di Antonio, M.; Matsis, S.; Riou, J.-F.; Rodriguez, R.; Balasubramanian, S. Pyridostatin Analogues Promote Telomere Dysfunction and Long-Term Growth Inhibition in Human Cancer Cells. *Org. Biomol. Chem.* **2012**, *10*, 6537–6546. [CrossRef] [PubMed]
15. Maizels, N.; Gray, L.T. The G4 Genome. *PLOS Genet.* **2013**, *9*, e1003468. [CrossRef]
16. Gowan, S.M.; Heald, R.; Stevens, M.F.G.; Kelland, L.R. Potent Inhibition of Telomerase by Small-Molecule Pentacyclic Acridines Capable of Interacting with G-Quadruplexes. *Mol. Pharmacol.* **2001**, *60*, 981–988. [CrossRef] [PubMed]
17. Berardinelli, F.; Tanori, M.; Muoio, D.; Buccarelli, M.; di Masi, A.; Leone, S.; Ricci-Vitiani, L.; Pallini, R.; Mancuso, M.; Antoccia, A. G-Quadruplex Ligand RHPS4 Radiosensitizes Glioblastoma Xenograft in Vivo through a Differential Targeting of Bulky Differentiated- and Stem-Cancer Cells. *J. Exp. Clin. Cancer Res.* **2019**, *38*, 311. [CrossRef]
18. Zizza, P.; Cingolani, C.; Artuso, S.; Salvati, E.; Rizzo, A.; d'Angelo, C.; Porru, M.; Pagano, B.; Amato, J.; Randazzo, A.; et al. Intragenic G-Quadruplex Structure Formed in the Human CD133 and Its Biological and Translational Relevance. *Nucleic Acids Res.* **2016**, *44*, 1579–1590. [CrossRef] [PubMed]
19. Mulholland, K.; Siddiquei, F.; Wu, C. Binding Modes and Pathway of RHPS4 to Human Telomeric G-Quadruplex and Duplex DNA Probed by All-Atom Molecular Dynamics Simulations with Explicit Solvent. *Phys. Chem. Chem. Phys.* **2017**, *19*, 18685–18694. [CrossRef] [PubMed]
20. Santos, T.; Salgado, G.F.; Cabrita, E.J.; Cruz, C. G-Quadruplexes and Their Ligands: Biophysical Methods to Unravel G-Quadruplex/Ligand Interactions. *Pharmaceuticals* **2021**, *14*, 769. [CrossRef]
21. Gavathiotis, E.; Searle, M.S. Structure of the Parallel-Stranded DNA Quadruplex d(TTAGGGT)4 Containing the Human Telomeric Repeat: Evidence for A-Tetrad Formation from NMR and Molecular Dynamics Simulations. *Org. Biomol. Chem.* **2003**, *1*, 1650–1656. [CrossRef] [PubMed]
22. Lagah, S.; Tan, I.-L.; Radhakrishnan, P.; Hirst, R.A.; Ward, J.H.; O'Callaghan, C.; Smith, S.J.; Stevens, M.F.G.; Grundy, R.G.; Rahman, R. RHPS4 G-Quadruplex Ligand Induces Anti-Proliferative Effects in Brain Tumor Cells. *PLoS ONE* **2014**, *9*, e86187. [CrossRef] [PubMed]
23. Kantidze, O.L.; Luzhin, A.V.; Nizovtseva, E.V.; Safina, A.; Valieva, M.; Golov, A.; Velichko, A.K.; Lyubitelev, A.V.; Feofanov, A.V.; Gurova, K.V.; et al. The Anti-Cancer Drugs Curaxins Target Spatial Genome Organization. *Nat. Commun.* **2019**, *10*, 1441. [CrossRef] [PubMed]
24. Lu, K.; Liu, C.; Liu, Y.; Luo, A.; Chen, J.; Lei, Z.; Kong, J.; Xiao, X.; Zhang, S.; Wang, Y.-Z.; et al. Curaxin-Induced DNA Topology Alterations Trigger the Distinct Binding Response of CTCF and FACT at the Single-Molecule Level. *Biochemistry* **2021**, *60*, 494–499. [CrossRef] [PubMed]
25. Dallavalle, S.; Mattio, L.M.; Artali, R.; Musso, L.; Aviñó, A.; Fàbrega, C.; Eritja, R.; Gargallo, R.; Mazzini, S. Exploring the Interaction of Curaxin CBL0137 with G-Quadruplex DNA Oligomers. *Int. J. Mol. Sci.* **2021**, *22*, 6476. [CrossRef] [PubMed]
26. Dallavalle, S.; Musso, L.; Artali, R.; Aviñó, A.; Scaglioni, L.; Eritja, R.; Gargallo, R.; Mazzini, S. G-Quadruplex Binding Properties of a Potent PARP-1 Inhibitor Derived from 7-Azaindole-1-Carboxamide. *Sci. Rep.* **2021**, *11*, 3869. [CrossRef] [PubMed]
27. Sengar, A.; Vandana, J.J.; Chambers, V.S.; di Antonio, M.; Winnerdy, F.R.; Balasubramanian, S.; Phan, A.T. Structure of a (3 + 1) Hybrid G-Quadruplex in the PARP1 Promoter. *Nucleic Acids Res.* **2019**, *47*, 1564–1572. [CrossRef] [PubMed]
28. Dyson, R.M.; Kaderli, S.; Lawrance, G.A.; Maeder, M. Second Order Global Analysis: The Evaluation of Series of Spectrophotometric Titrations for Improved Determination of Equilibrium Constants. *Anal. Chim. Acta* **1997**, *353*, 381–393. [CrossRef]
29. Gargallo, R. Hard/Soft Hybrid Modeling of Temperature-Induced Unfolding Processes Involving G-Quadruplex and i-Motif Nucleic Acid Structures. *Anal. Biochem.* **2014**, *466*, 4–15. [CrossRef] [PubMed]
30. De la Hoz, A.; Navarro, A.; Aviñó, A.; Eritja, R.; Gargallo, R. Studies on the Interactions of Ag(i) with DNA and Their Implication on the DNA-Templated Synthesis of Silver Nanoclusters and on the Interaction with Complementary DNA and RNA Sequences. *RSC Adv.* **2021**, *11*, 9029–9042. [CrossRef]
31. Di Fonzo, S.; Amato, J.; D'Aria, F.; Caterino, M.; D'Amico, F.; Gessini, A.; Brady, J.W.; Cesàro, A.; Pagano, B.; Giancola, C. Ligand Binding to G-Quadruplex DNA: New Insights from Ultraviolet Resonance Raman Spectroscopy. *Phys. Chem. Chem. Phys.* **2020**, *22*, 8128–8140. [CrossRef] [PubMed]
32. Marchand, A.; Granzhan, A.; Iida, K.; Tsushima, Y.; Ma, Y.; Nagasawa, K.; Teulade-Fichou, M.-P.; Gabelica, V. Ligand-Induced Conformational Changes with Cation Ejection upon Binding to Human Telomeric DNA G-Quadruplexes. *J. Am. Chem Soc.* **2015**, *137*, 750–756. [CrossRef]
33. Rodriguez, R.; Müller, S.; Yeoman, J.A.; Trentesaux, C.; Riou, J.-F.; Balasubramanian, S. A Novel Small Molecule That Alters Shelterin Integrity and Triggers a DNA-Damage Response at Telomeres. *J. Am. Chem Soc.* **2008**, *130*, 15758–15759. [CrossRef] [PubMed]

34. Dallavalle, S.; Princiotto, S.; Mattio, L.M.; Artali, R.; Musso, L.; Aviñó, A.; Eritja, R.; Pisano, C.; Gargallo, R.; Mazzini, S. Investigation of the Complexes Formed between PARP1 Inhibitors and PARP1 G-Quadruplex at the Gene Promoter Region. *Int. J. Mol. Sci.* **2021**, *22*, 8737. [CrossRef] [PubMed]
35. Musso, L.; Mazzini, S.; Rossini, A.; Castagnoli, L.; Scaglioni, L.; Artali, R.; di Nicola, M.; Zunino, F.; Dallavalle, S. C-MYC G-Quadruplex Binding by the RNA Polymerase I Inhibitor BMH-21 and Analogues Revealed by a Combined NMR and Biochemical Approach. *Biochim. Biophys. Acta BBA* **2018**, *1862*, 615–629. [CrossRef] [PubMed]
36. Morris, G.M.; Goodsell, D.S.; Halliday, R.S.; Huey, R.; Hart, W.E.; Belew, R.K.; Olson, A.J. Automated Docking Using a Lamarckian Genetic Algorithm and an Empirical Binding Free Energy Function. *J. Comput. Chem.* **1998**, *19*, 1639–1662. [CrossRef]
37. Morris, G.M.; Huey, R.; Lindstrom, W.; Sanner, M.F.; Belew, R.K.; Goodsell, D.S.; Olson, A.J. AutoDock4 and AutoDockTools4: Automated Docking with Selective Receptor Flexibility. *J. Comput. Chem.* **2009**, *30*, 2785–2791. [CrossRef]
38. Sanner, M.F. Python: A Programming Language for Software Integration and Development. *J. Mol. Graph. Model* **1999**, *17*, 57–61.
39. Gasteiger, J.; Marsili, M. Iterative Partial Equalization of Orbital Electronegativity—A Rapid Access to Atomic Charges. *Tetrahedron* **1980**, *36*, 3219–3228. [CrossRef]
40. Galindo-Murillo, R.; Robertson, J.C.; Zgarbová, M.; Šponer, J.; Otyepka, M.; Jurečka, P.; Cheatham, T.E. Assessing the Current State of Amber Force Field Modifications for DNA. *J. Chem. Theory Comput.* **2016**, *12*, 4114–4127. [CrossRef] [PubMed]
41. Ivani, I.; Dans, P.D.; Noy, A.; Pérez, A.; Faustino, I.; Hospital, A.; Walther, J.; Andrio, P.; Goñi, R.; Balaceanu, A.; et al. Parmbsc1: A Refined Force Field for DNA Simulations. *Nat. Methods* **2016**, *13*, 55–58. [CrossRef]
42. Lamoureux, G.; Roux, B. Modeling Induced Polarization with Classical Drude Oscillators: Theory and Molecular Dynamics Simulation Algorithm. *J. Chem. Phys.* **2003**, *119*, 3025–3039. [CrossRef]
43. Jiang, W.; Hardy, D.J.; Phillips, J.C.; MacKerell, A.D.; Schulten, K.; Roux, B. High-Performance Scalable Molecular Dynamics Simulations of a Polarizable Force Field Based on Classical Drude Oscillators in NAMD. *J. Phys. Chem. Lett.* **2011**, *2*, 87–92. [CrossRef] [PubMed]
44. Phillips, J.C.; Hardy, D.J.; Maia, J.D.C.; Stone, J.E.; Ribeiro, J.V.; Bernardi, R.C.; Buch, R.; Fiorin, G.; Hénin, J.; Jiang, W.; et al. Scalable Molecular Dynamics on CPU and GPU Architectures with NAMD. *J. Chem. Phys.* **2020**, *153*, 044130. [CrossRef] [PubMed]
45. Ryckaert, J.-P.; Ciccotti, G.; Berendsen, H.J.C. Numerical Integration of the Cartesian Equations of Motion of a System with Constraints: Molecular Dynamics of n-Alkanes. *J. Comput. Phys.* **1977**, *23*, 327–341. [CrossRef]
46. Miyamoto, S.; Kollman, P.A. Settle: An Analytical Version of the SHAKE and RATTLE Algorithm for Rigid Water Models. *J. Comput. Chem.* **1992**, *13*, 952–962. [CrossRef]
47. Darden, T.; York, D.; Pedersen, L. Particle Mesh Ewald: An N·log(N) Method for Ewald Sums in Large Systems. *J. Chem. Phys.* **1993**, *98*, 10089–10092. [CrossRef]
48. Essmann, U.; Perera, L.; Berkowitz, M.L.; Darden, T.; Lee, H.; Pedersen, L.G. A Smooth Particle Mesh Ewald Method. *J. Chem. Phys.* **1995**, *103*, 8577–8593. [CrossRef]
49. Berendsen, H.J.C.; Postma, J.P.M.; van Gunsteren, W.F.; DiNola, A.; Haak, J.R. Molecular Dynamics with Coupling to an External Bath. *J. Chem. Phys.* **1984**, *81*, 3684–3690. [CrossRef]
50. Goddard, T.D.; Huang, C.C.; Meng, E.C.; Pettersen, E.F.; Couch, G.S.; Morris, J.H.; Ferrin, T.E. UCSF ChimeraX: Meeting Modern Challenges in Visualization and Analysis. *Protein Sci.* **2018**, *27*, 14–25. [CrossRef] [PubMed]

Article

Sustainable Protocol for the Synthesis of 2′,3′-Dideoxynucleoside and 2′,3′-Didehydro-2′,3′-dideoxynucleoside Derivatives

Virginia Martín-Nieves [1], Yogesh S. Sanghvi [2], Susana Fernández [1,*] and Miguel Ferrero [1,*]

[1] Departamento de Química Orgánica e Inorgánica, Universidad de Oviedo, 33006 Oviedo, Spain; martinvirginia@uniovi.es
[2] Rasayan Inc., Encinitas, CA 92024, USA; ysanghvi@rasayan.us
* Correspondence: fernandezgsusana@uniovi.es (S.F.); mferrero@uniovi.es (M.F.); Tel.: +34-985-102-984 (S.F.); +34-985-105-013 (M.F.)

Citation: Martín-Nieves, V.; Sanghvi, Y.S.; Fernández, S.; Ferrero, M. Sustainable Protocol for the Synthesis of 2′,3′-Dideoxynucleoside and 2′,3′-Didehydro-2′,3′-dideoxynucleoside Derivatives. *Molecules* 2022, 27, 3993. https://doi.org/10.3390/molecules27133993

Academic Editors: Aldo Galeone and Jussara Amato

Received: 2 June 2022
Accepted: 20 June 2022
Published: 21 June 2022

Publisher's Note: MDPI stays neutral with regard to jurisdictional claims in published maps and institutional affiliations.

Copyright: © 2022 by the authors. Licensee MDPI, Basel, Switzerland. This article is an open access article distributed under the terms and conditions of the Creative Commons Attribution (CC BY) license (https://creativecommons.org/licenses/by/4.0/).

Abstract: An improved protocol for the transformation of ribonucleosides into 2′,3′-dideoxynucleoside and 2′,3′-didehydro-2′,3′-dideoxynucleoside derivatives, including the anti-HIV drugs stavudine (d4T), zalcitabine (ddC) and didanosine (ddI), was established. The process involves radical deoxygenation of xanthate using environmentally friendly and low-cost reagents. Bromoethane or 3-bromopropanenitrile was the alkylating agent of choice to prepare the ribonucleoside 2′,3′-bisxanthates. In the subsequent radical deoxygenation reaction, tris(trimethylsilyl)silane and 1,1′-azobis(cyclohexanecarbonitrile) were used to replace hazardous Bu_3SnH and AIBN, respectively. In addition, TBAF was substituted for camphorsulfonic acid in the deprotection step of the 5′-*O*-silyl ether group, and an enzyme (adenosine deaminase) was used to transform 2′,3′-dideoxyadenosine into 2′,3′-dideoxyinosine (ddI) in excellent yield.

Keywords: 2′,3′-dideoxynucleosides; 2′,3′-didehydro-2′,3′-dideoxynucleosides; synthesis; zalcitabine (ddC); didanosine (ddI); stavudine (d4T)

1. Introduction

Emerging viruses continue to be a global threat to human health. During the past 25 years, human immunodeficiency virus (HIV), the cause of AIDS, reached virtually every corner of the globe, with 680,000 dying of HIV-related illnesses worldwide in 2020 [1]. More than two-thirds of people infected with HIV live in Asia and Africa. Despite substantial progress in the development of anti-HIV drugs, only 20% of low- and middle-income countries in need of these drugs are receiving them. Among the different anti-HIV chemotherapeutic agents known, the Nucleoside Reverse Transcriptase Inhibitors (NRTI, Figure 1) represent an important class.

Since Mitsuya et al. [2] identified 3′-azido-2′,3′-dideoxythymidine (zidovudine, AZT) as a potent antiviral agent against HIV-1, other nucleoside derivatives showing activity against this virus, such as ddI, ddC, d4T, 3TC, FTC, ABV and TDF, have been successfully developed [3,4]. Most of these compounds are 2′,3′-dideoxynucleosides or 2′,3′-didehydro-2′,3′-dideoxynucleosides and are characterized by lacking hydroxyl groups at the 2′- and 3′-positions.

Various methodologies are reported in the literature for the synthesis of the title compounds. These protocols require formation of the glycosidic bonds [5–11], the Eastwood procedure [12,13], the Corey–Winter synthesis [14–18], the Barton–McCombie deoxygenation [16,19–22], the Garegg–Samuelsson reaction [23], photoinduced deoxygenations [24,25], reductive elimination [13,26–35], or metathesis reaction [36,37]. However, careful review of the literature indicated that the majority of these protocols are not amenable for large-scale production to meet the global demand of antiviral nucleosides. Particularly, some of the methods described involve difficult control of diastereoselectivity in glycosidic bond formation, reagents that are expensive or not environmentally friendly, or partial nucleoside decomposition with loss of the pyrimidine base.

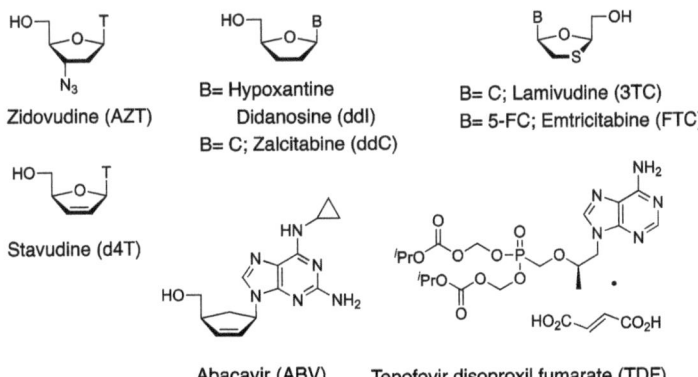

Figure 1. Several approved NRTIs against HIV.

Considering the ongoing challenge of HIV infections in underdeveloped countries, among NRTIs, ddI, ddC and d4T are the most affordable drugs for poor patient populations in Asia and Africa. Our objective is to develop improved protocols that are simple, inexpensive, safe and industrially benign for the large-scale syntheses of these three nucleoside derivatives and their analogs, with different heterocyclic bases. For that purpose, we develop a procedure that involves a Barton–McCombie deoxygenation and the use of commercial ribonucleosides as starting materials.

2. Results and Discussion

The selective removal of the hydroxyl groups at the 2′- and 3′-positions of the ribonucleoside requires appropriate protection of the 5′-OH group. Due to prior experience in our group [38,39], we decided to carry out the regioselective enzymatic acylation of the primary hydroxyl with acetonoxime levulinate as an acylating agent and *Candida antarctica* lipase B (CAL-B) as the catalyst. The reactions were performed in THF at 250 rpm, varying the number of equivalents of the acyl donor, the temperature and the substrate concentration, depending on the starting nucleoside (Scheme 1).

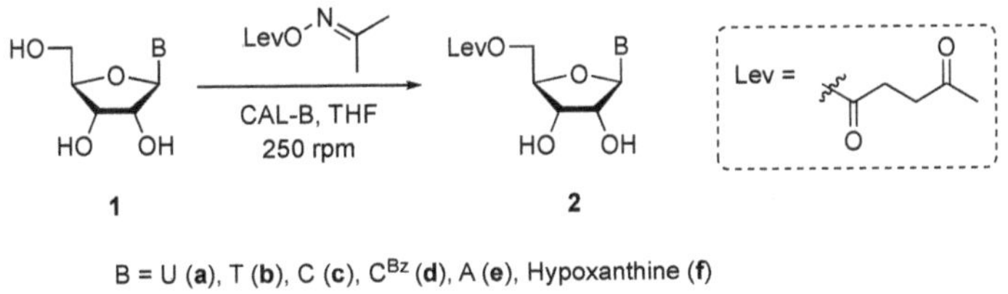

B = U (a), T (b), C (c), C^{Bz} (d), A (e), Hypoxanthine (f)

Scheme 1. Regioselective enzymatic acylation of 1.

Enzymatic acylation of β-D-uridine (**1a**) and β-D-5-methyluridine (**1b**) with 3 equiv of acetonoxime levulinate at 30 °C in the presence of CAL-B afforded the 5′-O-levulinyl esters **2a** and **2b** with excellent regioselectivity and high yields in short reaction times (entries 1 and 2, Table 1). However, the reaction with β-D-cytidine (**1c**) is slower, and complete conversion is not achieved, despite using long reaction times, 55 °C instead 30 °C, more dilute conditions, a large excess of acylating agent (9 vs. 3 equiv), and a higher ratio of **1c**:CAL-B, 1:2 (w/w). This resulted in the undesired acylation of the secondary hydroxyl group (entry 3, Table 1). The low reactivity was attributed to the poor solubility of the starting nucleoside in the reaction mixture. Next, the enzymatic acylation reaction of the base-protected cytidine was attempted. A complete conversion was observed when the same process was carried out with N^4-benzoyl-β-D-cytidine (**1d**), giving rise to the acylated derivative **2d**, with total selectivity and 93% yield (entry 4, Table 1). A moderate selectivity and absence of complete conversion was also observed when the substrate was adenosine (**1e**), which was attributed to the low solubility of this compound in the reaction medium (entry 5, Table 1). In the case of inosine (**1f**), 90 h of reaction time was needed to achieve complete conversion, and although the formation of other acylation products occurred in a low ratio (entry 6, Table 1), compound **2f** was obtained in low yield after column chromatography purification.

Table 1. Regioselective enzymatic acylation of ribonucleosides **1**.

Entry	Substrate	T (°C)	conc (M)	t (h)	1 (%) [a]	2 (%) [a,b]	Other Acylated Compounds (%) [a]
1	**1a** [c]	30	0.1	2	-	>97 (80)	-
2	**1b** [c]	30	0.1	2.5	-	>97 (78)	-
3	**1c** [d]	55	0.025	54	26	53 (50)	21
4	**1d** [d]	55	0.025	24	-	>97 (93)	-
5	**1e** [d]	55	0.025	48	10	70 (42)	20
6	**1f** [d]	55	0.025	90	-	87 (40)	13

[a] Based on ^{1}H-NMR signal integration. [b] Percentage of isolated yields are given in parenthesis. [c] 3 equiv of acetonoxime levulinate and ratio 1:CAL-B, 1:1 (w/w). [d] 9 equiv of acetonoxime levulinate and ratio 1:CAL-B, 1:2 (w/w).

Next, transformation of the 5′-O-levulinylribonucleoside **2a** into the corresponding bisxanthate was carried out by reaction with carbon disulfide followed by alkylation with bromoethane, a safer and cheaper reagent than other alkylating agents previously used, such as iodomethane or 3-bromopropanenitrile (Scheme 2) [16]. However, the desired bisxanthate **3a** was obtained in a low 25% yield because compound **4a**, resulting from the reaction at the primary hydroxyl, which was deprotected under the reaction conditions (NaOH 5 M), was formed as a by-product. Although different bases (inorganic: ^{t}BuOK, K_2CO_3; organic: DIPEA, DBU) were studied as alternatives, the appropriate conditions to carry out the reaction were not found, and the levulinyl group was not pursued as protecting group for the 5′-position.

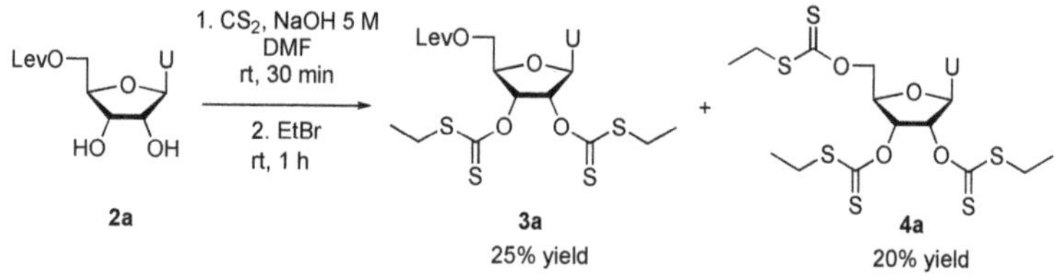

Scheme 2. Transformation of 5′-O-Lev-uridine into bisxanthates.

Therefore, we elected 5′-O-*tert*-butyldimethylsilyl (TBS) as the protecting group of choice due to low cost, high regioselectivity and stability during base treatment. Various ribonucleosides **1** were regioselectively protected at the primary hydroxyl as silyl ethers by treatment with TBSCl and imidazole in DMF for 12 h at room temperature (Scheme 3), furnishing the 5′-O-TBS protected nucleosides **5** in high to excellent yields (Table 2). TBS-protected nucleosides **5** were pure enough to carry forward into the next step without further purification by column chromatography. Next, the conversion of **5** to **6** was carefully optimized using the correct combination of the solvent, base, and reaction temperature. The ideal reaction condition calls for the reaction of **5** with CS_2 in the presence of 3 M aqueous NaOH solution and DMF as solvent for 30 min at 0 °C, and subsequent in situ alkylation with bromoethane for 20 min, affording bisxanthates **6a–f** in high yields. It is important to note that compounds **6** were isolated with suitable purity by thorough washing with heptane, avoiding chromatographic purification. We expect the two-step simple chromatography-free protocol for the synthesis of bisxanthates **6a–f** will be conducive for scale-up.

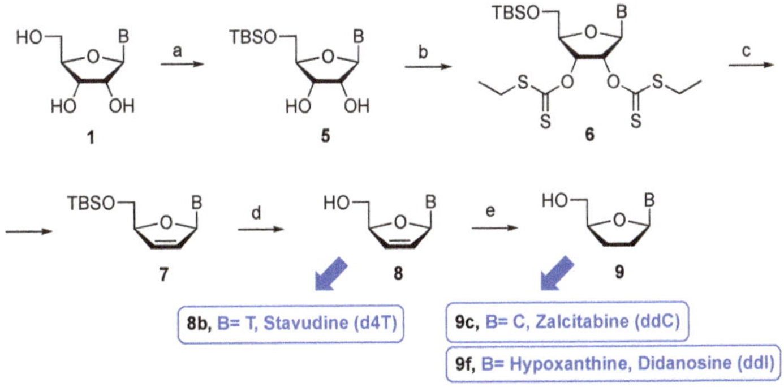

Scheme 3. Synthesis of 2′,3′-didehydro-2′,3′-dideoxynucleosides and 2′,3′-dideoxynucleosides. Synthesis of d4T, ddC and ddI. *Reagents and conditions*: (**a**) TBSCl, imidazole, DMF, rt, 12 h; (**b**) (1) CS_2, 3 M NaOH, DMF, 0 °C, 30 min; (2) EtBr, 0 °C → rt, 20 min; (**c**) Method A: Bu_3SnH, AIBN, MeCN, reflux, 1 h; Method B: $(Me_3Si)_3SiH$, ACHN, MeCN, reflux, 1 h (**7a,b,d,e**) or 6 h (**7f**); (**d**) Method A: TBAF, THF, 0 °C → rt, 1 h; Method B: (−)-CSA, MeOH, 0 °C → rt, 1 h; (**e**) H_2, 10% Pd-C, MeOH, rt, 2 h.

Table 2. Reaction yields of **5**, **6**, **7** and **9**.

B	1→5	5→6	6→7			7→8		8→9
			Bu_3SnH	$(Me_3Si)_3SiH$	TBAF	(−)-CSA		
a = U	93	82	60	65	95	92	82	
b = T	85	81	60	75	90	95	87	
c = C	91	75	35	ND	-	-	70 [a,b]	
d = C^{Bz}	80	72	60	40	90	ND	-	
e = A	85	90	60	77	95	ND	88	
f = Hypoxanthine	80	70	ND	80	75	ND	80	

[a] From **8d**. [b] See Scheme 4. ND, not the desired product. -, reaction not performed.

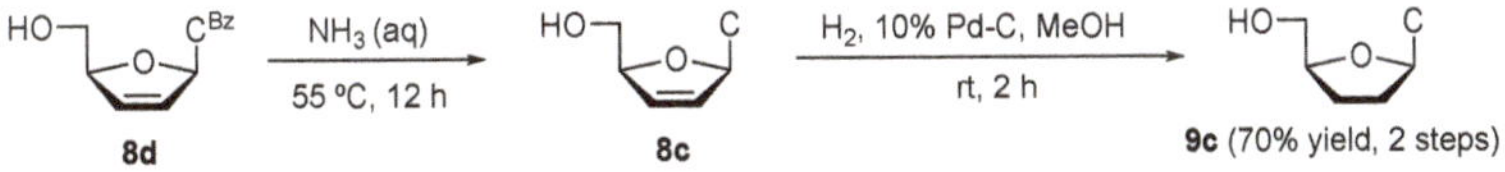

Scheme 4. Synthesis of Zalcitabine (ddC) from **8d**.

Next, we tested the reduction of Bisxanthates **6** using conventional conditions to ensure the formation of desired nucleosides **7**. Using tributyltin hydride (Bu_3SnH) and 2,2′-azobis(2-methylpropionitrile) (AIBN) in refluxing acetonitrile furnished **7a,b,d,e** in moderate yield (60%) and **7c** in low yield (35%) (Table 2). Interestingly, conversion of the hypoxanthine derivative **6f** resulted in a mixture of products difficult to separate and identify. Next, we sought to find a replacement for the traditional reducing agent Bu_3SnH, which is toxic, expensive and difficult to remove from the reaction mixture. We elected to use tris(trimethylsilyl)silane [$(Me_3Si)_3SiH$] [40,41] as a greener, non-toxic reagent for reduction. We also replaced hazardous AIBN with a safer radical initiator 1,1′-azobis(cyclohexanecarbonitrile) (ACHN), which has a longer half-life than AIBN. Under optimized reaction conditions, reduction of bisxanthates **6** with green reagents afforded improved yields for uracil, thymine and adenine derivatives furnishing **7a**, **7b** and **7e** in 65%, 75% and 77% yield, respectively. In the case of cytosine, better conversion was observed with the N-protected derivative. It is important to note that reaction of the hypoxanthine derivative **6f** with (Me$_3$Si)$_3$SiH and ACHN allowed the synthesis of 2′,3′-didehydro-2′,3′-dideoxynucleoside **7f** in 80% yield, while its synthesis with Bu$_3$SnH was not possible. Thus, the combination of [(Me$_3$Si)$_3$SiH] and ACHN represents a considerable improvement in the scalable green synthetic strategy proposed for the synthesis of these nucleoside analogs.

Compounds **7** were desilylated with tetrabutylammonium fluoride (TBAF) at room temperature to offer the 2′,3′-didehydro-2′,3′-dideoxynucleosides **8** in excellent yields. Nucleoside **8b** is the antiretroviral drug stavudine (d4T), establishing an efficient route of synthesis. The use of TBAF for deprotection of nucleosides during the final step results in trace contamination of the reagent. Therefore, we searched for an alternative TBS deprotection reagent that is easily removed. Camphorsulfonic acid [(−)-CSA] [42,43] emerged as a reagent of choice; it is an acid derived from camphor that has low sensitivity to air, is compatible with water, and is environmentally friendly. Treatment of **7** with (−)-CSA in MeOH leads to the 2′,3′-didehydro-2′,3′-dideoxynucleosides of uracil and thymine **8a** and **8b** with 92% and 95% yield, respectively. However, this protocol is not suitable for purine derivatives due to the cleavage of the glycosidic bond in the acidic reaction medium. Other TBS deprotection methods using povidone-iodine (PVP-1) [44] or phosphomolybdic acid [45] were not successful.

Hydrogenation of 2′,3′-didehydro-2′,3′-dideoxynucleosides **8** using palladium on carbon in methanol at room temperature afford the corresponding 2′,3′-dideoxynucleosides **9a,b,e,f** in high yields. The reaction of the N^4-benzoylcytidine derivative **8d** was carried out under similar conditions, but it resulted in the formation of a mixture of products. Therefore, we opted to reverse the sequence of the reactions, first carrying out the N-benzoyl deprotection by treating **8d** with an aqueous ammonia solution at 55 °C and then performing hydrogenation under the same conditions, isolating the drug zalcitabine (**9c**) with a 70% yield (Scheme 4).

Additionally, the drug didanosine (**9f**) was obtained via enzymatic deamination of adenosine analogue **9e** (Scheme 5) [46]. Treatment of **9e** with adenosine deaminase (ADA) in a 0.10 M phosphate buffer (pH 7) and 3% DMSO provides the 2′,3′-dideoxynucleoside **9f** in an almost quantitative yield (95%) after 3 h of reaction.

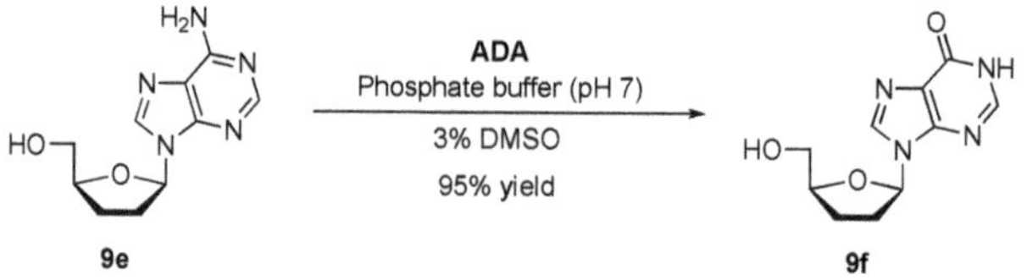

Scheme 5. Synthesis of Didanosine (ddI) through enzymatic deamination of **9e**.

The structure of the synthesized compounds was determined by NMR spectroscopy. The signals of the ^{1}H and ^{13}C-NMR spectra of the nucleoside derivatives are fully assigned on the basis of ^{1}H and ^{13}C chemical shifts, proton coupling constants, and two-dimensional ^{1}H-^{1}H (COSY) and ^{1}H-^{13}C spectra (HSQC and HMBC). As an illustrating example, the identification of zalcitabine (**9c**) was performed as follows. The protons H1′, H4′, H5′, H5 and H6 are assigned by ^{1}H-NMR. Subsequent analysis of the ^{1}H-^{13}C HSQC experiment leads to identification of the corresponding carbons. Several multiplets at 1.6–2.5 ppm in the ^{1}H-NMR spectrum are assigned, but not identified, to the hydrogens H2′ and H3′. In addition, the signals at 24.7 and 31.9 ppm in the ^{13}C-NMR spectrum are assigned to C2′ and C3′. A correlation cross-peak in the ^{1}H-^{13}C HMBC experiment between the H5′ protons and the carbon at 24.7 ppm allows the assignment of C3′. This has been corroborated by a correlation cross-peak between H1′ and C3′. Further analysis of the ^{1}H-^{13}C HSQC experiment leads to unambiguously identification of H2 and H3. Finally, a correlation cross-peak between H1' and the signal at 157.3 ppm in the ^{1}H-^{13}C HMBC experiment allows the assignment of C2, being the signal of the ^{13}C-NMR spectrum at 165.9 ppm, which does not appear in the DEPT-135 experiment, identified as C4. The COSY experiment validates the assignment made. It is worth mentioning the three-bond correlation of H1' with the two hydrogens H2', but not with H3', as well as the three-bond correlation of H4' with the two hydrogens of H3', but not with H2'.

3. Materials and Methods

3.1. General

All chemical reagents were purchased from Aldrich, Sigma, Merck, Acros or Alfa Aesar, and used without further purification. Thin-layer chromatography (TLC) was carried out on aluminum-backed Silica-Gel 60 F_{254} plates. The spots were visualized with UV light. Column chromatography was performed using Silica Gel (60 Å, 230 × 400 mesh).

Candida antarctica lipase type B (CAL-B, Novozyme 435, immobilized by adsorption in Lewatit, 9120 PLU/g) was purchased from Novozymes. Adenosine deaminase (ADA, 2–5 units/mg, intestinal bovine source, lyophilized) was purchased from Creative Enzymes.

NMR spectra were measured on Bruker DPX-300 (^{1}H 300.13 MHz and ^{13}C 75.5 MHz). High resolution mass spectra (HRMS) were recorded on a Bruker MicrOTOF-Q mass spectrometer under electron spray ionization (ESI). Melting points were recorded on a Gallemkamp apparatus with samples in open capillary tubes. Full analytical data for new compounds are available in the Supporting Information.

The structure of the synthesized compounds was determined by NMR spectroscopy. The signals of the ^{1}H and ^{13}C-NMR spectra are fully assigned on the basis of ^{1}H and ^{13}C chemical shifts, proton coupling constants, and two-dimensional ^{1}H-^{1}H (COSY) and ^{1}H-^{13}C spectra (HSQC and HMBC). Full NMR data are available in the Supporting Information. The level of purity is indicated by the inclusion of copies of ^{1}H, ^{13}C, DEPT and 2D NMR spectra.

3.2. General Procedure for Enzymatic Acylation of 1 Synthesis of 2

Anhydrous THF was added to an Erlenmeyer flask containing ribonucleoside **1** (0.2 mmol), acetonoxime levulinate and CAL-B (acylating agent equiv, enzyme ratio, concentration, temperature, and reaction time are indicated in Table 1) under nitrogen. The reaction was stirred at 250 rpm and followed by TLC (10% MeOH/CH$_2$Cl$_2$). Next, the enzyme was filtered and washed with CH$_2$Cl$_2$ and MeOH, and the solvents were removed under reduced pressure. The reaction crude was purified by column chromatography (gradient eluent: 2–5% MeOH/CH$_2$Cl$_2$), obtaining the corresponding acylated ribonucleosides **2a–f** (yields are indicated in Table 1).

5'-*O*-Levulinyl-β-D-uridine (**2a**). White solid, mp: 60–62 °C. R_f: 0.32 (10% MeOH/CH$_2$Cl$_2$). HRMS (ESI$^+$, *m/z*): Calcd. for C$_{14}$H$_{19}$N$_2$O$_8$ [M + H]$^+$: 343.1136. Found: 343.1131.

5'-*O*-Levulinyl-β-D-5-methyluridine (**2b**). White solid, mp: 134–136 °C. R_f: 0.33 (10% MeOH/CH$_2$Cl$_2$). HRMS (ESI$^+$, *m/z*): Calcd. for C$_{15}$H$_{21}$N$_2$O$_8$ [M + H]$^+$: 357.1292. Found: 357.1279.

5'-*O*-Levulinyl-β-D-cytidine (**2c**). White solid, mp: 53–55 °C. R_f: 0.35 (20% MeOH/CH$_2$Cl$_2$). HRMS (ESI$^+$, *m/z*): Calcd. for C$_{14}$H$_{20}$N$_3$O$_7$ [M + H]$^+$: 342.1296. Found: 342.1295.

N^4-Benzoyl-5'-*O*-levulinyl-β-D-cytidine (**2d**). White solid, mp: 193–195 °C. R_f: 0.47 (10% MeOH/CH$_2$Cl$_2$). HRMS (ESI$^+$, *m/z*): Calcd. for C$_{21}$H$_{24}$N$_3$O$_8$ [M + H]$^+$: 446.1563. Found: 446.1564.

5'-*O*-Levulinyl-β-D-adenosine (**2e**). White solid, mp: 116 °C (decompose). R_f: 0.26 (10% MeOH/CH$_2$Cl$_2$). HRMS (ESI$^+$, *m/z*): Calcd. for C$_{15}$H$_{20}$N$_5$O$_6$ [M + H]$^+$: 366.1408. Found: 366.1406.

5'-*O*-Levulinyl-β-D-inosine (**2f**). White solid, mp: 54–56 °C. R_f: 0.44 (20% MeOH/CH$_2$Cl$_2$). HRMS (ESI$^+$, *m/z*): Calcd. for C$_{15}$H$_{19}$N$_4$O$_7$ [M + H]$^+$: 367.1248. Found: 367.1253.

3.3. Synthesis of 5

To a solution of ribonucleoside **1** (0.4 M for **1a,b** and 0.2 M for **1c–f**) in anhydrous DMF were added imidazole (2.4 equiv) and TBSCl (1.2 equiv). The mixture was stirred at rt for 12 h. Then, the residue was poured into EtOAc and washed with water. The organic phase was dried, filtered and evaporated under reduced pressure. Compounds **5** were obtained with sufficient purity for the next step and the following yields: 93% for **5a**, 85% for **5b**, 91% for **5c**, 80% for **5d**, 85% for **5e** and 80% for **5f**. If desired, a chromatographic column could be performed (gradient eluent: 5–10% MeOH/CH_2Cl_2).

5′-O-(*tert*-Butyldimethylsilyl)-β-D-uridine (**5a**). White solid, mp: 94–96 °C. R_f: 0.41 (10% MeOH/CH_2Cl_2). HRMS (ESI$^+$, m/z): Calcd. for $C_{15}H_{27}N_2O_6Si$ [M + H]$^+$: 359.16329. Found: 359.16332.

5′-O-(*tert*-Butyldimethylsilyl)-β-D-5-methyluridine (**5b**). White solid, mp: 197–198 °C. R_f: 0.35 (10% MeOH/CH_2Cl_2). HRMS (ESI$^+$, m/z): Calcd. for $C_{16}H_{29}N_2O_6Si$ [M + H]$^+$: 373.1789. Found: 373.1790.

5′-O-(*tert*-Butyldimethylsilyl)-β-D-cytidine (**5c**). Colorless foam. R_f: 0.52 (10% MeOH/CH_2Cl_2). HRMS (ESI$^+$, m/z): Calcd. for $C_{15}H_{28}N_3O_5Si$ [M + H]$^+$: 358.1798. Found: 358.1791.

N^4-Benzoyl-5′-O-(*tert*-butyldimethylsilyl)-β-D-cytidine (**5d**). White solid, mp: 86–88 °C. R_f: 0.47 (10% MeOH/CH_2Cl_2). HRMS (ESI$^+$, m/z): Calcd. for $C_{22}H_{31}N_3O_6Si$ [M + H]$^+$: 462.2055. Found: 462.2048.

5′-O-(*tert*-Butyldimethylsilyl)-β-D-adenosine (**5e**). White solid, mp: 178–179 °C. R_f: 0.33 (10% MeOH/CH_2Cl_2). HRMS (ESI$^+$, m/z): Calcd. for $C_{16}H_{28}N_5O_4Si$ [M + H]$^+$: 382.1911. Found: 382.1902.

5′-O-(*tert*-Butyldimethylsilyl)-β-D-inosine (**5f**). White solid, mp: 229–230 °C. R_f: 0.17 (10% MeOH/CH_2Cl_2). HRMS (ESI$^+$, m/z): Calcd. for $C_{16}H_{27}N_4O_5Si$ [M + H]$^+$: 383.1745. Found: 383.1743.

3.4. Synthesis of 6

To a solution of 5′-O-silyl protected ribonucleosides **5** and CS_2 (7 equiv) in DMF (0.4 M) at 0 °C, an aqueous 3 M NaOH solution (3 equiv) was added dropwise. After being stirred for 30 min at this temperature, bromoethane (15 equiv) was added dropwise, and stirring continued for 20 min at rt. Then, the residue was poured into EtOAc and washed with water. The organic phase was dried, filtered, and evaporated under reduced pressure. The resulting solid was thoroughly washed with heptane to afford compounds **6** with suitable purity, avoiding chromatographic purification. Yields: 82% for **6a**, 81% for **6b**, 75% for **6c**, 72% for **6d**, 90% for **6e** and 70% for **6f**.

5′-O-(*tert*-Butyldimethylsilyl)-2′,3′-bis-O-[(ethylthio)thiocarbonyl]-β-D-uridine (**6a**). White solid, mp: 102–104 °C. R_f: 0.45 (40% EtOAc/hexane). HRMS (ESI$^+$, m/z): Calcd. for $C_{21}H_{35}N_2O_6S_4Si$ [M + H]$^+$: 567.1142. Found: 567.1133.

5′-O-(*tert*-Butyldimethylsilyl)-2′,3′-bis-O-[(ethylthio)thiocarbonyl]-β-D-5-methyluridine (**6b**). White solid, mp: 131–132 °C. R_f: 0.50 (40% EtOAc/hexane). HRMS (ESI$^+$, m/z): Calcd. for $C_{22}H_{37}N_2O_6S_4Si$ [M + H]$^+$: 581.1298. Found: 581.1292.

5′-O-(*tert*-Butyldimethylsilyl)-2′,3′-bis-O-[(ethylthio)thiocarbonyl]-β-D-cytidine (**6c**). White solid, mp: 99–101 °C. R_f: 0.35 (10% MeOH/CH_2Cl_2). HRMS (ESI$^+$, m/z): Calcd. for $C_{21}H_{36}N_3O_5S_4Si$ [M + H]$^+$: 566.1302. Found: 566.1295.

N^4-Benzoyl-5′-O-(*tert*-butyldimethylsilyl)-2′,3′-bis-O-[(ethylthio)thiocarbonyl]-β-D-cytidine (**6d**). White solid, mp: 140–141 °C. R_f: 0.25 (40% EtOAc/hexane). HRMS (ESI$^+$, m/z): Calcd. for $C_{28}H_{40}N_3O_6S_4Si$ [M + H]$^+$: 670.1564. Found: 670.1558.

5′-O-(*tert*-Butyldimethylsilyl)-2′,3′-bis-O-[(ethylthio)thiocarbonyl]-β-D-adenosine (**6e**). White solid, mp: 164–165 °C. R_f: 0.18 (50% EtOAc/hexane). HRMS (ESI$^+$, m/z): Calcd. for $C_{22}H_{36}N_5O_4S_4Si$ [M + H]$^+$: 590.1414. Found: 590.1409.

5′-O-(*tert*-Butyldimethylsilyl)-2′,3′-bis-O-[(ethylthio)thiocarbonyl]-β-D-inosine (**6f**). White solid, mp: 201–203 °C. R_f: 0.36 (10% MeOH/CH_2Cl_2). HRMS (ESI$^+$, m/z): Calcd. for $C_{22}H_{36}N_4O_5S_4Si$ [M + H]$^+$: 591.1254. Found: 591.1239.

3.5. Synthesis of 7

3.5.1. Method A: Bu$_3$SnH

To a solution of **6** in anhydrous MeCN (0.13 M) at reflux was added dropwise a solution of Bu$_3$SnH (4 equiv) and AIBN (0.4 equiv) in anhydrous MeCN (0.5 M). After being stirred for 1 h at this temperature, the solvent was removed under vacuum, and the residue was purified by column chromatography (gradient eluents: 40–50% EtOAc/hexane for **7a,b**; 70% EtOAc/hexane-EtOAc for **7d**; 2–5% MeOH/CH$_2$Cl$_2$ for **7c,e**) to afford **7a**, **7b**, **7d** and **7e** in 60% yield and **7c** in 35% yield.

3.5.2. Method B: (Me$_3$Si)$_3$SiH

To a solution of **6** in anhydrous MeCN (0.13 M) at reflux, was added dropwise a solution of (Me$_3$Si)$_3$SiH (4 equiv) and 1,1'-azobis(cyclohexanecarbonitrile) (0.4 equiv) in anhydrous MeCN (0.5 M). The mixture was stirred for 1 h (**6a–e**) or 6 h (**6f**) at this temperature. Next, the solvent was removed under vacuum, and the residue was purified by column chromatography (gradient eluents: 40–50% EtOAc/hexane for **7a,b**; 70% EtOAc/hexane-EtOAc for **7d**; 2–5% MeOH/CH$_2$Cl$_2$ for **7e,f**) to afford **7** (65% for **7a**, 75% for **7b**, 40% for **7d**, 77% for **7e** and 80% yield for **7f**).

5'-O-(*tert*-Butyldimethylsilyl)-2',3'-didehydro-2',3'-dideoxy-β-D-uridine (**7a**). White solid, mp: 166–168 °C. R_f: 0.16 (40% EtOAc/hexane). HRMS (ESI$^+$, *m/z*): Calcd. for C$_{15}$H$_{25}$N$_2$O$_4$Si [M + H]$^+$: 325.1578. Found: 325.1573.

5'-O-(*tert*-Butyldimethylsilyl)-2',3'-didehydro-3'-deoxy-β-D-5-thymidine (**7b**). White solid, mp: 169–171 °C. R_f: 0.29 (40% EtOAc/hexane). HRMS (ESI$^+$, *m/z*): Calcd. for C$_{16}$H$_{27}$N$_2$O$_4$Si [M + H]$^+$: 339.1735. Found: 339.1729.

5'-O-(*tert*-Butyldimethylsilyl)-2',3'-didehydro-2',3'-dideoxy-β-D-cytidine (**7c**). White solid, mp: 176–178 °C. R_f: 0.52 (10% MeOH/CH$_2$Cl$_2$). HRMS (ESI$^+$, *m/z*): Calcd. for C$_{15}$H$_{26}$N$_3$O$_3$Si [M + H]$^+$: 324.1738. Found: 324.1743.

N^4-Benzoyl-5'-O-(*tert*-butyldimethylsilyl)-2',3'-didehydro-2',3'-dideoxy-β-D-cytidine (**7d**). White solid, mp: 137–138 °C. R_f: 0.19 (40% EtOAc/hexane). HRMS (ESI$^+$, *m/z*): Calcd. for C$_{22}$H$_{30}$N$_3$O$_4$Si [M + H]$^+$: 428.2000. Found: 428.1993.

5'-O-(*tert*-Butyldimethylsilyl)-2',3'-didehydro-2',3'-dideoxy-β-D-adenosine (**7e**). White solid, mp: 118–120 °C. R_f: 0.48 (10% MeOH/CH$_2$Cl$_2$). HRMS (ESI$^+$, *m/z*): Calcd. for C$_{16}$H$_{26}$N$_5$O$_2$Si [M + H]$^+$: 348.1850. Found: 348.1848.

5'-O-(*tert*-Butyldimethylsilyl)-2',3'-didehydro-2',3'-dideoxy-β-D-inosine (**7f**). White solid, mp: 178–180 °C. R_f: 0.36 (10% MeOH/CH$_2$Cl$_2$). HRMS (ESI$^+$, *m/z*): Calcd. for C$_{16}$H$_{25}$N$_4$O$_3$Si [M + H]$^+$: 349.1690. Found: 349.1690.

3.6. Synthesis of 8

3.6.1. Method A: TBAF

TBAF (2 equiv, 1.0 M in THF) was added dropwise to a stirred solution of **7** (1 equiv) in anhydrous THF (0.1 M) at 0 °C. After 5 min, the ice bath was removed, and the reaction mixture was stirred at rt for 1 h. Next, the solvent was removed under vacuum, and the residue was purified by column chromatography (5% MeOH/CH$_2$Cl$_2$ for **8a,b,d**; 15% MeOH/CH$_2$Cl$_2$ for **8e,f**) to afford **8a,e** in 95%, **8b,d** in 90%, and **8f** in 75% yields.

3.6.2. Method B: (−)-CSA

(−)-CSA (1 equiv) was added to a solution of **7** in anhydrous MeOH (0.1 M) at 0 °C, and the reaction was stirred at rt for 1 h. Solid NaHCO$_3$ was then added, and the mixture was stirred for a further 5 min. Next, the solvent was removed under vacuum, and the residue was purified by column chromatography (5% MeOH/CH$_2$Cl$_2$) to afford **8a** in 92% and **8b** in 95% yields.

2',3'-Didehydro-2',3'-dideoxy-β-D-uridine (**8a**). White solid, mp: 154–155 °C. R_f: 0.40 (10% MeOH/CH$_2$Cl$_2$). HRMS (ESI$^+$, *m/z*): Calcd. for C$_9$H$_{10}$N$_2$NaO$_4$ [M + Na]$^+$: 233.0533. Found: 233.0537.

2′,3′-Didehydro-3′-deoxy-β-D-5-thymidine (**8b**). White solid, mp: 165–166 °C. R_f: 0.42 (10% MeOH/CH$_2$Cl$_2$). HRMS (ESI$^+$, m/z): Calcd. for C$_{10}$H$_{13}$N$_2$O$_4$ [M + H]$^+$: 225.0870. Found: 225.0873.

N^4-Benzoyl-2′,3′-didehydro-2′,3′-dideoxy-β-D-cytidine (**8d**). White solid, mp: 280 °C (decompose). R_f: 0.66 (10% MeOH/CH$_2$Cl$_2$). HRMS (ESI$^+$, m/z): Calcd. for C$_{16}$H$_{16}$N$_3$O$_4$ [M + H]$^+$: 314.1135. Found: 314.1140.

2′,3′-Didehydro-2′,3′-dideoxy-β-D-adenosine (**8e**). White solid, mp: 185–186 °C. R_f: 0.24 (10% MeOH/CH$_2$Cl$_2$). HRMS (ESI$^+$, m/z): Calcd. for C$_{10}$H$_{12}$N$_5$O$_2$ [M + H]$^+$: 234.0986. Found: 234.0984.

2′,3′-Didehydro-2′,3′-dideoxy-β-D-inosine (**8f**). White solid, mp: >300 °C. R_f: 0.19 (10% MeOH/CH$_2$Cl$_2$). HRMS (ESI$^+$, m/z): Calcd. for C$_{10}$H$_{11}$N$_4$O$_3$ [M + H]$^+$: 235.0826. Found: 235.0826.

3.7. Synthesis of 9

A flask containing **8** and 10% Pd/C (0.1 equiv) was exposed to a positive pressure of hydrogen gas (balloon). Anhydrous MeOH (0.02M) was added, and the mixture was stirred vigorously for 2 h under a hydrogen atmosphere. The suspension was filtered on Celite®® and washed with MeOH, and the solvent was removed under vacuum. The crude was purified by column chromatography (gradient eluent 2–10% MeOH/CH$_2$Cl$_2$) to afford **9a** in 82%, **9b** in 87%, **9e** in 88%, and **9f** in 80% yields.

2′,3′-Dideoxy-β-D-uridine (**9a**). White solid, mp: 116–117 °C. R_f: 0.42 (10% MeOH/CH$_2$Cl$_2$). HRMS (ESI$^+$, m/z): Calcd. for C$_9$H$_{13}$N$_2$O$_4$ [M + H]$^+$: 213.0870. Found: 213.0875.

3′-Deoxy-β-D-5-thymidine (**9b**). White solid, mp: 155–156 °C. R_f: 0.44 (10% MeOH/CH$_2$Cl$_2$). HRMS (ESI$^+$, m/z): Calcd. for C$_{10}$H$_{15}$N$_2$O$_4$ [M + H]$^+$: 227.1026. Found: 227.1032.

2′,3′-Dideoxy-β-D-adenosine (**9e**). White solid, mp: 186–188 °C. R_f: 0.33 (10% MeOH/CH$_2$Cl$_2$). HRMS (ESI$^+$, m/z): Calcd. for C$_{10}$H$_{14}$N$_5$O$_2$ [M + H]$^+$: 236.1142. Found: 236.1148.

2′,3′-Dideoxy-β-D-inosine (**9f**). White solid, mp: 160–163 °C. R_f: 0.28 (10% MeOH/CH$_2$Cl$_2$). HRMS (ESI$^+$, m/z): Calcd. for C$_{10}$H$_{13}$N$_4$O$_3$ [M + H]$^+$: 237.0982. Found: 237.0989.

3.7.1. Synthesis of Zalcitabine (9c)

A suspension of **8d** (50 mg, 0.16 mmol) in an aqueous 32% NH$_3$ solution (2.5 mL) was stirred at 55 °C for 12 h. The solvent was removed under vacuum. Then, a mixture of the resulting residue and 10% Pd/C (17 mg) was exposed to a positive pressure of hydrogen gas (balloon). Anhydrous MeOH (8 mL) was added, and the mixture was stirred vigorously for 2 h under a hydrogen atmosphere. The suspension was filtered on Celite®® and washed with MeOH, and the solvent was removed under vacuum. The crude was purified by column chromatography (20% MeOH/CH$_2$Cl$_2$) previously packed with silica gel using a 10% Et$_3$N solution in MeOH:CH$_2$Cl$_2$ (2:8, v:v). Compound **9c** was isolated in 70% yield.

2′,3′-Dideoxy-β-D-cytidine (**9c**). White solid, mp: 208–210 °C. R_f: 0.27 (20% MeOH/ CH$_2$Cl$_2$). HRMS (ESI$^+$, m/z): Calcd. for C$_9$H$_{14}$N$_3$O$_3$ [M + H]$^+$: 212.1030. Found: 212.1034.

3.7.2. Synthesis of Didanosine (9f)

To a suspension of **9e** (40 mg, 0.17 mmol) in a phosphate buffer solution pH 7 (0.8 mL) and 3% of DMSO, to promote dissolution, 2 mg of adenosine deaminase dissolved in the same buffer (0.2 mL) was added. The reaction was stirred at 250 rpm and 30 °C for 3 h. The crude was purified by column chromatography (10% MeOH/CH$_2$Cl$_2$) to afford **9f** in 95% yield.

4. Conclusions

We report an economical and green synthesis of 2′,3′-dideoxynucleoside and 2′,3′-didehydro-2′,3′-dideoxynucleoside derivatives of uracil, thymine, cytosine, adenine and hypoxanthine through deoxygenation of the corresponding 2′,3′-O-bisxanthate ribonucleosides. This protocol involves the use of tris(trimethylsilyl)silane [(Me$_3$Si)$_3$SiH] instead of Bu$_3$SnH, which is toxic, expensive and difficult to remove from the reaction mixture, as a radical-based reducing agent. We also replaced potentially explosive AIBN with 1,1′-azobis(cyclohexanecarbonitrile) (ACHN) as a safer alternative. In addition, for the deprotection of silyl ethers at the 5′-position of the nucleosides, we were able to substitute TBAF for camphorsulfonic acid as a more sustainable reagent, in pyrimidine derivatives. The use of $(Me_3Si)_3SiH$ in the deoxygenation of bisxanthate hypoxanthine derivative allows easy access to 2′,3′-didehydro-2′,3′-dideoxyinosine, an antiviral agent. As an alternative synthesis, this nucleoside was also obtained in excellent yield via enzymatic deamination of 2′,3′-dideoxyadenosine with adenosine deaminase. It is important to emphasize that these protocols may have potential industrial application for the synthesis of three of the most demanding anti-HIV drugs: stavudine (d4T), zalcitabine (ddC) and didanosine (ddI).

Supplementary Materials: The following supporting information can be downloaded at: https://www.mdpi.com/article/10.3390/molecules27133993/s1, ^{1}H and ^{13}C-NMR data with their assignment for all compounds. Level of purity is indicated by the inclusion of copies of ^{1}H, ^{13}C, and DEPT NMR spectra; in addition, some 2D NMR experiments are shown, which were used to assign the peaks.

Author Contributions: Conceptualization, M.F., S.F. and Y.S.S.; methodology, M.F., S.F. and Y.S.S.; validation, M.F. and S.F.; investigation, V.M.-N.; resources, M.F., S.F. and Y.S.S.; data curation, M.F. and S.F.; writing—original draft preparation, M.F., S.F., Y.S.S. and V.M.-N.; writing—review and editing, M.F., S.F. and Y.S.S.; visualization, M.F. and V.M.-N.; supervision, M.F., S.F. and Y.S.S.; project administration, M.F. and S.F.; funding acquisition, M.F. All authors have read and agreed to the published version of the manuscript.

Funding: This research was funded by Principado de Asturias, project number SV-PA-21-AYUD-2021-51542, and the APC was waived.

Institutional Review Board Statement: Not applicable.

Informed Consent Statement: Not applicable.

Data Availability Statement: Not applicable.

Acknowledgments: V.M.-N. thanks FICYT (Asturias) for a predoctoral Severo Ochoa fellowship.

Conflicts of Interest: The authors declare no conflict of interest. The funders had no role in the design of the study; in the collection, analyses, or interpretation of data; in the writing of the manuscript, or in the decision to publish the results.

Sample Availability: Not applicable.

References

1. World Health Organization. Number of Deaths due to HIV/AIDS. Available online: https://www.who.int/data/gho/data/indicators/indicator-details/GHO/number-of-deaths-due-to-hiv-aids (accessed on 19 June 2022).
2. Mitsuya, H.; Weinhold, K.; Furman, P.A.; St. Clair, M.H.; Lehrman, S.N.; Gallo, R.C.; Bolognesi, D.; Barry, D.W.; Broder, S. 3′-Azido-3′-deoxythymidine (BW A509U): An antiviral agent that inhibits the infectivity and cytopathic effect of human T-lymphotropic virus type III/lymphadenopathy-associated virus in vitro. *Proc. Natl. Acad. Sci. USA* **1985**, *82*, 7096–7100. [CrossRef] [PubMed]
3. Yates, M.K.; Seley-Radtke, K.L. The evolution of antiviral nucleoside analogues: A review for chemists and non-chemists. Part II: Complex modifications to the nucleoside scaffold. *Antivir. Res.* **2019**, *162*, 5–21. [CrossRef] [PubMed]
4. Seley-Radtke, K.L.; Yates, M.K. The evolution of nucleoside analogue antivirals: A review for chemists and non-chemists. Part 1: Early structural modifications to the nucleoside scaffold. *Antivir. Res.* **2018**, *154*, 66–86. [CrossRef] [PubMed]
5. Wilson, L.J.; Liotta, D. A general method for controlling glycosylation stereochemistry in the synthesis of 2′-deoxyribose nucleosides. *Tetrahedron Lett.* **1990**, *31*, 1815–1818. [CrossRef]
6. Chu, C.K.; Babu, J.R.; Beach, J.W.; Ahn, S.K.; Huang, H.; Jeong, L.S.; Lee, S.J. A highly stereoselective glycosylation of 2-(phenylselenenyl)-2,3-dideoxyribose derivative with thymine: Synthesis of 3′-deoxy-2′,3′-didehydrothymidine and 3′-deoxythymidine. *J. Org. Chem.* **1990**, *55*, 1418–1420. [CrossRef]
7. Beach, J.W.; Kim, H.O.; Jeong, L.S.; Nampalli, S.; Islam, Q.; Ahn, S.K.; Babu, J.R.; Chu, C.K. A highly stereoselective synthesis of anti-HIV 2′,3′-dideoxy- and 2′,3′-didehydro-2′,3′-dideoxynucleosides. *J. Org. Chem.* **1992**, *57*, 3887–3894. [CrossRef]
8. McDonald, F.E.; Gleason, M.M. Asymmetric Syntheses of Stavudine (d4T) and Cordycepin by Cycloisomerization of Alkynyl Alcohols to Endocyclic Enol Ethers. *Angew. Chem. Int. Ed. Engl.* **1995**, *34*, 350–352. [CrossRef]

9. Diaz, Y.; El-Laghdach, A.; Matheu, M.S.; Castillon, S. Stereoselective Synthesis of 2′,3′-Dideoxynucleosides by Addition of Selenium Electrophiles to Glycals. A Formal Synthesis of D4T from 2-Deoxyribose. *J. Org. Chem.* **1997**, *62*, 1501–1505. [CrossRef]
10. Chiacchio, U.; Rescifina, A.; Iannazzo, D.; Romeo, G. Stereoselective Synthesis of 2′-Amino-2′,3′-dideoxynucleosides by Nitrone 1,3-Dipolar Cycloaddition: A New Efficient Entry Toward d4T and Its 2-Methyl Analogue. *J. Org. Chem.* **1999**, *64*, 28–36. [CrossRef]
11. Álvarez de Cienfuegos, L.; Mota, A.J.; Rodríguez, C.; Robles, R. Highly efficient synthesis of 2′,3′-didehydro-2′,3′-dideoxy-β-nucleosides through a sulfur-mediated reductive 2′,3′-trans-elimination. From iodomethylcyclopropanes to thiirane analogs. *Tetrahedron Lett.* **2005**, *46*, 469–473. [CrossRef]
12. Shiragamai, H.; Irie, Y.; Shirae, H.; Yokozeki, K.; Yasuda, N. Synthesis of 2′, 3′-dideoxyuridine via deoxygenation of 2′, 3′-O-(methoxymethylene) uridine. *J. Org. Chem.* **1988**, *53*, 5170–5173. [CrossRef]
13. Mansuri, M.M.; Starrett, J.E.; Wos, J.A.; Tortolani, D.R.; Brodfuehrer, P.R.; Howell, H.G.; Martin, J.C. Preparation of 1-(2,3-dideoxy-.beta.-D-glycero-pent-2-enofuranosyl)thymine (d4T) and 2′,3′-dideoxyadenosine (ddA): General methods for the synthesis of 2′,3′-olefinic and 2′,3′-dideoxy nucleoside analogs active against HIV. *J. Org. Chem.* **1989**, *54*, 4780–4785. [CrossRef]
14. Corey, E.J.; Winter, R.A.E. A new, stereospecific olefin synthesis from 1, 2-diols. *J. Am. Chem. Soc.* **1963**, *85*, 2677–2678. [CrossRef]
15. Corey, E.J.; Hopkins, P.B. A mild procedure for the conversion of 1,2-diols to olefins. *Tetrahedron Lett.* **1982**, *23*, 1979–1982. [CrossRef]
16. Chu, C.K.; Bhadti, V.S.; Doboszewski, B.; Gu, Z.P.; Kosugi, Y.; Pullaiah, K.C.; Van Roey, P. General syntheses of 2′,3′-dideoxynucleosides and 2′,3′-didehydro-2′,3′-dideoxynucleosides. *J. Org. Chem.* **1989**, *54*, 2217–2225. [CrossRef]
17. Dudycz, L.W. Synthesis of 2′,3′-Dideoxyuridine Via the Corey-Winter Reaction. *Nucleosides Nucleotides* **1989**, *8*, 35–41. [CrossRef]
18. Manchand, P.S.; Belica, P.S.; Holman, M.J.; Huang, T.N.; Maehr, H.; Tam, S.Y.K.; Yang, R.T. Syntheses of the anti-AIDS drug 2′,3′-dideoxycytidine from cytidine. *J. Org. Chem.* **1992**, *57*, 3473–3478. [CrossRef]
19. Barton, D.H.R.; Jang, D.O.; Jaszberenyi, J.C. Towards dideoxynucleosides: The silicon approach. *Tetrahedron Lett.* **1991**, *32*, 2569–2572. [CrossRef]
20. Barton, D.H.R.; Jang, D.O.; Jaszberenyi, J.C. Radical mono- and dideoxygenations with the triethylsilane + benzoyl peroxide system. *Tetrahedron Lett.* **1991**, *32*, 7187–7190. [CrossRef]
21. Jang, D.O.; Cho, D.H. Radical deoxygenation of alcohols and vicinal diols with N-ethylpiperidine hypophosphite in water. *Tetrahedron Lett.* **2002**, *43*, 5921–5924. [CrossRef]
22. Oba, M.; Suyama, M.; Shimamura, A.; Nishiyama, K. Radical-based transformation of vicinal diols to olefins via thioxocarbamate derivatives: A simple approach to 2′,3′-didehydro-2′,3′-dideoxynucleosides. *Tetrahedron Lett.* **2003**, *44*, 4027–4029. [CrossRef]
23. Luzzio, F.A.; Menes, M.E. A Facile Route to Pyrimidine-Based Nucleoside Olefins: Application to the Synthesis of d4T (Stavudine). *J. Org. Chem.* **1994**, *59*, 7267–7272. [CrossRef]
24. Saito, I.; Ikehira, H.; Kasatani, R.; Watanabe, M.; Matsuura, T. Photoinduced reactions. 167. Selective deoxygenation of secondary alcohols by photosensitized electron-transfer reaction. A general procedure for deoxygenation of ribonucleosides. *J. Am. Chem. Soc.* **1986**, *108*, 3115–3117. [CrossRef]
25. Shen, B.; Bedore, M.W.; Sniady, A.; Jamison, T.F. Continuous flow photocatalysis enhanced using an aluminum mirror: Rapid and selective synthesis of 2′-deoxy and 2′,3′-dideoxynucleosides. *Chem. Commun.* **2012**, *48*, 7444–7446. [CrossRef]
26. Greenberg, S.; Moffatt, J.G. Reactions of 2-acyloxyisobutyryl halides with nucleosides. I. Reactions of model diols and of uridine. *J. Am. Chem. Soc.* **1973**, *95*, 4016–4025. [CrossRef]
27. Russell, A.F.; Greenberg, S.; Moffatt, J.G. Reactions of 2-acyloxyisobutyryl halides with nucleosides. II. Reactions of adenosine. *J. Am. Chem. Soc.* **1973**, *95*, 4025–4030. [CrossRef]
28. Jain, T.C.; Jenkins, I.D.; Russell, A.F.; Verheyden, J.P.H.; Moffatt, J.G. Reactions of 2-Acyloxyisobutyryl Halides with Nucleosides. 1V.l A Facile Synthesis of 2,3-Unsaturated Nucleosides Using Chromous Acetate. *J. Org. Chem.* **1974**, *39*, 30–34. [CrossRef]
29. Mengel, R.; Seifert, J.M. Über einen neuen Zugang zu 2′,3′-ungesättigten Nucleosiden—Eine milde Umwandlung vicinaler cis-Diole in Olefine. *Tetrahedron Lett.* **1977**, *48*, 4203–4206. [CrossRef]
30. Robins, M.J.; Wilson, J.S.; Madej, D.; Low, N.H.; Hansske, F.; Wnuk, S.F. Nucleic Acid-Related Compounds. 88. Efficient Conversions of Ribonucleosides into Their 2′,3′-Anhydro, 2′(and 3′)-Deoxy, 2′,3′-Didehydro-2′,3′-dideoxy, and 2′,3′-Dideoxynucleoside Analogs. *J. Org. Chem.* **1995**, *60*, 7902–7908. [CrossRef]
31. Chen, B.C.; Quinlan, S.L.; Stark, D.R.; Reid, J.G.; Audia, V.H.; George, J.G.; Eisenreich, E.; Brundidge, S.P.; Racha, S.; Spector, R.H. 5′-Benzoyl-2′α-bromo-3′-O-methanesulfonylthymidine: A superior nucleoside for the synthesis of the anti-AIDS drug D4T (Stavudine). *Tetrahedron Lett.* **1995**, *36*, 7957–7960. [CrossRef]
32. Shiragami, H.; Ineyama, T.; Uchida, Y.; Izawa, K. Synthesis of 1-(2,3-Dideoxy-β-D-glycero-pent-2-enofuranosyl)thymine (d4T; Stavudine) from 5-Methyluridine. *Nucleosides Nucleotides* **1996**, *15*, 47–58. [CrossRef]
33. Chen, B.C.; Quinlan, S.L.; Reid, J.G.; Spector, R.H. A new thymine free synthesis of the anti-AIDS drug d4T via regio/stereo controlled β-elimination of bromoacetates. *Tetrahedron Lett.* **1998**, *39*, 729–732. [CrossRef]
34. Guo, Z.; Sanghvi, Y.S.; Brammer, L.E., Jr.; Hudlicky, T. Synthesis of 2′, 3′-Dideoxy-2′, 3′-didehydro Nucleosides via a Serendipitous Route. *Nucleosides Nucleotides Nucleic Acids* **2001**, *20*, 1263–1266. [CrossRef]
35. Sagandira, C.R.; Akwi, F.M.; Sagandira, M.B.; Watts, P. Multistep Continuous Flow Synthesis of Stavudine. *J. Org. Chem.* **2021**, *86*, 13934–13942. [CrossRef]

36. Gillaizeau, I.; Lagoja, I.M.; Nolan, S.P.; Aucagne, V.; Rozenski, J.; Herdewijn, P.; Agrofoglio, L.A. Straightforward Synthesis of Labeled and Unlabeled Pyrimidine d4Ns via 2′,3′-Diyne *seco* Analogues through Olefin Metathesis Reactions. *Eur. J. Org. Chem.* **2003**, *2003*, 666–671. [CrossRef]
37. Ewing, D.F.; Glaçon, V.; Mackenzie, G.; Postelb, D.; Len, C. Synthesis of acyclic bis-vinyl pyrimidines: A general route to d4T via metathesis. *Tetrahedron* **2003**, *59*, 941–945. [CrossRef]
38. García, J.; Fernández, S.; Ferrero, M.; Sanghvi, Y.S.; Gotor, V. Novel enzymatic synthesis of levulinyl protected nucleosides useful for solution phase synthesis of oligonucleotides. *Tetrahedron Assymetry* **2003**, *14*, 3533–3540. [CrossRef]
39. Martínez-Montero, S.; Fernández, S.; Sanghvi, Y.S.; Gotor, V.; Ferrero, M. Enzymatic Parallel Kinetic Resolution of Mixtures of D/L 2′-Deoxy and Ribonucleosides: An Approach for the Isolation of β-L-Nucleosides. *J. Org. Chem.* **2010**, *75*, 6605–6613. [CrossRef]
40. Chatgilialoglu, C.; Griller, D.; Lesage, M. Tris(trimethylsilyl)silane. A new reducing agent. *J. Org. Chem.* **1988**, *53*, 3641–3642. [CrossRef]
41. Chatgilialoglu, C. (Me$_3$Si)$_3$SiH: Twenty Years After Its Discovery as a Radical-Based Reducing Agent. *Chem. Eur. J.* **2008**, *14*, 2310–2320. [CrossRef]
42. Brahmachari, G.; Nurjamal, K.; Karmakar, I.; Mandal, M. Camphor-10-Sulfonic Acid (CSA): A Water Compatible Organocatalyst in Organic Transformations. *Curr. Organocatal.* **2018**, *5*, 165–181. [CrossRef]
43. Fehr, M.; Appl, A.; Esdaile, D.J.; Naumann, S.; Schulz, M.; Dahms, I. D-10-camphorsulfonic acid: Safety evaluation. *Mutat. Res. Genet. Toxicol. Environ. Mutagen.* **2020**, *858–860*, 503257. [CrossRef]
44. Lu, G.; Wang, D.; Ren, J.; Ke, Y.; Zeng, B.-B. Catalytic removal of tert-butyldimethylsilyl (TBS) ether by PVP-I. *Tetrahedron Lett.* **2019**, *60*, 150831. [CrossRef]
45. Kumar, G.D.K.; Baskaran, S. A Facile, Catalytic, and Environmentally Benign Method for Selective Deprotection of tert-Butyldimethylsilyl Ether Mediated by Phosphomolybdic Acid Supported on Silica Gel. *J. Org. Chem.* **2005**, *70*, 4520–4523. [CrossRef]
46. Santaniello, E.; Ciuffreda, P.; Alessandrini, L. Synthesis of Modified Purine Nucleosides and Related Compounds Mediated by Adenosine Deaminase (ADA) and Adenylate Deaminase (AMPDA). *Synthesis* **2005**, *2005*, 509–526. [CrossRef]

Article

Properties of Parallel Tetramolecular G-Quadruplex Carrying *N*-Acetylgalactosamine as Potential Enhancer for Oligonucleotide Delivery to Hepatocytes

Anna Clua [1,2], Santiago Grijalvo [1,2], Namrata Erande [3], Swati Gupta [3], Kristina Yucius [3], Raimundo Gargallo [4], Stefania Mazzini [5], Muthiah Manoharan [3] and Ramon Eritja [1,2,*]

1. Institute for Advanced Chemistry of Catalonia (IQAC-CSIC), Jordi Girona 18-26, E-08034 Barcelona, Spain; acvtnt@cid.csic.es (A.C.); sgrgma@cid.csic.es (S.G.)
2. Networking Center on Bioengineering, Biomaterials and Nanomedicine (CIBER-BBN), E-08034 Barcelona, Spain
3. Alnylam Pharmaceuticals, 300 Third Street, Cambridge, MA 02142, USA; nd.erande@gmail.com (N.E.); sgupta@alnylam.com (S.G.); kyucius@alnylam.com (K.Y.); mmanoharan@alnylam.com (M.M.)
4. Department of Chemical Engineering and Analytical Chemistry, University of Barcelona, Martí i Franquès 1-11, E-08028 Barcelona, Spain; raimon_gargallo@ub.edu
5. DEFENS-Dipartimento Di Scienze per Gli Alimenti, la Nutrizione e l'Ambiente, Università degli Studi di Milano, Via Celoria, 2, 20133 Milan, Italy; stefania.mazzini@unimi.it
* Correspondence: recgma@cid.csic.es; Tel.: +34-934-006-145

Citation: Clua, A.; Grijalvo, S.; Erande, N.; Gupta, S.; Yucius, K.; Gargallo, R.; Mazzini, S.; Manoharan, M.; Eritja, R. Properties of Parallel Tetramolecular G-Quadruplex Carrying *N*-Acetylgalactosamine as Potential Enhancer for Oligonucleotide Delivery to Hepatocytes. *Molecules* **2022**, *27*, 3944. https://doi.org/10.3390/molecules27123944

Academic Editor: Harri Lönnberg

Received: 13 May 2022
Accepted: 17 June 2022
Published: 20 June 2022
Corrected: 23 December 2022

Publisher's Note: MDPI stays neutral with regard to jurisdictional claims in published maps and institutional affiliations.

Copyright: © 2022 by the authors. Licensee MDPI, Basel, Switzerland. This article is an open access article distributed under the terms and conditions of the Creative Commons Attribution (CC BY) license (https://creativecommons.org/licenses/by/4.0/).

Abstract: The development of oligonucleotide conjugates for in vivo targeting is one of the most exciting areas for oligonucleotide therapeutics. A major breakthrough in this field was the development of multifunctional GalNAc-oligonucleotides with high affinity to asialoglycoprotein receptors (ASGPR) that directed therapeutic oligonucleotides to hepatocytes. In the present study, we explore the use of G-rich sequences functionalized with one unit of GalNAc at the 3′-end for the formation of tetrameric GalNAc nanostructures upon formation of a parallel G-quadruplex. These compounds are expected to facilitate the synthetic protocols by providing the multifunctionality needed for the binding to ASGPR. To this end, several G-rich oligonucleotides carrying a TGGGGGGT sequence at the 3′-end functionalized with one molecule of *N*-acetylgalactosamine (GalNAc) were synthesized together with appropriate control sequences. The formation of a self-assembled parallel G-quadruplex was confirmed through various biophysical techniques such as circular dichroism, nuclear magnetic resonance, polyacrylamide electrophoresis and denaturation curves. Binding experiments to ASGPR show that the size and the relative position of the therapeutic cargo are critical for the binding of these nanostructures. The biological properties of the resulting parallel G-quadruplex were evaluated demonstrating the absence of the toxicity in cell lines. The internalization preferences of GalNAc-quadruplexes to hepatic cells were also demonstrated as well as the enhancement of the luciferase inhibition using the luciferase assay in HepG2 cell lines versus HeLa cells. All together, we demonstrate that tetramerization of G-rich oligonucleotide is a novel and simple route to obtain the beneficial effects of multivalent *N*-acetylgalactosamine functionalization.

Keywords: G-quadruplex; *N*-acetylgalactosamine; antisense; oligonucleotide conjugates; asialoglycoprotein receptor; luciferase gene; gapmers

1. Introduction

The development of oligonucleotides for therapeutic use has entered an expansive phase with an arsenal of more than a dozen drugs based on synthetic oligonucleotides [1,2]. This is especially evident in the last three years, with an increase of new oligonucleotide approvals per year being one of the most prominent biomolecules approved for human use [3]. Most of these therapeutic drugs are antisense oligonucleotides [4] or siRNA duplexes [5] that target overexpressed messenger RNA (mRNA). Others are aptamers that target overexpressed proteins, such as pegaptanib that target vascular endothelial growth factor (VEGF) [6]. However, soon it is expected that new oligonucleotides will be used for modulating other important RNA targets such as microRNA [7] and long non coding (lnc) RNA [8].

In order to achieve the desired therapeutic action, oligonucleotides should penetrate inside the cells and bind readily to their complementary sequence before exonuclease activity degrades them. A large research activity is being made to design modified oligonucleotides with enhanced nuclease resistance and cellular uptake as well as higher selectivity to target tissues [1,5]. However, the delivery of modified DNA and RNA is the major bottleneck in the development of nucleic acids as drugs [9]. The importance of this issue can be demonstrated in the newly approved siRNAs, Givosiran [10] and Inclisiran [11], which are functionalized with triantennary N-acetylgalactosamine (GalNAc) allowing the drug to maintain their silencing properties for 6 months [11]. In these conjugates, the 3′ terminus of the siRNA sense strand is linked to three molecules of GalNAc that have a high binding affinity to the asialoglycoprotein receptor [12], or by successive additions of a GalNAc monomer [13,14]. The asialoglycoprotein receptor is mainly found in hepatocytes exhibiting a high affinity for galactose glycoproteins and trivalent GalNAc oligonucleotides with a rapid internalization rate via the clathrin-mediated pathway [12,15].

Another potential interest of oligonucleotides is the possibility of using their singular non-canonical structures for protein binding, especially guanine-rich oligonucleotide that may form G-quadruplex, as some of them have antiviral, anticancer and anticoagulant properties [16–18]. The antiviral properties usually result from the binding of the quadruplex to viral proteins [19]. These aptamers have also been used for targeted drug delivery as they have affinity to proteins that are overexpressed in cancer cells, such as AS1411 that binds nucleolin [18,20]. Recently, antisense oligonucleotides were attached to simple G-rich sequences to form parallel G-quadruplex structures. These simple DNA scaffolds were efficient in delivering the antisense oligonucleotide cargo inside the cells without the presence of cationic lipids as transfecting reagents [21,22]. Moreover, parallel G-quadruplexes have been used for increased delivery of antiproliferative floxuridine oligonucleotides upon tetramerization of G-rich sequences [23], as some cancer cells have abundant G-quadruplex binding proteins.

In the present study, we explore the use of G-rich sequences functionalized with one unit of GalNAc at the 3′-end for the formation of tetrameric GalNAc nanostructures upon formation of parallel G-quadruplex. This study was aimed to provide an alternative method to obtain multifunctional GalNAc-oligonucleotides with a more simplified synthetic method (Scheme 1, Figure S1). To this end, we prepare G-rich oligonucleotides carrying a TGGGGGGT sequence at the 3′-end functionalized with one molecule of N-acetylgalactosamine (GalNAc). These compounds are easier to synthesize than the "standard" triantennary GalNAc. We aim to study if the association of four strands produces G-quadruplex tetrameric GalNAc derivatives that may provide a similar affinity to ASGPR than the "standard" triantennary-GalNAc. The formation of self-assembled parallel G-quadruplexes is studied through various biophysical techniques such as circular dichroism, nuclear magnetic resonance, polyacrylamide electrophoresis and denaturation curves. Binding experiments to ASGPR show that the size and the relative position of the therapeutic cargo are important for the binding of these nanostructures to ASGPR. All together we demonstrate that tetramerization of G-rich sequences is a novel strategy to facilitate specific cellular uptake of short single-stranded oligonucleotides to hepatocytes including antisense gapmers directed against luciferase.

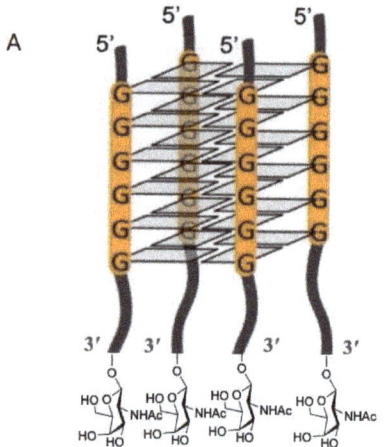

Scheme 1. Chemical schemes of potential multifunctional GalNAc derivatives including (**A**) tetrameric G-quadruplexes studied in this work and (**B**) the triantennary GalNAc.

2. Results

2.1. Oligonucleotides

Tables 1 and 2 show the oligonucleotides prepared in this study indicating the backbone and the presence or absence of N-acetylgalactosamine modification at the 3′-position. The first group of derivatives is derived from a 17 mer (TTGGGGGGTACAGTGCA) sequence (Table 1). This sequence is present in the HIV-1 genome [24,25]. Lipid-modified oligonucleotide derivatives or lipoquads have antiviral properties (both against HIV [26] and HCV [17]) as inhibitors of viral entry. These sequences contain natural phosphodiester backbone (H2–H4). Two different GalNAc derivatives with different length were studied (L193 (longer), and L235 (shorter), Scheme 2). The synthesis of the appropriate solid supports functionalized with these GalNAc derivatives has been reported [13]. Some of these oligonucleotides (H5, H6) were fluorescein labeled to visualize cellular uptake by flow cytometry. Control oligonucleotides include unmodified control sequence (H1) and A-rich analogues (H3, H5). Finally, some oligonucleotides contained two phosphorothioate linkages at the terminal 3′ and 5′-ends (H8–H11). The A-rich oligonucleotide H12 carrying the "classical" triantennary GalNAc ligand L96 was also made for comparison purposes during the evaluation of the binding of G-rich oligonucleotides to the asialoglycoprotein (ASGPR) receptor.

The second group of oligonucleotides combined a gapmer sequence against *Renilla* luciferase (5′-<u>CsGsUsUs</u>TsCsCsTsTsTsGsTsTsCs<u>UsGsGsA</u>-3′) [27] carrying 2′-O-methyl-RNA modifications at the last four positions of the 3′- and 5′-ends (underlined) and phosphorothioate linkages with a G-quadruplex forming TG$_6$T sequence. These include two different spatial distributions. In one oligonucleotide, gapmer was located at the 5′-end and TG$_6$T sequence was near the 3′-end (L3), and in another oligonucleotide, TG_6T sequence was located at the 5′-end and the gapmer sequences were located near the 3′-end (L4). In addition, we prepared a fluorescein labeled oligonucleotide (L7), an oligonucleotide carrying a scrambled sequence instead of the luciferase sequence (L6), unmodified antisense (L1) and gapmer (L2) antisense control sequences and control T-rich oligonucleotides unable to form G-quadruplex (L5, L8). All these sequences except L8 contained phosphorothioate linkages in the antisense/gapmer moiety, but they contained the natural phosphodiester linkages in the TG_6T or T_4 sequence.

Table 1. Oligonucleotide derivatives from 17 mer antiviral sequence.

N	Name	Sequence (5′-3′)	Backbone	GalNAc
H1	control	TTGGGGGGTACAGTGCA	PO	-
H2	G-PO-HCV-L235	TTGGGGGGTACAGTGCA-L235	PO	L235
H3	A-PO-HCV-L235	TTGAAAGGTACAGTGCA-L235	PO	L235
H4	G-PO-HCV-L193	TTGGGGGGTACAGTGCA-L193	PO	L193
H5	A-PO-HCV-L193	TTGAAAGGTACAGTGCA-L193	PO	L193
H6	FAM-G-PO-HCV-L235	FAM-TTGGGGGGTACAGTGCA-L235	PO	L235
H7	FAM-G-PO-HCV-L193	FAM-TTGGGGGGTACAGTGCA-L193	PO	L193
H8	A-PS-HCV-L235	TsTsGAAAGGTACAGsTsGsCsA-L235	PO/PS	L235
H9	A-PS-HCV-L193	TsTsGAAAGGTACAGsTsGsCsA-L193	PO/PS	L193
H10	G-PS-HCV-L235	TsTsGGGGGGTACAGsTsGsCsA-L235	PO/PS	L235
H11	G-PS-HCV-L193	TsTsGGGGGGTACAGsTsGsCsA-L193	PO/PS	L193
H12	A-PO-HCV-L96	TTGAAAGGTACAGTGCA-L96	PO	L96

s indicates the position of the phosphorothioate linkages.

Table 2. Oligonucleotide derivatives carrying anti-luciferase sequence.

N	Name	Sequence (5′-3′)	Backbone	GalNAc
L1	ASO control	CsGsTsTsTsCsCsTsTsTsGsTsTsCsTsGsGsA	PS	-
L2	Gapmer control	CsGsUsUsTsCsCsTsTsTsGsTsTsCsUsGsGsA	PS	-
L3	Luc-G₆-GalNAc	CsGsUsUsTsCsCsTsTsTsGsTsTsCsUsGsGsATGGGGGGT-L193	PO,PS	L193
L4	G₆-Luc-GalNAc	TGGGGGGTCsGsUsUsTsCsCsTsTsTsGsTsTsCsUsGsGsAs-L193	PO,PS	L193
L5	Luc-T₄-GalNAc	CsGsUsUsTsCsCsTsTsTsGsTsTsCsUsGsGsATTTT-L193	PO,PS	L193
L6	Scr-Luc-G₆-GalNAc	CsUsGsUsCsTsGsAsCsGsTsTsCsTsUsUsGsUsTGGGGGGT-L193	PO,PS	L193
L7	FAM-Luc-G₆-GalNAc	FAM-sCsUsGsUsTsCsCsTsTsTsGsTsTsCsUsGsGsAsTTGGGGGGT-L193	PO,PS	L193
L8	T₈	TTTTTTTT	PO	-

s indicates the position of the phosphorothioate linkages. Underline nucleotides are 2-O-methyl-RNA derivatives.

A third group of oligonucleotides (Table S2) were derived from the therapeutic siRNA against the mouse transthyretin (mTTR) gene [12]. In this case, they were a pair of RNA strands that were highly modified with 2′-O-methyl-RNA and 2′-fluoro-RNA nucleotides as well as two phosphorothioate linkages at the two last 5′-terminal positions. The TG_6T sequence was added to the 3′-position of the sense strand ending with the L193 GalNAc ligand in oligonucleotide R2. Oligonucleotide R1 was a control sense sequence carrying the sense strand and the triantennary GalNAc (L96) ligand. The oligonucleotide R3 was the antisense strand complementary to R1 and R2 oligonucleotides to obtain the double-stranded form.

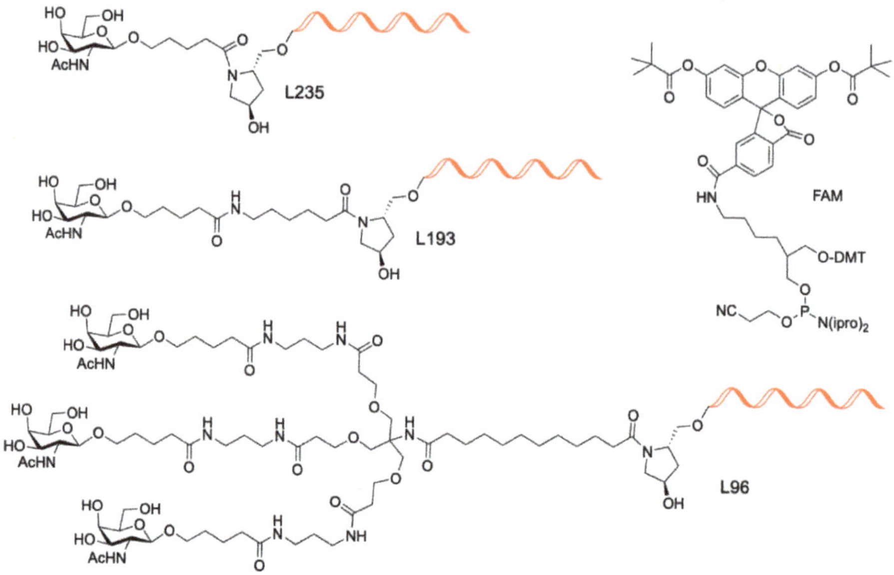

Scheme 2. Chemical structures of the GalNAc and FAM ligands used in this work.

2.2. Characterization of G-Quadruplex Formation

First of all, the formation of a parallel G-quadruplex was studied by biophysical techniques, such as circular dichroism (CD), gel electrophoresis and nuclear magnetic resonance (NMR).

2.2.1. Characterization of G-Quadruplex Formation by CD

Figure 1A shows the CD spectra of the G–quadruplex sequence H2 (G-PO-HCV-L235) compared with the A-rich sequence H3 (A-PO-HCV-L235) from 205 to 310 nm. The CD spectra of control oligonucleotides TG_6T (tetrameric G-quadruplex) and T_8 (single stranded) were added for comparison. The G-rich oligonucleotides (H2 and TG_6T) showed intense positive bands at around 210 and 261 nm characteristic of a parallel G-quadruplex [28], while the A-rich sequence (H3) and T_8 presented a maximum at 274 nm. In addition the CD spectrum of the G-rich sequences H2, H4, H10 and H11 maintain the same shape unaltered up to 90 °C (Figure S2) indicating that the parallel G-quadruplex structure is thermally stable at all studied temperatures. This thermal stability is in agreement with the reported values for the well-studied tetrameric parallel G-quadruplex obtained with TG_nT oligonucleotides [21–23,29]. TG_4T melting can only be observed in sodium buffers, but not in potassium buffers [21–23,29]. The melting temperatures of the six tetrads quadruplex (TG_6T derivatives) cannot be measured because they do not melt even at temperatures higher than 80 °C [29]. For example, TG_4T melts at 59.4 °C, but TG_6T cannot be measured (>80 °C) in 10 mM sodium cacodylate buffer with 0.15 M NaCl (pH 7.2) [21,22]. For the modified TG_6T oligonucleotides carrying an antisense oligonucleotide with lipids, or floxuridine or positively charged ligands, we cannot observe any melting behavior as G-quadruplex were stable up to 80 °C [21–23].

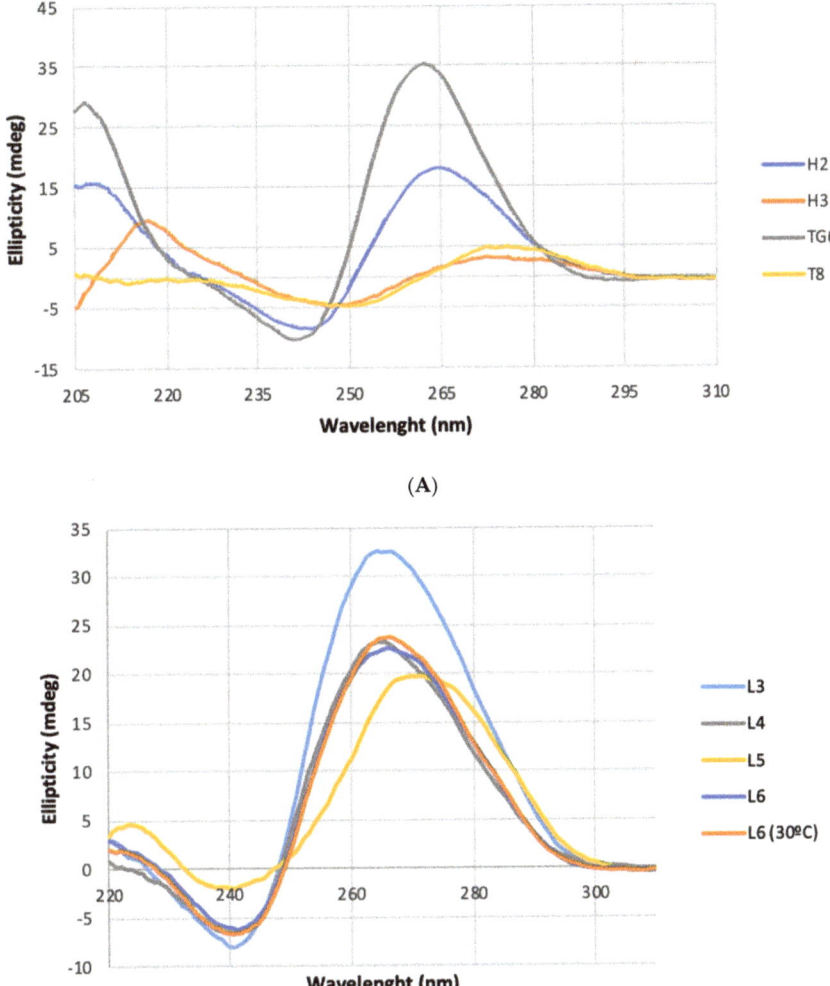

Figure 1. (**A**) CD spectra of G-rich sequence (G-PO-HCV-L235, H2) versus A-rich sequence (A-PO-HCV-L235, H3). The CD spectra of TG$_6$T and T$_8$ are included for comparison. Buffer conditions PBS 1×. (**B**) CD spectra of G-rich sequences carrying the antiluciferase sequence (L3, L4, L6) versus the control T-rich sequence (L5) Buffer conditions PBS 1×.

CD spectra of the rest of the G-rich oligonucleotides show similar positive bands at 265 nm and negative bands near 240 nm (Figures 1B, S2 and S3) characteristic of a parallel G-quadruplex. Oligonucleotide L5 carrying T's instead of G's showed a maximum at around 272 nm (Figures 1B and S3).

2.2.2. Characterization of G-Quadruplex Formation by NMR

Oligonucleotide H2 (G-PO-HCV-L235) was dissolved in PBS buffer, pH 7.4 and the resulting solution was heated at 90 °C and cooled down to room temperature. Several NMR spectra were taken at different times. Figure 2a shows the NMR spectra of the imino protons region. In all of them, the presence of the imino protons characteristic of a

G-quadruplex formation can be observed. The presence of these signals is observed just dissolving the sample. The formation of a G-tetrad gives rise to characteristic guanine imino protons (H1), which exhibit their chemical shifts within the range of 10–12 ppm, as compared to 13–14 ppm for those involved in Watson–Crick base pairing. Guanine imino protons in a G-quadruplex also exchange more slowly with solvent than the counterparts in a Watson–Crick duplex. The imino protons of guanines in the center G-tetrad exchange very slowly with the solvent and remain detected long after dissolving the sample in D_2O solution [30]. Annealing (heating and cooling) the solution shows the presence of imino protons all the time, indicating that the formation of the G-quadruplex in the short 17 mers is very fast, as they do not change with the time (Figure 2). The guanine protons displayed the characteristic sequential imino H1/H1 and H1/H8 NOE interactions [31], thus confirming the formation quadruplex structure (Figure 2b).

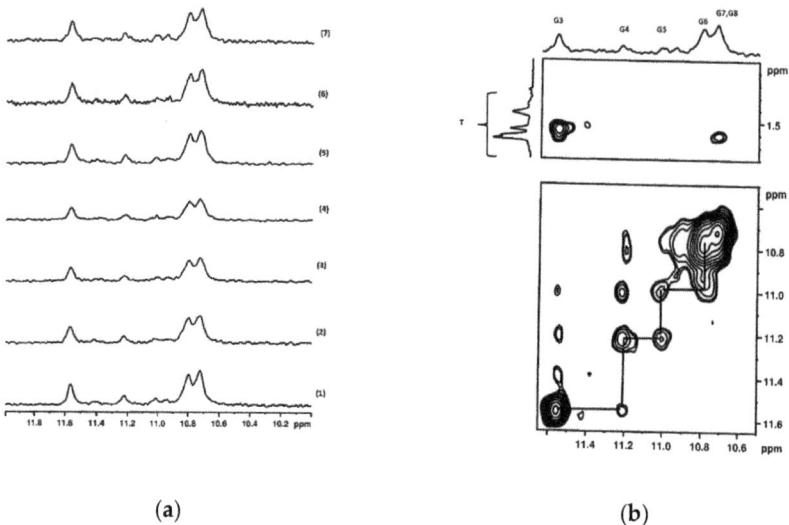

(a) (b)

Figure 2. (a) Imino proton NMR spectra of oligonucleotide H2 (G-PO-HCV-L235). (1) Sample before heating. (2) Sample after heating at 90 °C for 1 min and move the sample to the magnet for NMR acquisition (approx. at 50 °C) (3) after 13 min, (4) after 30 min, (5) after 50 min, (6) after 65 min, and (7) after 5 months. (b) Imino and methyl protons region of 2D-NOESY spectrum of H2 (G-PO-HCV-L235) at 25 °C in H_2O/D_2O (9:1), PBS buffer (pH 7.4).

A similar experiment was done with the longer oligonucleotide L3, but in this case the signals at the imino protons regions were too small to analyze (data not shown).

2.2.3. Characterization of G-Quadruplex Formation by Gel Electrophoresis

Next, the G-quadruplex sequences were analyzed by native gel electrophoresis. Figure 3 shows the presence of single bands in the lanes with G-rich sequences that are retarded from the single-stranded control sequences in agreement with G-quadruplex formation (Figure 3A,B).

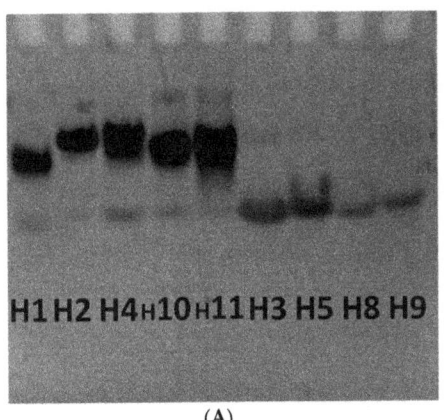 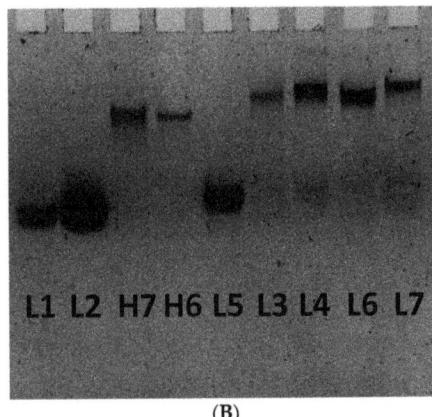

Figure 3. Native 12% PAGE of TBE 1X supplemented with 100 mM KCl using the same running buffer at 150 Watts maintaining a 20 °C for 4–5 h; stained with SYBR Green. (**A**) Short HIV G-rich sequences (H1, H2, H4, H10, H11) versus A-rich control sequences (H3, H5, H8, H9). (**B**) G-rich sequences carrying antiluciferase sequence (L3, L4, L6, L7) versus control single stranded sequences (L1, L2, L5) compared with short G-rich sequences (H6, H7).

2.3. ASGPR Receptor Binding Studies

Next, the affinity binding of G-rich oligonucleotides to asialoglycoprotein receptors (ASGPR) were analyzed by a flow cytometry based competitive binding assay using mouse hepatocytes [12,14] and determining the inhibitory constant (KI). For comparison the oligonucleotide H12 (A-PO-HCV-L96) carrying the triantennary GalNAc was included in addition to G-rich oligonucleotides (G-PO-HCV-L193 (H4), G-PO-HCV-L235 (H2), G-PS-HCV-L193 (H10) and G-PS-HCV-L193 (H11) and a control A-rich sequence that is not able to form tetrameric G-quadruplex (A-PO-HCV-L193, H5). Figure 4 shows the displacement curves of all the GalNAc oligonucleotides from where a KI is estimated (Figure 4 and Table 3). The best affinity was found for the triantennary GalNAc (A-PO-L96, 10.4 nM) followed by the G-rich oligonucleotides carrying the shorter GalNAc linker (L235) (31.3 and 39.9 nM). Then, the G-rich oligonucleotides carrying the longer GalNAc linker (L193) followed with a KI of 42.8 and 47.2 nM. In both cases the oligonucleotide carrying phosphorothioate linkages had a slight better affinity. The affinity to ASGPR of the control single-stranded A-rich sequence was not possible to determine at concentrations tested. These results demonstrate the benefit of tetramerization by increasing ASGPR binding close to the affinity values observed for the triantennary GalNAc using G-rich oligonucleotides carrying a single monomeric GalNAc ligand.

Table 3. Binding affinity to ASGPR using a mouse primary hepatocytes fluorescence-based assay [12,14].

N	Name	KI (nM)	Stdev
H4	G-PO-HCV-L193	47.2	7.3
H2	G-PO-HCV-L235	39.9	3.2
H11	G-PS-HCV-L193	42.8	7.6
H10	G-PS-HCV-L235	31.3	7.2
H12	A-PO-HCV-L96	10.4	3.2
H5	A-PO-HCV-L193	N/D [1]	N/D [1]

[1] N/D = not determined at concentrations tested.

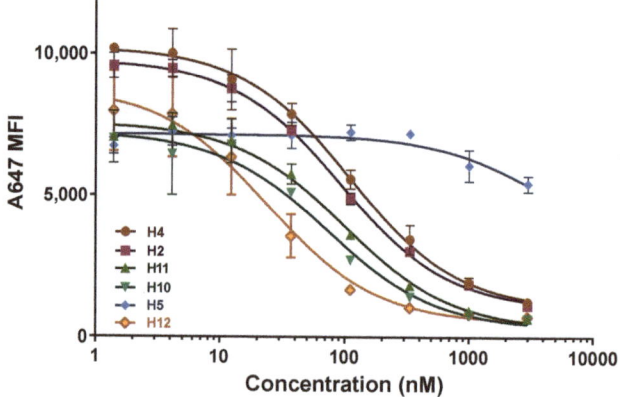

Figure 4. Affinity binding to ASGPR in primary mouse hepatocytes. Oligonucleotides derived from 17 mer antiviral sequence.

The binding affinities to ASGPR of GalNAc-functionalized gapmers and RNA oligonucleotides linked to a TG_6T sequence are shown in Figure 5 and Table 4. In this case we observed a loss of the affinity to ASGPR. Single-stranded and double stranded RNA carrying the TG_6T sequence and one single monomeric GalNAc ligand have very low affinity to ASGPR (Sense-TG_6T-L193 696 nM, Duplex-TG_6T-L193 N/D). Gapmers (L3 and L4) have slightly better affinity especially L4 (G_6-Luc-L193, 121 nM), but still far from the RNA carrying a triantennary GalNAc ligand R1 (sense-L96, 8.8 nM). There is a clear loss in binding affinity when the G-quadruplex sequence is next to the GalNAc ligand (L3, Luc-G6-L193, 357.8 nM) becoming of the same magnitude that the T-rich sequence that cannot form a quadruplex (L5, Luc-T_4-L193, 300.7 nM). These results may indicate that when the oligonucleotide attached to the TG_6T sequence becomes longer and complex the tetramerization becomes more difficult and the G-quadruplex structure dissociates rapidly to trimeric, dimeric and monomeric species that have less affinity for ASGPR. Although the gel retardation data show G-quadruplex formation, these quadruplexes may undergo to dissociation upon dilution.

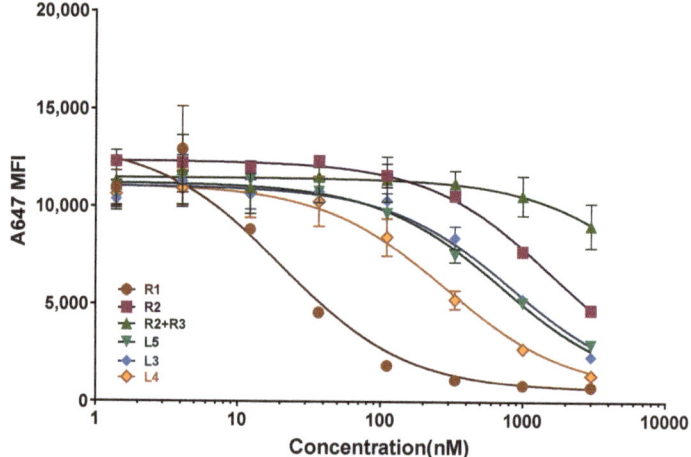

Figure 5. Affinity binding to ASGPR in primary mouse hepatocytes of gapmers and RNA derivatives carrying the anti-luciferase and anti-mTTR (Table S2) sequences.

Table 4. Binding affinity to ASGPR using a mouse primary hepatocytes fluorescence-based assay [12,14].

N	Motif	KI (nM)	Stdev
R1	Sense-L96	8.8	2.2
R2	Sense-TG$_6$T-L193	696.0	37.2
R2 + R3	Duplex-TG$_6$T-L193	N/D [1]	N/D [1]
L5	Luc-T$_4$-L193	300.7	23.9
L3	Luc-G$_6$-L193	357.8	48.5
L4	G$_6$-Luc-L193	121.1	17.6

[1] N/D = not determined at concentrations tested.

2.4. Toxicity Assays

Prior the evaluation of internalization and antisense inhibition of luciferase, we evaluate the potential toxicity of GalNAc-oligonucleotides by using MTT assay at 60 nM, 120 nM and 300 nM oligonucleotide concentration after 24 h. Results are shown in Figure S5. Most of the compounds were not toxic, as all the MTT values were 80% or higher in HeLa cells (cervix cancer cells) and HepG2 (hepatic cells).

2.5. Analysis of the Stability of Oligonucleotides towards Snake Venom Phosphodiesterase and 10% FBS

Selected G-rich oligonucleotides (H6, H7, L4 and L7) were incubated with a solution of snake venom phosphodiesterase and a 10% FBS solution (experimental conditions used in cell studies). Results are shown in supplementary materials (Figure S4A,B). H6 (FAM-G-PO-HCV-L235) and H7 (FAM-G-PO-HCV-L193) G-rich oligonucleotides incubated with snake venom phosphodiesterase show some stability as the spots corresponding to the full length H6 and H7 are clearly visible, but disappeared between 8–24 h. In 10% FBS two groups of spots are seen: one having the mobility of the spots seen in the treatment snake venom phosphodiesterase and a retarded band that may correspond to the oligonucleotide complexed with serum proteins that are unchanged over the time of incubation (Figure S4A).

In the snake venom phosphodiesterase treatment of L4 (G$_6$-Luc-GalNAc) and L7 (FAM-Luc-G6-GalNAc), the spots corresponding to the full length L4 and L7 are clearly visible after 24 h confirming the stability of these highly modified oligonucleotides (gapmers with full phosphorothioate linkages). In 10% FBS, only a main retarded band is seen that may correspond to the oligonucleotide complexed with serum proteins that are unchanged over the time of incubation (Figure S4B). These data indicate that these oligonucleotides are more stable than the H series because they carry phosphorothioate and 2′-O-methyl residues and these modifications increase binding of oligonucleotides to serum proteins that may protect oligonucleotides from degradation.

2.6. Internalization Assay

The internalization of fluoresceine (FAM)-labelled G-rich GalNAc-oligonucleotides (H6, H7 and L7) were measured with a cell cytometer. Results are shown in Figure 6. Oligonucleotide L7 is the oligonucleotide whose internalization is higher in both HeLa and HepG2 cell lines. In fact, the internalization magnitude in HepG2 cells of the G-quadruplex compound is by a factor of 3 higher than in HeLa cells at same conditions so, this secondary structure push up the internalization process in HepG2 cells.

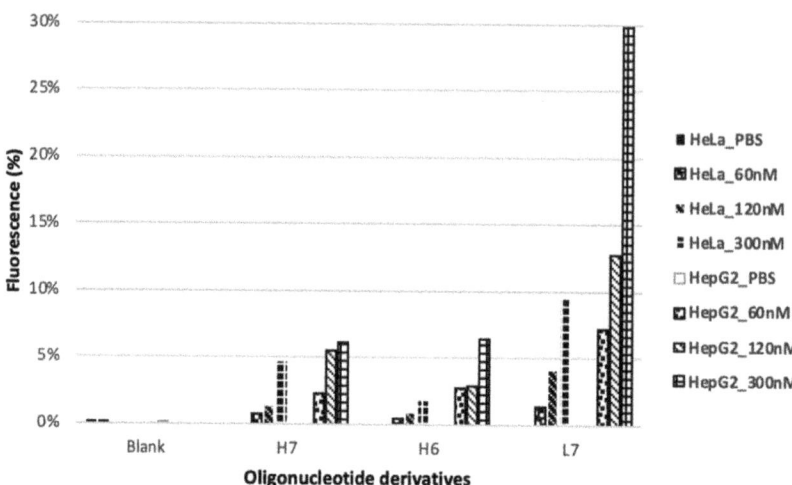

Figure 6. Cytometry assay evaluating the internalization of H7 (L193), H6 (L235), L7 (FAM-G$_6$-Luc, FAM derivative of L3) at 60, 120 and 300 nM in HeLa and HepG2 cell lines.

Independently analyzing the three compounds, on the one hand, oligonucleotide H7 in HeLa cells go from 1% green fluorescence intensity internalization to around 5% at 300 nM. On the other hand, in the HepG2 cell line, the green fluorescence intensity is slightly higher by going from around 2% to 6%. Oligonucleotide H6, a shorter linker than H7, it slims the internalization down in HeLa cells, but in HepG2 cells it can be seen a blip between low concentrations and the higher concentration of the compost. Finally, looking deeper on the results obtained with the FAM-G$_6$-Luc (L7) compound, the differences between the two cell lines are very large. In HeLa cells, percentage of internalization goes from 3% at 60 nM to 10% at 300 nM. Then, in HepG2 cells goes from 7% to 30%.

2.7. Antisense Studies

Next the luciferase silencing activity of antisense oligonucleotides was evaluated in HeLa and HepG2 cell lines. G-rich oligonucleotides carrying the antiluciferase gapmer sequence L3 (Luc-G$_6$-L193) and L4 (G$_6$-Luc-L193) were compared to T-rich oligonucleotide L5 (Luc-T$_4$-L193) and the scrambled G-rich sequence L6 (Scr-Luc-G$_6$-L193). As a positive control we used L1 (ASO) and L2 (Gapmer) transfected with Lipofectamine 2000. In the following graphs (Figure 7), the luciferase inhibition assay is represented in gymnotic conditions for the GalNAc-oligonucleotides. In HeLa cells (Figure 7a), the luciferase inhibitory effect of the G-rich oligonucleotides (L3 and L4) is low compared with the ASO and Gapmer controls transfected with Lipofectamine that inhibits around 70%. In Hela cells the product that has the best inhibitory properties is the L5 that is not able to form a G-quadruplex. This is in agreement with the lack of ASGPR as cellular uptake that is more favorable for the smallest oligonucleotide (L5) without the possibility of G-quadruplex tetramerization.

On the contrary the analysis of the luciferase assay results in HepG2 cell line (Figure 7b) shows the reverse situation. In this case the G-rich oligonucleotide L3 (Luc-G$_6$-L193) is the most efficient antisense luciferase inhibitor, followed by L4 (G$_6$-Luc-L193) that near 50% luciferase inhibition at 300 nM in gymnotic conditions reaching a very significant difference ($p < 0.001$) at 120 nM. In HepG2 cells, the smaller oligonucleotide has low inhibitory properties presenting an erratic shape from 60 nM to 300 nM concentration.

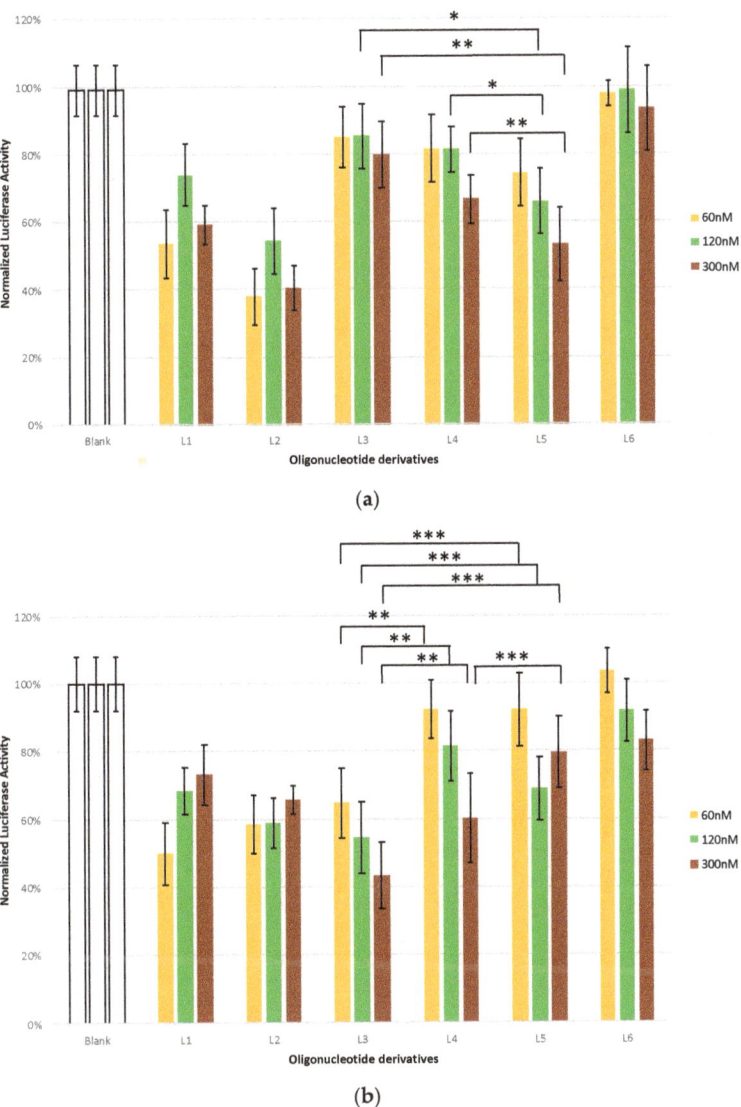

Figure 7. Luciferase inhibitory assay of [L1–L5] in HeLa (**a**) and HepG2 (**b**) cell lines, using L6 as a scrambled negative control and L1 (ASO) and L2 (Gapmer) transfected with lipofectamine as the positive control. Oligonucleotides carrying GalNAc are administrated without transfecting agent. * corresponds to t-test results with * $p \leq 0.1$, ** $p \leq 0.05$, *** $p \leq 0.01$ significantly different within each indicated pair.

Comparing the inhibitory data in HepG2 cells with the results on ASGPR affinity data in mouse primary hepatocytes, a reverse effect is observed. Specifically, L3 (Luc-G_6-L193) has higher inhibitory properties in HepG2 cells than L4 (G_6-Luc-L193), but a lower ASGPR affinity in mouse primary hepatocytes. On the other, the luciferase inhibition data are in agreement with the cytometry assays that evaluate the internalization of L7 (FAM-Luc-G_6-L193) that is the fluorescent version of L3. Figure 6 clearly demonstrated the high internalization efficacy of L7 in HepG2 cells in agreement of the high luciferase inhibitory properties of L3.

3. Discussion

The last five years have witnessed the approval of several oligonucleotides for human use [1,2]. A major success in this field is the development of the active targeting system based in the use of the triantennary GalNAc ligand directly conjugated to the nucleic acid part [32]. This ligand directs the therapeutic oligonucleotides to hepatocytes [12]. For these reasons, there is an interest for the synthesis of these valuable oligonucleotides. It has been proposed that a potential simplification on the synthesis of tri-GalNAc oligonucleotides is the stepwise addition of three monomeric GalNAc units at the 3'-end [13,14]. In this work, we have studied an alternative method aiming to simplify the synthetic process that consists in the use of the tetramerization of individual monomeric GalNAc-oligonucleotides by adding a short G-rich sequence (TGGGGGT) that form a parallel G-quadruplex. Monomeric GalNAc-oligonucleotides are easier to prepare than the triantennary GalNAc and the addition of eight extra deoxynucleotides will still be a clear simplification of the preparation of these valuable oligonucleotide conjugates. We hypothesized that the formation of the tetramer will have a similar effect to the addition of a triantennary GalNAc derivative. In addition, the presence of the G-quadruplex may also have a beneficial effect in cellular uptake as it has been hypothesized that G-quadruplex may bind to serum proteins and to certain protein membrane receptors overexpressed in cancer cells [18–20].

For these reasons we have previously studied the gene inhibitory properties of antisense oligonucleotides linked to lipid-modified G-quadruplex forming oligonucleotides [21,22]. Analysis of gene expression showed that the formation of self-assembled G-quadruplex nanostructures did not disrupt the antisense mechanism and flow cytometry analyses confirmed that antisense G-quadruplex nanostructures were efficiently taken up by cancer cells. In addition we found that some lipid–functionalized G-quadruplexes are effective antiviral compounds against Hepatitis C [17]. Moreover, floxuridine oligomers linked to parallel G-quadruplex were able to deliver fluoropyrimidines to cancer cells acting as prodrugs [23].

In this work, we studied the generation of parallel G-quadruplex functionalized with a monomeric GalNAc unit at the 3'- end and their potential in the selective targeting of antisense oligonucleotide against *Renilla* luciferase gene. To this end, two types of sequences were selected. First, we incorporated a GalNAc unit at the 3'- end of the 17 mer HIV-1 sequence [17] instead of the lipid moiety. The formation of a self-assembled parallel G-quadruplex was confirmed through various biophysical techniques such as circular dichroism, nuclear magnetic resonance, polyacrylamide electrophoresis and denaturation curves. In this work, we did measure the NMR spectra of the GalNAc-17 mer before and after annealing, observing that the NMR signals assigned to the imino protons of the G-quadruplex were present all the time, indicating the presence of quadruplex in all conditions. This is contrasted with the slow formation kinetics of parallel G-quadruplexes described by several authors [17,29]. The kinetics of tetramolecular quadruplex formation has mainly been studied in TG_4T oligonucleotides [29]. These studies have also shown that association is slow and concentration-dependent as well as some other factors [29]. In addition, it has been described that the association rate can be accelerated by the presence of 8-modified guanine derivatives such as 8-bromo-G [33] or 8-amino-G [34]. There are no studies with TG_6T in NMR experimental conditions because it is harder to quantify the kinetic parameters, as TG_6T does not dissociate at 90 °C [29]. However, Mergny et al. described that the longer the G-tract, the faster the association and that the addition of one tetrad from TG_4T introduces a 10-fold larger association rate [29]. One should expect that the association rates of TG_6T will become 100-fold larger than TG_4T. In addition, we have to consider that NMR experiments use relatively high concentrations compared with UV studies. The observed acceleration of quadruplex formation observed in our NMR experiments may also respond to the high concentration used in the experimental conditions, or because of the higher stability of the TG$_6$T quadruplex. Finally, the effect of the presence of the extra nucleotides or the linkers with GalNAc has never been studied. Therefore, our study may indicate that there is an interesting story to be revealed. From the practical view in the biological experiments, we prepared a concentrated solution a couple of days before to make sure that all the quadruplex is formed without taking into account the faster association rates of the G-rich oligonucleotides presented in this work.

As expected for the formation of tetramers, the affinity binding of GalNAc-G-rich-17 mers to ASGPR (KI: 31.3–47.2 nM) were found to have a lower affinity than a single-stranded oligonucleotide functionalized with the triantennary GalNAc (A-PO-L96, KI 10.4 nM), but higher than the affinity to ASGPR of a control single-stranded A-rich sequence that was not possible to determine at concentrations tested. This result confirmed that tetramerization of monofunctionalized GalNAc G-rich sequences does indeed increase the affinity to ASGPR most probably by emulating the rapid internalization via the clathrin-mediated pathway described for the multifunctional triantennary oligonucleotides [12,15]. We have prepared two different GalNAc derivatives with different length (L193 and L235) as well as oligonucleotides carrying two phosphorothioate linkages at the terminal 3′ and 5′- ends, but the analysis of the ASGPR binding affinity resulted in values of the same range (KI: 31.3–47.2 nM). Although a slight higher affinity is observed for oligonucleotides carrying phosphorothioate linkages (KI: 31.3 and 42.8 vs. 39.9 and 47.2 nM) and carrying the shorter linker L235 (31.3 and 39.9 vs. 42.8 and 47.2).

Next we prepared a series of oligonucleotides that combined a gapmer sequence against *Renilla* luciferase carrying 2′-O-methyl-RNA modifications at the last four positions of the 3′- and 5′-ends and phosphorothioate linkages with a G-quadruplex forming TG$_6$T sequence with phosphodiester linkages. In these cases, the characterization of the G-quadruplex is more challenging due to the complexity of the molecules, but the analysis of polyacrylamide electrophoresis clearly showed the presence of the G-quadruplex as a major species. Unfortunately, the affinity binding of GalNAc-G-rich oligonucleotides to ASGPR decrease more than one order of magnitude (KI: 8.8 vs. 121 and 357 nM). In addition, duplex formation abolishes binding of GalNAc-G-rich oligonucleotides to ASGPR, precluding the potential use of tetramerization for GalNAc active hepatocyte targeting for siRNA and other double-stranded oligonucleotides such as microRNA mimetics.

The biological properties of the hybrid gapmer/G-quadruplex molecules were evaluated in two cell lines (HepG2, HeLa) demonstrating the absence of toxicity. The analysis of the stability of hybrid gapmer/G-quadruplex molecules towards degradation in 10% PBS indicates that these oligonucleotides are more stable than the H series because they carry phosphorothioate and 2′-O-methyl residues and these modifications increase binding of oligonucleotides to serum proteins that may protect oligonucleotides from degradation.

The internalization preferences of GalNAc-quadruplexes to hepatic cells were also demonstrated and specifically the cellular uptake of oligonucleotide carrying the antisense sequence (FAM-Luc-G$_6$-GalNAc, L7) and the G-rich sequence goes from 9% in HeLa cells to 30% in HepG2 cells (at 300 nM concentration). Consequently, to the good internalization properties of the gapmer-G-quadruplex hybrid molecules L3 and L4, we observed an enhancement of the luciferase inhibition in HepG2 cell lines versus HeLa cells, L3 being the more active gapmer in HepG2 cells.

All together, we have demonstrated the possibility of obtaining multifunctional nanoassemblies by simple hybridization of monofunctionalized oligonucleotides. This approach has been extensively studied in the bibliography and holds promise for future developments [35]. For example, three oligonucleotides designed to form a triplex were functionalized with a short coiled peptide that interacts between them, stabilizing the triplex structure [36]. Similarly, G-quadruplex formation has been shown to direct the assembly of two peptide strands generating two-loop structures on top of the G-quadruplex. This approach can be used with homo and hetero peptide sequences [37]. G-rich oligonucleotides designed to form parallel G-quadruplex functionalized with hydrophobic group are able to tetramerize resulting in multifunctionalized G-quadruplex with affinity to viral proteins [17,38] and/or cell membranes [21–23]. The strategy here described can be extended for the hepatic delivery of antiproliferative and antiviral nucleosides by the prodrug strategy described recently [23].

4. Materials and Methods

Lipofectamine 2000 was purchased from Invitrogen. Dulbecco's Modified Eagle's Medium (DMEM) was supplemented with 10% heat-inactive fetal serum bovine (FBS). DMEM, PBS buffer and distilled water (DNAse/RNAse free) were purchased from Gibco (Waltham, MA, USA). Luciferase assay kits were acquired from Promega (Madison, WI, USA). Luminescence was measured in a Promega Glomax Multidetection System instrument. Flow cytometer analyses were carried out in a Guava® easyCyte. Luciferase plasmids (p-GL3 and p-RL) were extracted of growing up E. coli transfected previously with the corresponding plasmid and purified with Qiagen Giga plasmid purification kit purchased from Qiagen.

4.1. Synthesis of GalNac Oligonucleotide Conjugates and Controls

Control oligonucleotides were synthesized on an Applied Biosystems DNA synthesizer using solid-phase phosphoramidite chemistry. These oligonucleotides were 5'-TG$_6$T-3', 5'-T$_8$-3' the anti-luciferase phosphorothioate oligonucleotide 5'-CsGsTsTsTsCsCsTsTsTsGs TsTsCsTsGsGsA-3', and the gapmer sequence 5'-<u>CsGsUsUs</u>TsCsCsTsTsGsTsTsCs<u>UsGsGsA</u>-3' where the underlined nucleotides are 2'-O-methyl-RNA and contain phosphorothioate linkages. The antisense phosphorothioate and the gapmer oligonucleotides are complementary to the mRNA of the *Renilla* luciferase gene which target to a predominant accessible site between 20 and 40 nt of the luciferase gene [27].

Oligonucleotides carrying N-acetylgalactosamine (GalNAc) shown in Tables 1 and 2 have been synthesized using solid-phase phosphoramidite chemistry. For the introduction of the GalNAc at the 3'-end two special solid supports were prepared functionalized with O-tetraacetyl-N-acetyl galactosamine connected to 4-hydroxyprolinol through alkyl chains of different lengths (L235 and L193) (Scheme 1). The synthesis of solid supports functionalized with these GalNAc derivatives has been reported in detail previously [13]. All oligonucleotides were purified by HPLC, and the major peak was characterized by mass spectrometry (Table S1).

4.2. G-Quadruplex Formation

All G-rich and control oligonucleotides were treated in the same conditions to reduce the differences between samples. First, oligonucleotides were quantified, aliquoted appropriately for the different experiments and concentrated to dryness. Oligonucleotides used in CD, gel electrophoresis and in cell experiments were dissolved in 100 mM PBS (pH 7.4). In all cases, solutions were annealed by heating at 93 °C for 2 min in a thermo-block and slow cooling down to room temperature (2–3 days). The resulting oligonucleotide solutions were stored at 4 °C until used.

4.3. Circular Dichroism

CD spectra of the annealed solutions were registered between either from 205 to 320 nm or from 220 to 320 nm at 20 °C in PBS 1× buffer. CD thermal denaturation experiment of the G-quadruplex-forming oligonucleotide (G-PO-HCV-L235, H2) (4–5 µM) was performed in a range of temperature between 20 and 85 °C using a heating rate of 0.8 °C min^{-1} and monitoring the ellipticity at 263 nm.

4.4. Nuclear Magnetic Resonance (NMR) Experiments

The NMR sample of H2 (G-PO-HCV-L235, 2 mg in 0.55 mL, 0.63 mM) was prepared in of PBS buffer (0.01 M phosphate buffer, 0.0027 M KCl and 0.137 M NaCl) (H$_2$O:D$_2$O 9:1), pH 7.4. The oligonucleotide was heated to 90 °C for 1 min and then cooled at room temperature overnight. ^{1}H-NMR spectra were acquired at 25 °C at different times after the heating with a Bruker AV600, 600 MHz spectrometer, equipped with a TXI probe with z-gradient, and processed with TOPSPIN 2.1 software. Two-dimensional NOESY spectra were acquired with mixing time of 300 ms and 2D TOCSY spectra with mixing time of 60 ms.

4.5. Polyacrylamide Electrophoresis Assays

A native polyacrylamide gel electrophoresis (PAGE) was carried out using 12% (v/v) acrylamide to characterize the mobility of the assembled oligonucleotides. Samples were dissolved in PBS and the gel was run in 1× TBE buffer supplemented with 100 mM KCl at 150 V (12% PAGE) for approximately 4–5 h maintaining a fixed temperature of 20 °C. SYBR green (20 µL in 200 mL 1× TBE) was used to stain the DNA bands and then images were taken using Fujifilm LAS-1000 Intelligent Dark Box II as well as IR LAS-1000 Lite v1.2. The ladder used was a solution containing Bromophenol Blue and Xylene Cyanol for visual tracking of oligonucleotides migration during the electrophoretic process.

4.6. Affinity Binding to ASGPR Receptors in Primary Mouse Hepatocytes

ASGPR competitive binding was evaluated as previously described [14]. Briefly, 20 nM triantennary GalNAc-conjugated, Alexa647-labeled siRNA, described previously [12], and 3 µM to 1.4 nM of relevant oligonucleotide were premixed and then co-incubated with 1×10^5 of viable plateable primary mouse (CD-1) cryopreserved hepatocytes (ThermoFisher Scientific, Waltham, MA, USA) in Dulbecco's Modified Eagle Medium (DMEM) with 2% bovine serum albumin (BSA). Samples were incubated at 4 °C for 15 min, then washed twice with 2% BSA in Dulbecco's Phosphate-Buffered Saline with magnesium and calcium (DPBS). Cells were suspended in a solution of 2% BSA in DPBS with 2 µg/mL propidium iodide and analyzed on a BD LSRII flow cytometer. Hepatocytes were gated by size using forward scatter and side scatter, and dead cells stained with propidium iodide were excluded from analysis. Median fluorescence intensity of the Triantennary GalNAc-conjugated, Alexa647-labeled siRNA was quantified (A647 MFI). Data were analyzed using FlowJo and KI was calculated in GraphPad Prism using a derived Michaelis-Menten equation for competitive inhibition.

4.7. MTT Assays

HeLa and HepG2 cells were regularly passaged to maintain exponential growth and proprieties. Cells were seeded for 8000 cells into each well of a 96-well cell culture plate. Then, cells were incubated overnight at 37 °C and 5% CO_2 in DMEM supplemented with 10% FBS. After that, the oligonucleotide conjugates were added and the cells incubated for 24 h at increasing concentrations (60, 120 and 300 nM) in 200 mL of new DMEM (10% FBS). PBS buffer was used as a control. Before that time, growth medium was removed and cells were washed with PBS (200 mL), so all compounds not internalized were removed and 200 mL of fresh DMEM (10% FBS) was added to incubated the cells again for 12 h more at 37 °C. Next day, using a MTT dye solution (25 mL; 5 mg mL^{-1}) was added per well and cells were incubated for two additional hours. Medium was removed and DMSO (100 mL) was added to dissolve formazan crystals formed and the absorbance of the solution was measured in a plate reader. Data was analyzed in a excel sheet. The MTT assay was repeated 12 times in six independent experiments in Hela cells and 10 times in 5 independent experiments in HepG2 cells.

4.8. Flow Cytometry

HeLa and HepG2 cells (80,000 and 100,000 cells per well, respectively) were seeded on 24-well cell culture plate. Then, they were incubated overnight at 37 °C and 5% CO_2 in DMEM supplemented with 10% FBS. Next day, fluorescein-labelled oligonucleotides were added and the cells were incubated for 24 h. After that time, cells were washed with PBS (500 µL) and harvested with trypsin (200 mL) at 37 °C for some minutes. DMEM (800 µL) was added and cells were centrifuged (3.0 rcf 8 min). DMEM supernatants were removed and pellets were washed with 800 µL of PBS to analyze them in a cytometer instrument. For each sample, 5000 events were collected in a selected gate (R1) that corresponds to the cell population (each line was analyzed with a different template). Guava soft software Incite surface was used to analyze the relationship between fluorescently labeled and unlabeled cell populations and quantify the percentages of each type of cells.

4.9. Luciferase Assays

HeLa and HepG2 cells were seeded in the same 24-well plate as the previous experiment, but in a confluence of 100,000 and 200,000 cells per well, respectively. After overnight incubation at 37 °C and 5% CO_2 in DMEM + 10% FBS, cells were transfected with a solution of two luciferase plasmids *Renilla* and Firefly luciferase (pRL (10 ng/µL) and pGL3 (100 ng/µL)) in Optimem buffer (without plasma proteins) using lipofectamine 2000 to internalize the plasmids into the different cell lines. After 4 h of incubation, cells were thoroughly washed, and the incubation with the Luc-oligonucleotides at 60–120–300 nM was done without lipofectamine, so differences between cells will affect to the internalization of the drug. All these experiments were done in a final volume of 600 µL. An ASO as control at the same concentration was used for comparison purposes. Transfections were performed in triplicate and for 24 h. After this time, cells were washed again to eliminate all the Luc-compound not internalized and frozen. Finally, lysates were analyzed by comparing the luminescence of both Firefly and *Renilla* luciferase protein according to manufacturer's protocol. The luciferase assay was repeated 30 times in 12 independent experiments in Hela cells and 30 times in 8 independent experiments in HepG2 cells.

4.10. Analysis of the Stability of Oligonucleotides towards Snake Venom Phosphodiesterase and 10% FBS

Briefly, 10 µL of the oligonucleotide at 5 µM were dissolved in 100 µL of a mixture of Tris·HCl buffer (100 mM), $MgCl_2$ solution (100 mM) and 1 µL of phosphodiesterase I from *Crotalus adamanteus* venom (USB) was added and incubated in 37 °C. Then, 5 µL of the solution was removed (every certain time) and added to 15 µL of urea solution (8M) and heated to 85 °C for 5 min and stored at the freeze. Finally, the different samples were analyzed by denaturing (8M urea) 12% polyacrylamide gel electrophoresis. Similarly, 10 µL of the oligonucleotide at 5 µM were dissolved in 10% FBS and incubated at 37 °C. As before, 5 µL of the solution was removed and added to 15 µL of urea solution (8M) and heated to 85 °C for 5 min and stored at the freeze. Then, the different samples were analyzed in denaturing (8M urea) 12% polyacrylamide gel electrophoresis.

5. Conclusions

There is a large interest in the development of defined DNA nanostructures carrying multifunctional ligands that are recognized by membrane receptors to achieve more efficient delivery systems for the fast-growing field of therapeutic oligonucleotides. To this end, we addressed the potential use of the tetramerization of G-rich sequences to build novel methods to achieve the multivalency of the well-known GalNAc ligands using simple oligonucleotides carrying a single GalNAc ligand. To this end, several G-rich oligonucleotides carrying the TGGGGGT sequence were prepared carrying one GalNAc molecule per oligonucleotide. The synthesis of these monofunctionalized GalNAc-oligonucleotides is easier than the standard triantennary GalNAc derivatives due to the complexity of the synthesis of the triantennary GalNAc ligand. We studied three potential tetrameric G-quadruplexes carrying four GalNAc molecules by tetramerization of monofunctionalized GalNAc G-rich oligonucleotide: I. Tetravalent GalNAc linked to the HVC sequence (H series, Table 1). II. Tetravalent GalNAc linked to an antisense oligonucleotide against luciferase gene (Gapmer, L series, Table 2). III. Tetravalent GalNAc linked to a siRNA (R series, Table S2). The study show that structures I and II can be observed and may be responsible for an increased affinity to ASGPR and an increased antisense activity in hepatocytes while structure III most probably is not formed (Figure S1). The formation of a self-assembled parallel G-quadruplex in structures I and II was confirmed through various biophysical techniques. Binding experiments to asialoglycoprotein receptors (ASGPR) show that structure I achieve the best affinity binding to ASGPR and clearly better than single-stranded control oligonucleotides that cannot form G-quadruplex demonstrating that powerful advantages of multifunctionality. Binding experiments to ASGPR in structures II show a less efficient ASGPR binding, but still better than single-stranded control oligonucleotides. In addition some differences are observed depending of the relative position of the therapeutic cargo. The internalization preferences of GalNAc-quadruplexes to hepatic cells were also demonstrated by cell cytometry. As well as the enhancement of the luciferase inhibition under gymnotic conditions using the luciferase assay in the HepG2 cell line. All together demonstrates that tetramerization of G-rich oligonucleotides can be used to obtain the beneficial effects of multivalent GalNAc functionalization although in long oligonucleotides as well as duplex siRNA tetramerization is less efficient probably by competing duplex formation.

Supplementary Materials: The following supporting information can be downloaded at: https://www.mdpi.com/article/10.3390/molecules27123944/s1, Figure S1. Schematic representation of the potential tetrameric G-quadruplexes. Table S1: Sequences and mass spectra of oligonucleotide derivatives prepared in this work. Table S2: Oligonucleotide RNA derivatives carrying anti-mTTR siRNA sequence. Figure S2: Melting curves on Gquadruplex sequences. Figure S3: CD spectra of G-rich and control oligonucleotides. Figure S4: Analysis of stability of G-rich oligonucleotides to FBS and phosphodiesterase. Figure S5: MTT assay in HeLa (a), and HepG2 (b) cells.

Author Contributions: Conceptualization, R.E., M.M., S.G. (Santiago Grijalvo) and A.C.; methodology, A.C., S.G. (Santiago Grijalvo), N.E., S.G. (Swati Gupta), K.Y., R.G., S.M., M.M. and R.E.; validation, A.C., S.G. (Santiago Grijalvo), N.E., S.G. (Swati Gupta), K.Y., R.G., S.M., M.M. and R.E.; formal analysis, R.E., M.M., S.G. (Santiago Grijalvo) and A.C.; investigation, A.C., S.G. (Santiago Grijalvo), N.E., S.G. (Swati Gupta), K.Y., R.G., S.M., M.M. and R.E.; resources, R.E., M.M. and S.G. (Santiago Grijalvo) data curation, A.C. and N.E.; writing—original draft preparation, R.E., M.M., S.G. (Santiago Grijalvo) and A.C.; writing—review and editing, A.C., S.G. (Santiago Grijalvo), N.E., S.G. (Swati Gupta), K.Y., R.G., S.M., M.M. and R.E.; visualization, R.E., S.G. (Santiago Grijalvo) and A.C.; supervision, R.E., M.M. and S.G. (Santiago Grijalvo); project administration, R.E.; funding acquisition, M.M. and R.E. All authors have read and agreed to the published version of the manuscript.

Funding: This research was funded by Spanish Ministerio de Ciencia e Innovación (MICINN) (Projects CTQ2017-84415-R, PID2019-107158GB-I100, PID2020-118145RB-I100), CIBER-BBN grant number CB06/01/0019, PIANO DI SOSTEGNO ALLA RICERCA 2020—Linea 2 azione B (DEFENS) and a predoctoral contract grant (PRE2018-084056) to A.C. CIBER-BBN is an initiative funded by the VI National R + D + I Plan 2008–2011, Iniciativa Ingenio 2010, Consolider Program, CIBER Actions and financed by the Instituto de Salud Carlos III with assistance from the European Regional Development. The APC was funded by MICINN, project PID2020-118145RB-I100.

Institutional Review Board Statement: Not applicable.

Informed Consent Statement: Not applicable.

Data Availability Statement: The data presented in this study are available on request from the corresponding author.

Acknowledgments: This work has been partially done at the NANBIOSIS Unit U29; Oligonucleotide synthesis platform, we thank Drs. Anna Aviñó, Carme Fàbrega, Sébastien Lyonnais and Giles Mirambeau for helpful discussions and support.

Conflicts of Interest: The authors declare no conflict of interest.

Sample Availability: Samples of the oligonucleotides prepared in this work are available from the authors upon request to the corresponding author.

References

1. Khvorova, A.; Watts, J.K. The chemical evolution of oligonucleotide therapies of clinical utility. *Nat. Biotech.* **2017**, *35*, 238–248. [CrossRef] [PubMed]
2. Jorge, A.F.; Grijalvo, S.; Fàbrega, C.; Aviñó, A.; Eritja, R. Advances in therapeutic oligonucleotide chemistry. In *Nucleic Acids Chemistry. Modifications and Conjugates for Biomedicine and Nanotechnology*; Eritja, R., Ed.; De Gruyter: Berlin, Germany, 2021; pp. 273–329.
3. Al Shaer, D.; Al Musaimi, O.; Albericio, F.; de la Torre, B.G. 2019 FDA TIDES (peptides and oligonucleotides) harvest. *Pharmaceuticals* **2020**, *13*, 40. [CrossRef] [PubMed]
4. Crooke, S.T.; Liang, X.H.; Baker, B.F.; Crooke, R.M. Antisense technology: A review. *J. Biol. Chem.* **2021**, *296*, 100416. [CrossRef] [PubMed]
5. Egli, M.; Manoharan, M. Re-engineering RNA molecules into therapeutic agents. *Acc. Chem. Res.* **2019**, *52*, 1036–1047. [CrossRef]
6. Gragoudas, E.S.; Adamis, A.P.; Cunningham, E.T., Jr.; Feinsod, M.; Guyer, D.R. Pegaptanib for neovascular age-related macular degeneration. *N. Engl. J. Med.* **2004**, *351*, 2805–2816. [CrossRef]
7. Chakraborty, C.; Sharma, A.R.; Sharma, G.; Doss, G.P.; Lee, S.S. Therapeutic miRNA and siRNA moving from bench to clinic as next generation on medicine. *Mol. Nucleic Acids* **2017**, *8*, 132–143. [CrossRef]
8. Mendell, J.T. Targeting a long noncoding RNA in breast cancer. *N. Engl. J. Med.* **2016**, *374*, 2287–2289. [CrossRef]
9. Dong, Y.; Siegwart, D.J.; Anderson, D.G. Strategies, design, and chemistry in siRNA delivery systems. *Adv. Drug Deliv. Rev.* **2019**, *144*, 133–147. [CrossRef]
10. Balwani, M.; Sardh, E.; Ventura, P.; Aguilera Peiró, P.; Rees, D.C.; Stölzel, U.; Bissell, D.M.; Bonkovsky, H.L.; Windyga, J.; Anderson, K.E.; et al. Phase 3 trial of RNAi therapeutic Givosiran for acute intermittent porphyria. *N. Engl. J. Med.* **2020**, *382*, 2289–2301. [CrossRef]
11. Ray, K.K.; Wright, R.S.; Kallend, D.; Koenig, W.; Leiter, L.A.; Raal, F.J.; Bisch, J.A.; Richardson, T.; Jaros, M.; Wijngaard, P.L.J.; et al. Two phase 3 trials of inclisiran in patients with elevated LDL cholesterol. *N. Engl. J. Med.* **2020**, *382*, 1507–1519. [CrossRef]
12. Nair, J.K.; Willoughby, J.L.S.; Chan, A.; Charisse, K.; Alam, R.; Wang, Q.; Hoekstra, M.; Kandasamy, P.; Kel'in, A.V.; Milstein, S.; et al. Multivalent N-acetylgalactosamine-conjugated siRNA localizes in hepatocytes and elicits robust RNAi-mediated gene silencing. *J. Am. Chem. Soc.* **2014**, *136*, 16958–16961. [CrossRef] [PubMed]
13. Rajeev, K.G.; Nair, J.K.; Jayaraman, M.; Charisse, K.; Taneja, N.; O'Shea, J.; Willoughby, J.L.S.; Yucius, K.; Nguyen, T.; Shulga-Morskaya, S.; et al. Hepatocyte-specific delivery of siRNAs conjugated to novel non-nucleosidic trivalent N-acetylgalactosamine elicits robust gene silencing in vivo. *ChemBioChem* **2015**, *16*, 903–908. [CrossRef] [PubMed]
14. Matsuda, S.; Keiser, K.; Nair, J.K.; Charisse, K.; Manoharan, R.M.; Kretschmer, P.; Peng, C.G.; Kel'in, A.V.; Kandasamy, P.; Willoughby, J.L.S.; et al. siRNA conjugates carrying sequentially assembled trivalent N-acetylgalactosamine linked through nucleosides elicit robust gene silencing in vivo in hepatocytes. *ACS Chem. Biol.* **2015**, *10*, 1181–1187. [CrossRef] [PubMed]
15. Janas, M.M.; Schlegel, M.K.; Harbison, C.E.; Yilmaz, V.O.; Jiang, Y.; Parmar, R.; Zlatev, I.; Castoreno, A.; Xu, H.; Shulga-Morskaya, S.; et al. Selection of GalNAc-conjugated siRNAs with limited off-target-driven rat hepatotoxicity. *Nat. Commun.* **2018**, *9*, 723. [CrossRef]
16. Gatto, B.; Palimbo, M.; Sissi, C. Nucleic acid aptamers based on the G-quadruplex structure: Therapeutic and diagnostic potential. *Curr. Med. Chem.* **2009**, *16*, 1248–1265. [CrossRef]
17. Koutsoudakis, G.; Paris de León, A.; Herrera, C.; Dorner, M.; Pérez-Vilaró, G.; Lyonnais, S.; Grijalvo, S.; Eritja, R.; Meyerhans, A.; Mirambeau, G.; et al. Oligonucleotide-lipid conjugates forming G-quadruplex structures are potent and pangenotypic hepatitis C virus entry inhibitors in vitro and ex vivo. *Antimicrob. Agents Chemother.* **2017**, *61*, e02354-16. [CrossRef]
18. Bates, P.J.; Reyes-Reyes, E.M.; Malik, M.T.; Murphy, E.M.; O'Toole, M.G.; Trent, J.O. G-quadruplex oligonucleotide AS1411 as a cancer-targeting agent: Uses and mechanisms. *Biophys. Biochip. Acta* **2017**, *1861*, 1414–1428. [CrossRef]
19. Platella, C.; Riccardi, C.; Montesarchio, D.; Roviello, G.N.; Musumeci, D. G-quadruplex-based aptamers against protein targets in therapy and diagnostics. *Biophys. Biochim. Acta* **2017**, *1861*, 1429–1447. [CrossRef]
20. Riccardi, C.; Fàbrega, C.; Grijalvo, S.; Vitiello, G.; D'Errico, G.; Eritja, R.; Montesarchio, D. AS1411-decorated niosomes as effective nanocarriers for Ru(III)-based drugs in anticancer strategies. *J. Mat. Chem. B* **2018**, *6*, 5368–5384. [CrossRef]
21. Grijalvo, S.; Alagia, A.; Gargallo, R.; Eritja, R. Cellular uptake studies of antisense oligonucleotides using g-quadruplex-nanostructures: The effect of cationic residue in the biophysical and biological properties. *RSC Adv.* **2016**, *6*, 76099–76109. [CrossRef]
22. Grijalvo, S.; Clua, A.; Eres, M.; Gargallo, R.; Eritja, R. Tuning G-quadruplex nanostructures with lipids. Towards designing hybrid scaffolds for oligonucleotide delivery. *Int. J. Mol. Sci.* **2021**, *22*, 121. [CrossRef] [PubMed]
23. Clua, A.; Fàbrega, C.; García-Chica, J.; Grijalvo, S.; Eritja, R. Parallel G-quadruplex structures increase cellular uptake and cytotoxicity of 5-fluoro-2'-deoxyuridine oligomers in 5-fluorouracil resistant cells. *Molecules* **2021**, *26*, 1741. [CrossRef] [PubMed]
24. Lyonnais, S.; Hounsou, C.; Teulade-Fichou, M.P.; Jeusset, J.; Le Cam, E.; Mirambeau, G. G-quartets assembly within a G-rich DNA flap. A possible event at the center of the HIV-1 genome. *Nucleic Acids Res.* **2002**, *30*, 5276–5283. [CrossRef]
25. Lyonnais, S.; Gorelick, R.J.; Mergny, J.L.; Le Cam, E.; Mirambeau, G. G-quartets direct assembly of HIV-1 nucleocapsid protein along single-stranded DNA. *Nucleic Acids Res.* **2003**, *31*, 5754–5763. [CrossRef] [PubMed]

26. Lyonnais, S.; Grijalvo, S.; Alvarez-Fernández, C.; Fleta, E.; Martínez, J.; Meyerhans, A.; Sánchez-Palomino, S.; Mirambeau, G.; Eritja, R. Lipid-oligonucleotide conjugates forming G-quadruplex (lipoquads) as potent inhibitors of HIV entry. *Proceedings* **2017**, *1*, 670.
27. Zhang, H.-Y.; Mao, J.; Zhou, D.; Xu, D.; Thonberg, H.; Liang, Z.; Wahlestedt, C. mRNA accessible site tagging (MAST): A novel high throughput method for selecting effective antisense oligonucleotides. *Nucleic Acid Res.* **2003**, *31*, e72. [CrossRef]
28. Malgowska, M.; Gudanis, D.; Teubert, A.; Dominiak, G.; Gdaniec, Z. How to study G-quadruplex structures. *BioTechnol. J. Biotechnol. Comput. Biol. Bionanotechnol.* **2012**, *93*, 381–390. [CrossRef]
29. Mergny, J.L.; De Cian, A.; Ghelab, A.; Sacca, B.; Lacroix, L. Kinetics of tetramolecular quadruplexes. *Nucleic Acids Res.* **2005**, *33*, 81–94. [CrossRef]
30. Adrian, M.; Heddi, B.; Phan, A.T. NMR spectroscopy of G-quadruplex. *Methods* **2012**, *57*, 11–24. [CrossRef]
31. Lin, C.; Dickerhoff, J.; Yang, D. NMR studies of G-quadruplex structures and G-quadruplex-interactive compounds. *Methods Mol. Biol.* **2019**, *2035*, 157–176.
32. Paunovska, K.; Loughrey, D.; Dahlman, J.E. Drug delivery systems for RNA therapeutics. *Nat. Rev. Genet.* **2022**, *23*, 265–280. [CrossRef] [PubMed]
33. Gros, J.; Rosu, F.; Amrane, S.; De Cian, A.; Gabelica, V.; Lacroix, L.; Mergny, J.L. Guanines are a quartet's best friend: Impact of base substitutions on the kinetics and stability of tetramolecular quadruplexes. *Nucleic Acids Res.* **2007**, *35*, 3064–3075. [CrossRef] [PubMed]
34. Gros, J.; Aviñó, A.; Lopez de la Osa, J.; González, C.; Lacroix, L.; Pérez, A.; Orozco, M.; Eritja, R.; Mergny, J.L. 8-Aminoguanine accelerates tetramolecular G-quadruplex formation. *Chem. Comm.* **2008**, 2926–2928. [CrossRef] [PubMed]
35. Lacroix, A.; Sleiman, H.F. DNA nanostructures: Current challenges and opportunities for cellular delivery. *ACS Nano* **2021**, *15*, 3631–3645. [CrossRef] [PubMed]
36. Lou, C.; Christensen, N.J.; Martos-Maldonado, M.C.; Midtgaard, S.R.; Ejlersen, M.; Thulstrup, P.W.; Sørensen, K.K.; Jensen, K.J.; Wengel, J. Folding topology of a short coiled-coil peptide structure templated by an oligonucleotide triplex. *Chem. Eur. J.* **2017**, *23*, 9297–9305. [CrossRef]
37. Ghosh, P.S.; Hamilton, A.D. Noncovalent template-assisted mimicry of multiloop protein surfaces: Assembling discontinuous and functional domains. *J. Am. Chem. Soc.* **2012**, *134*, 13208–13211. [CrossRef]
38. Musumeci, D.; Montesarchio, D. Synthesis of a cholesterol-HEG phosphoramidite derivative and its application to lipid-conjugates of the anti-HIV 5′TGGGAG3′ Hotoda's sequence. *Molecules* **2012**, *17*, 12378–12392. [CrossRef]

Article

Exploring the Parallel G-Quadruplex Nucleic Acid World: A Spectroscopic and Computational Investigation on the Binding of the c-myc Oncogene NHE III1 Region by the Phytochemical Polydatin

Francesca Greco [1,†], Domenica Musumeci [2,3,†], Nicola Borbone [1,4], Andrea Patrizia Falanga [1], Stefano D'Errico [1], Monica Terracciano [1], Ilaria Piccialli [5], Giovanni Nicola Roviello [2,*] and Giorgia Oliviero [6]

[1] Department of Pharmacy, University of Naples Federico II, Via Domenico Montesano 49, 80131 Naples, Italy; francesca.greco@unina.it (F.G.); nicola.borbone@unina.it (N.B.); andreapatrizia.falanga@unina.it (A.P.F.); stefano.derrico@unina.it (S.D.); monica.terracciano@unina.it (M.T.)
[2] Institute of Biostructures and Bioimaging, Italian National Council for Research (IBB-CNR), Area di Ricerca Site and Headquarters-Via Pietro Castellino 111, 80131 Naples, Italy; domenica.musumeci@unina.it
[3] Department of Chemistry, University of Naples Federico II, Via Vicinale Cupa Cintia 21, 80126 Naples, Italy
[4] ISBE-IT, University of Naples Federico II, 80138 Naples, Italy
[5] Division of Pharmacology, Department of Neuroscience, Reproductive and Odontostomatological Sciences, University of Naples Federico II, Via Sergio Pansini 5, 80131 Naples, Italy; ilaria.piccialli@unina.it
[6] Department of Molecular Medicine and Medical Biotechnologies, University of Naples Federico II, Via Sergio Pansini 5, 80131 Naples, Italy; golivier@unina.it
* Correspondence: giroviel@unina.it
† These authors contributed equally to this work.

Citation: Greco, F.; Musumeci, D.; Borbone, N.; Falanga, A.P.; D'Errico, S.; Terracciano, M.; Piccialli, I.; Roviello, G.N.; Oliviero, G. Exploring the Parallel G-Quadruplex Nucleic Acid World: A Spectroscopic and Computational Investigation on the Binding of the c-myc Oncogene NHE III1 Region by the Phytochemical Polydatin. *Molecules* **2022**, *27*, 2997. https://doi.org/10.3390/molecules27092997

Academic Editors: Ramon Eritja, Daniela Montesarchio and Montserrat Terrazas

Received: 11 March 2022
Accepted: 5 May 2022
Published: 7 May 2022

Publisher's Note: MDPI stays neutral with regard to jurisdictional claims in published maps and institutional affiliations.

Copyright: © 2022 by the authors. Licensee MDPI, Basel, Switzerland. This article is an open access article distributed under the terms and conditions of the Creative Commons Attribution (CC BY) license (https://creativecommons.org/licenses/by/4.0/).

Abstract: Trans-polydatin (tPD), the 3-β-D-glucoside of the well-known nutraceutical trans-resveratrol, is a natural polyphenol with documented anti-cancer, anti-inflammatory, cardioprotective, and immunoregulatory effects. Considering the anticancer activity of tPD, in this work, we aimed to explore the binding properties of this natural compound with the G-quadruplex (G4) structure formed by the Pu22 [d(TGAGGGTGGGTAGGGTGGGTAA)] DNA sequence by exploiting CD spectroscopy and molecular docking simulations. Pu22 is a mutated and shorter analog of the G4-forming sequence known as Pu27 located in the promoter of the c-myc oncogene, whose overexpression triggers the metabolic changes responsible for cancer cells transformation. The binding of tPD with the parallel Pu22 G4 was confirmed by CD spectroscopy, which showed significant changes in the CD spectrum of the DNA and a slight thermal stabilization of the G4 structure. To gain a deeper insight into the structural features of the tPD-Pu22 complex, we performed an in silico molecular docking study, which indicated that the interaction of tPD with Pu22 G4 may involve partial end-stacking to the terminal G-quartet and H-bonding interactions between the sugar moiety of the ligand and deoxynucleotides not included in the G-tetrads. Finally, we compared the experimental CD profiles of Pu22 G4 with the corresponding theoretical output obtained using DichroCalc, a web-based server normally used for the prediction of proteins' CD spectra starting from their ".pdb" file. The results indicated a good agreement between the predicted and the experimental CD spectra in terms of the spectral bands' profile even if with a slight bathochromic shift in the positive band, suggesting the utility of this predictive tool for G4 DNA CD investigations.

Keywords: Pu22; G-quadruplex; c-myc; phytochemicals; circular dichroism; in silico simulations; molecular docking; CD prediction

1. Introduction

Among the noncanonical secondary structures of DNA, G-quadruplex (G4) is an appealing therapeutic target being found in specific regions of the genome such as telomeres and the regulatory regions of many oncogenes including c-kit, c-myc, and bcl-2 [1–14]. Remarkably, the promoter region of c-myc—an oncogene over-expressed in the majority of solid tumors and closely associated with cancer cell apoptosis, proliferation, invasion, cell-cycle arrest, and metastasis—can form a parallel G4 structure via Hoogsteen hydrogen bonds under specific conditions, and has been proposed as an effective target for antitumor drugs [15–23]. Particularly, it was found that molecules capable of binding and stabilizing this type of G4 downregulate the expression of c-myc, finally resulting in the apoptosis of cancer cells with great benefit in anticancer therapy [24–26].

Trans-polydatin (tPD, Figure 1a), the 3-β-D-glucoside of the well-known nutraceutical trans-resveratrol [27], is a natural polyphenol with documented anti-cancer, anti-inflammatory, cardioprotective, and immunoregulatory effects [28,29]. In a recent work, the G4-binding of tPD was explored toward three cancer-related G-rich DNA sequences, including c-myc, in comparison with a model duplex [30]. Interestingly, tPD displayed a clear binding ability with all the G4s and a higher ability, with respect to its aglycone derivative trans-resveratrol, to discriminate G4 over duplex DNA. Moreover, in vitro assays on melanoma cells proved that tPD significantly reduced telomerase activity, and inhibited cancer cell proliferation [30]. However, the adopted experimental conditions did not allow the detection of any significant conformational changes of the analyzed G4 DNA upon binding with tPD. Moreover, it was not possible to estimate the thermal stability of both c-myc and its complex with tPD, as needed for evaluating any stabilizing or destabilizing effects of the polyphenol on the G4-folded c-myc promoter [30]. On the other hand, other studies clearly indicated that the anticancer effects (including inhibition of cell proliferation and metastasis) of tPD took place through suppressing the c-myc expression, as proven in a model of human cervical cancer [31]. Therefore, conscious of the role of G4-structure binding and stabilization by ligands in c-myc deregulation [32], we decided to examine in more detail the molecular recognition of c-myc G4 by tPD through an approach differing from that previously reported in the literature from both experimental and in silico perspectives. To this aim, the interaction of tPD with c-myc DNA was studied in the present work focusing on the G4 structure formed by the Pu22 region having the sequence 5'-TGAGGGTGGGTAGGGTGGGTAA-3', a mutated and shorter analog of the sequence known as Pu27 located in the promoter of the c-myc oncogene and associated with the regulation of promoter activity and gene transcription.

Circular dichroism (CD) spectra of Pu22, either unliganded or in complex with the tPD, were recorded at variable temperatures in a buffer containing a lower concentration of potassium ions than previously reported [30]. CD spectroscopy is a technique typically employed to verify the formation of several secondary structures of nucleic acids and their analogs [33–36], including the G4 structure in G-rich DNA sequences [37–41], which also allows one to determine whether the denaturing temperature of a DNA secondary structure is affected by potential ligands [42]. Being aware of the utility of molecular docking in identifying DNA ligands [43,44] through verification of the favored binding sites in a complex, and in the estimation of the binding affinity, we decided to further characterize the molecular recognition of the G4 by tPD, by docking experiments of the tested polyphenol against the c-myc G4. The CD spectrum of the parallel G4 structure of Pu22 was further predicted by DichroCalc software [45], with the aim to verify whether the experimental profile could be reproduced by simulation as described below.

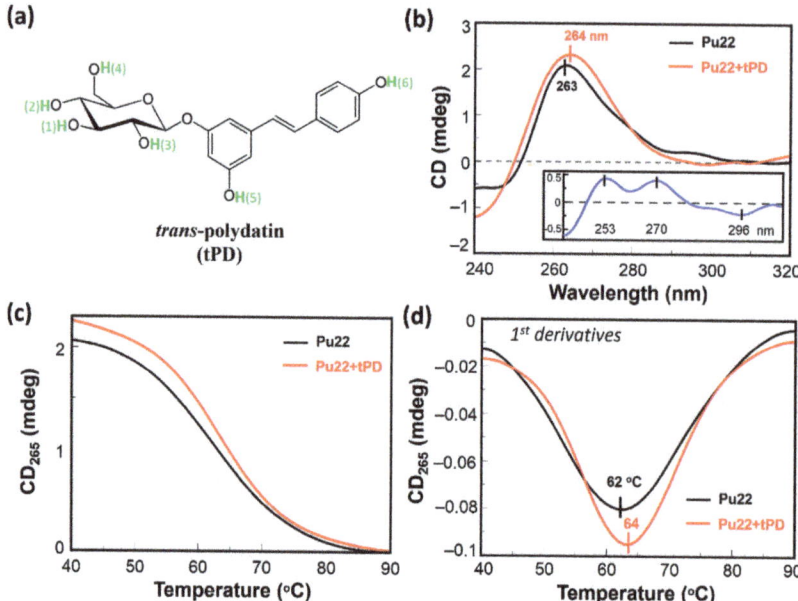

Figure 1. (a) Chemical structure of tPD; some atoms are numbered as in the docking program. (b) CD spectra of Pu22 2.5 µM (black) and its complex with tPD (red) at 40 °C. Inset shows the "difference" CD spectrum (tPD-Pu22 (mdeg)–Pu22 (mdeg)). (c) CD thermal denaturation curves (CD_{265} (mdeg) vs. T (°C)) and (d) their first derivatives vs. T plots for Pu22 (2.5 µM, black) and its complex with tPD (red). All experiments were run in PBS, pH 7.4 (optical path length = 0.1 cm).

2. Results

The effective binding of polydatin to Pu22 had been unequivocally shown by some of us by using various techniques including fluorescence [30]; however, with our work, we aimed to explore some biophysical characteristics of the complex, such as its thermal stability, and give more insights into the molecular aspects of the interaction by using in silico approaches. Our combined experimental and computational work started with the examination of the CD spectral features of Pu22 DNA and its complex with tPD. Moreover, a thermal denaturation study was conducted with both unliganded Pu22 and tPD-Pu22. The observations from CD spectroscopy were then interpreted in the light of the docking experiments performed by us on tPD-Pu22, but also on (tPD-Pu22)-Pu22, (Pu22)$_2$, and tPD-(Pu22)$_2$ molecular systems.

2.1. CD Spectroscopic Analysis of the Binding of Pu22 by tPD

With the aim to shed light on the possible mechanisms underlying the previously reported anticancer activity of tPD [31], we evaluated the potential of this polyphenol in binding Pu22. In our CD study, we observed a spectrum for Pu22 corresponding to a G4 with parallel topology, as identified by the characteristic positive band at ~265 nm and the negative one at 240 nm (Figure 1b, black line) [46]. In the presence of the polyphenol, we observed an increase in the positive CD signal at 263 nm accompanied with a 1 nm red-shift in the band maximum, and a concomitant reduction in the CD minimum at 240 nm (Figure 1b, red line). In addition, some differences in the CD spectra were evidenced in the 280–300 nm region. Overall, in the studied conditions, tPD induced a greater degree of structuration in the Pu22 G-quadruplex, as evidenced by the "difference" CD spectrum obtained by subtracting the CD spectrum of the Pu22 G4 to that of the tPD-Pu22 complex (inset of Figure 1b).

Then, we studied the effect of tPD on the stability of this G4 DNA by recording, for Pu22 and its mixture with the polyphenol, the CD values at 265 nm as a function of temperature (Figure 1c). We found a slight thermal stabilization in the presence of tPD, detectable by the increased value of the G4 melting temperature (T_m = 64 °C) with respect to the unliganded Pu22 G4 reference (T_m = 62 °C), leading to a ΔT_m of +2 °C (Figure 1d and Table 1). Furthermore, the overall variation in the CD signal at the λ_{max} upon heating, i.e., between 40 (folded state) and 90 °C (unfolded), for Pu22 alone or in complex with tPD, was 1.99 and 2.23, respectively, again confirming a higher structuration degree of the quadruplex when bound by the ligand. Specifically, the highest difference in the ΔCD for the two systems was evidenced between 40 and 50 °C (Figure 2, black-line dashed squares, and Table 1). Some differences between the two systems were also detected in the CD spectra recorded at the various temperatures in the 280–300 nm spectral region (Figure 2a,c).

Table 1. Summary of the CD and CD melting data for Pu22 and the complex tPD-Pu22. ΔT_m is the variation in the melting temperature of the complex with respect to the Pu22 reference; $\Delta CD_{max\ 40-90}$ is the difference in the CD value at the λ_{max} between 40 (folded state) and 90 °C (unfolded), whereas $\Delta CD_{max\ 40-50}$ is the one between 40 and 50 °C.

Entry	ΔT_m * (°C)	$\Delta CD_{max\ 40-90}$ °C (mdeg)	$\Delta CD_{max\ 40-50}$ °C (mdeg)
Pu22	0	1.99	0.54
tPD-Pu22	+2	2.23	0.15

* T_m Pu22 = 62 °C.

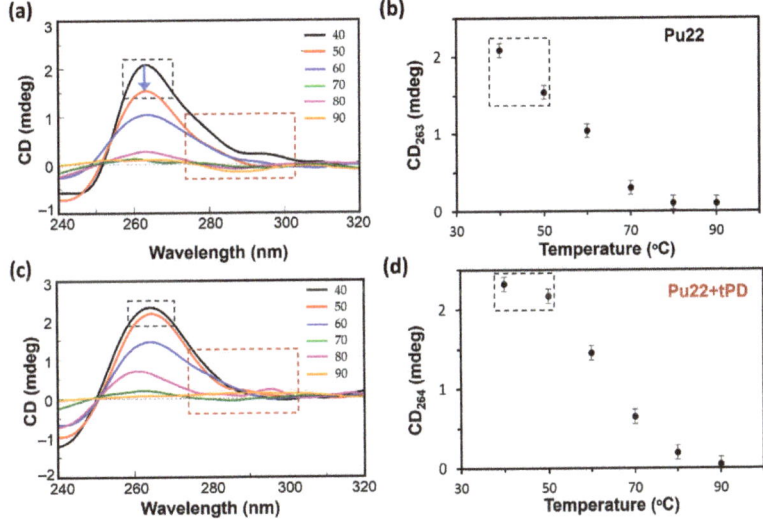

Figure 2. CD spectra of Pu22 (2.5 µM) (a) and its complex with tPD (c) recorded in the 40–90 °C temperature range. Plots of the CD signal at λ_{max} (in mdeg) vs. temperature (in °C) for Pu22 (b) and its complex with tPD (d) derived from panels a and c. All experiments were run in PBS, pH 7.4 (optical path length = 0.1 cm).

2.2. Molecular Docking Studies

The importance of phytochemicals in drug discovery [47] prompted the scientific community to investigate the potential of a plethora of natural products in anticancer strategies by using in silico approaches for a rapid screening or to corroborate and describe at a molecular level in vitro observations. In this context, we used herein in silico methods, and more specifically molecular docking, in analogy to other recent literature examples using polyphenols as anticancer drug candidates [48,49], to deeper analyze the interaction between tPD and the target Pu22 G4, whose sequence is located in a regulatory region of the c-myc oncogene. More in detail, we exploited the Hdock software [50,51] for the computational studies involving DNA. Hdock is used for both macromolecule–macromolecule [50] and small molecule–macromolecule [52] dockings, including those involving DNA and RNA G4s [53,54]. It is worth noting that the docking software provides dimensionless scores (Hdock scores) that are correlated to binding affinities [55]. This allows the comparison to made of the binding affinity of ligands for a given target by simply comparing their docking scores, with the most negative values being associated with the highest affinity ligands [55]. We found by Hdock docking that tPD bound the G4 target in proximity of the G4, G8, G13, and G17 nucleotides (Figure 3, Table 2). Comparing the Hdock scores for the top-1 poses, we can predict that the ligand bound Pu22 with a lower affinity than its aglycone resveratrol (tRES, Table 2), as experimentally shown in the literature [30].

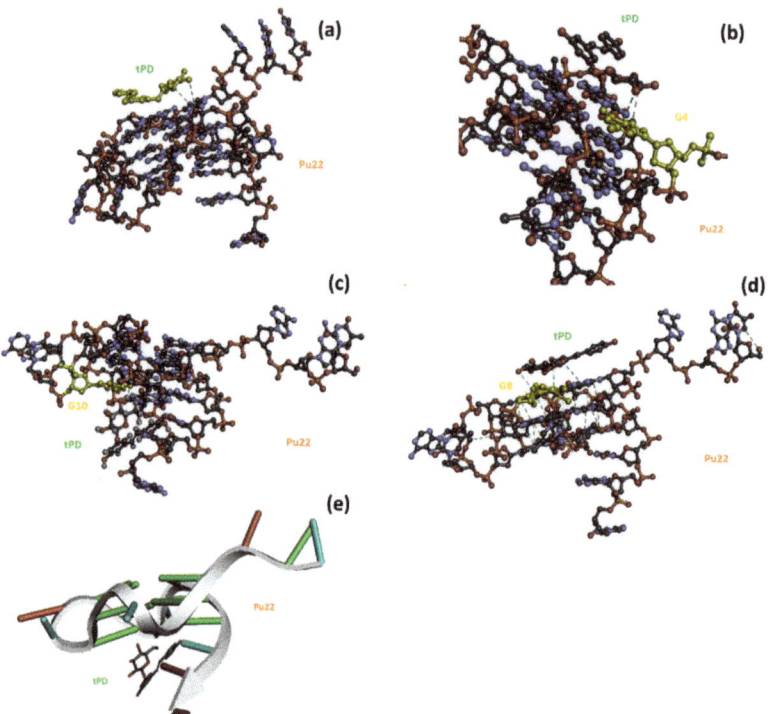

Figure 3. The docked structures of the tPD-Pu22, with the Pu22 PDB ID: 6AU4, corresponding to the top-1–3 ranked poses: (**a,b**) pose 1; (**c**) pose 2; (**d**) pose 3. Note how in poses 1 and 3, tPD seems to interact by end-stacking and H-bondings with the nucleotides represented in yellow in panels (**b,d**). Panel (**e**) reports a different depiction of pose 2 in which the backbone of Pu22 is represented as a white arrow and the base pairs as ladders for clarity.

Table 2. HDOCK docking scores (for the top-ranked pose and mean value from the top-1–3 poses). The interface nucleotide residues within 5.0 Å from the ligand in the top-1–3 complexes are reported in the last column.

Complex	HDOCK Score * Top-1 Ranked Pose	HDOCK Score Mean Value (Top-1–3 Poses) ± SD	Interface Residues
tPD/Pu22	−112.1	−111.7 ± 0.3	G4, G6, G8, G10, G13, G15, T16, G17, G19, T20, A21
tRES/Pu22	−120.6	−112.9 ± 7.3	G6, T7, G10, G15, T16, G19, T20, A21
tPD/(Pu22)$_2$	−103.2	−102.7 ± 0.5	G6, G10, T11, G15, T16, G19, T20, A21, G'14, G'15, T'16, G'17, G'18, G'19, T'20, A'21

* The docking energy scores.

The interactions that emerged by analyzing the top-1–3 poses are held by H-bondings with aromatic rings involving the tPD H1 (Figure 1a) and the guanine residues 4 (3.15 Å, ligand H1–G-ring; π donor H-bond) and 8 (3.05 Å, ligand H1–G ring; π donor H-bond), respectively, in poses 1 and 3 (Figure 3b,d). In pose 2, a H-bond between ligand H2 (Figure 1a) and the O6 (2.14 Å) of guanine residue 10 was also detected (Figure 3c). Interestingly, unlike pose 2 (Figure 3c,e), poses 1 and 3 show the tPD aromatic moieties laying almost parallel to the terminal quartets of the quadruplex (Figure 3), thus suggesting a partial end-stacking interaction of the polyphenol to the G4.

The dimerization of Pu22 G4 was described in the literature under some conditions; for example, a quadruplex dimer was clearly evidenced in the solid state [56], whereas in solution, this G4 is present mainly as a monomer [57]. Nonetheless, Jana and Weisz [58] using nondenaturing polyacrylamide gel electrophoresis showed that in solution, *MYC*-Δ1,6 and, albeit to a much lesser extent, Pu22 (indicated by them as "*MYC*-Δ1,6[1.2.1]", carrying two G replacements by T with respect to *MYC*-Δ1,6) presented dimeric forms corresponding to slower migrating bands, more evident in the former case and somewhat faint, but still detectable, in the case of Pu22 [58]. Similarly, the electrophoretic assays of Moriyama et al. showed for Pu22 (indicated by them as c-myc) a main band and two slower migrating bands [59]. The presence of dimeric Pu22 in solution was suggested also by size exclusion chromatography (SEC), revealing two main SEC peaks for the Pu22 solution that led to the hypothesis of the coexistence of monomeric and dimeric forms in solution [60]. Therefore, we hypothesize that Pu22 in solution is found mainly as a monomer, which justifies its usage in biomolecular studies as a model of G4 DNA not prone to undesirable multimerization, but also, albeit at a much lesser degree, as a dimer. G4 DNA dimer binding by ligands could, in principle, alter the monomer-dimer equilibrium, and importantly, some ligands can induce dimerization of truncated parallel c-myc G-quadruplexes [61].

With all the above considerations in mind, we decided to explore by molecular docking also the propensity of tPD to bind the (Pu22)$_2$ dimer model. We found for the top-ranked pose, as well as poses 1–3 of this docking, less negative Hdock scores (−103.2 and −102.7 ± 0.5, respectively; Table 2) with respect to those found in the case of the docking of the same ligand with the monomeric G4 (−112.1 and −111.7 ± 0.3), suggesting a slightly higher affinity of tPD for the most abundant monomeric form of the Pu22 G4 structure.

We also performed DNA–DNA dockings to explore the dimerization of Pu22 G-quadruplex to obtain (Pu22)$_2$ and the effects of tPD on this process. To this scope, in the first case, we docked Pu22 G4 to a second Pu22 G4 unit, set as the target (Figure 4a), while in the second docking, we used the pre-docked tPD-Pu22 G4 for docking to a second Pu22 G4 unit (Figure 4b). We found that tPD-Pu22 G4 binds Pu22 G4 with an affinity 1.3 times lower than that showed by unliganded Pu22 G4 with the same target (Hdock scores (mean of top-1–3 values): −460.1 ± 14.1 vs. −601.3 ± 7.3, respectively). In other terms, it seems that tPD hinders Pu22 G4 dimerization that, in its absence, is more favored (Figure 4a,b), and binds the Pu22 G4 monomer with slightly higher affinity than the dimeric $(Pu22)_2$ G4 (Figure 4b,d).

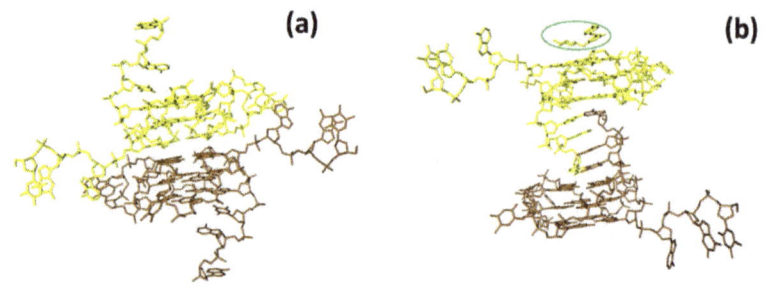

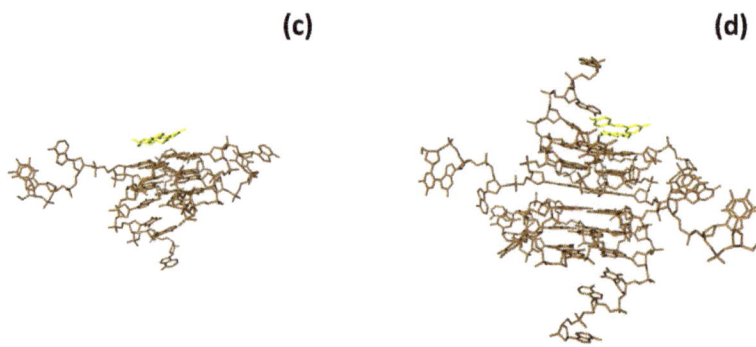

Figure 4. Docking of Pu22 (**a**) or tPD-Pu22 (**b**) to another Pu22 unit. Docking of tPD to Pu22 monomer (**c**) or dimer (**d**). Hdock scores (mean of top-1–3 values) were also indicated.

Remarkably, the dimeric form of Pu22 G4 with tPD (Figures 4b and 5a) was predicted to show considerable structural differences with respect to the unliganded $(Pu22)_2$ G4 dimer (Figure 4a). In this regard, it is worth noting how 18 hydrophobic/π–π stacking Pu22-Pu22 intermolecular interactions (pink, Figure 5b) along with six intermolecular H-bonds (green) are predicted to sustain the trimeric complex structure.

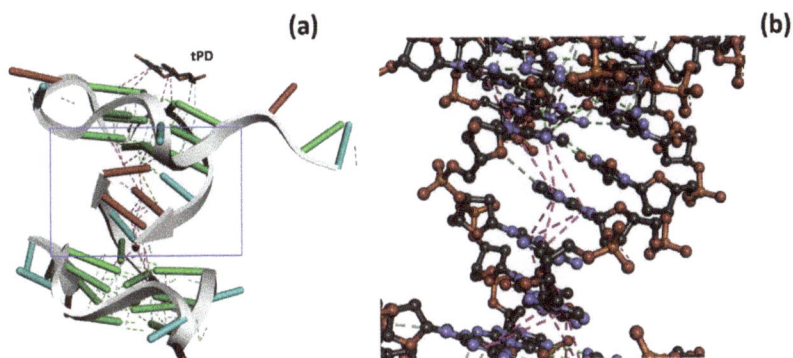

Figure 5. (a) Detailed pose view of the trimeric complex (tPD-Pu22)-Pu22 of Figure 4b; tPD structure is indicated. (b) Enlargement of the area delimited by the blue rectangle.

2.3. CD Predictions and Comparison with Experimental Spectroscopic Data

Furthermore, the solution NMR structure of the monomeric model of Pu22 G4 formed in human c-myc promoter [57] was used to simulate its CD spectrum by DichroCalc [45]. This software is routinely used for obtaining simulated CD spectra of proteins starting from their PDB structure files. In our approach, we applied the method to the prediction of the spectroscopic profile of the G4 structure object of our study. In particular, a positive band at 268 nm and a negative one at 244 were predicted by DichroCalc (Figure 6a), which were, to some extent, in analogy to what we experimentally found by CD (Figure 6b) and was previously described in the literature for the parallel G4 structure of Pu22, though with a bathochromic shift in the bands by about 5 nm.

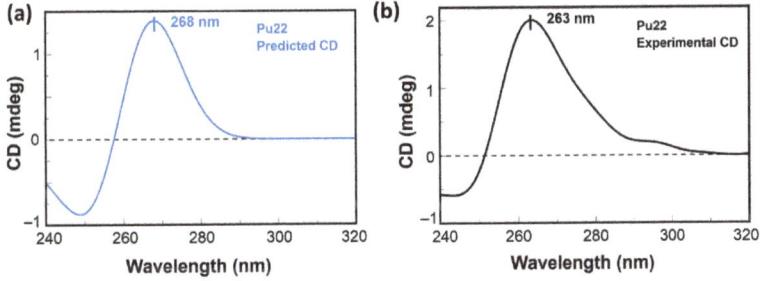

Figure 6. Theoretical CD spectrum of Pu22 G4 (a) as simulated by DichroCalc [45] using the PDB ID 1XAV, compared with the experimental counterpart (b) obtained for Pu22 at 2.5 µM in PBS.

3. Discussion

With this investigation, we aimed to give more insights into the interaction of the natural polyphenol tPD with the G4-forming DNA model of the c-myc promoter Pu22, as the anticancer effects of this phytochemical compound were previously associated to c-myc deregulation [31]. Specifically, a possible mechanism of anticancer activity could be the stabilization of a G4 structure within a regulatory region of this oncogene. Previous attempts [30] in this regard failed to show any stabilization effects of tPD due to the experimental conditions used and notably because of the particularly K^+-rich buffer [30]. In this work, we decided to substitute the previously used buffer with PBS, which corresponds to an overall 4.5 mM K^+ concentration. The binding of tPD with the parallel Pu22 G4 was confirmed by CD spectroscopy, which showed changes in the CD spectrum of this DNA secondary structure under our experimental conditions, especially in the characteristic positive band centered at 263 nm (Figure 1b). The overall variation in the CD spectrum of

Pu22 when bound by tPD was significant and evidenced by the "difference" CD spectrum (inset of Figure 1b). The thermal denaturation profiles in PBS revealed for both Pu22 and tPD-Pu22 sigmoidal shapes with transition midpoint temperatures (T_ms) of 62 and 64 °C, respectively (Figure 1c,d, Table 1), denoting a stabilization effect of tPD on the G4. Furthermore, by examining the variations in the CD curves upon heating, we observed that tPD in the complex slowed down the unfolding process of the G4 structure, especially in the 40–50 °C range. Then, to give a tentative interpretation of the experimental findings, we conducted a molecular docking study on different systems including Pu22 monomeric and dimeric G4s and tPD. The most interesting docking results revealed that tPD may bind the monomeric G4 c-myc model (PDB ID: 6AU4 [56]) used in our CD experiments by partial stacking to the terminal G-quartet of the 22-mer sequence Pu22 (Figure 3). The binding involves a region similar to that described previously [30] for the 24-mer G4 structure (PDB ID: 2A5P), in the proximity of nucleotides common to both computational studies, such as G13 [30] (Table 2). The tPD-Pu22 complex is held also by H-bonding interactions with the aromatic rings [62] between the tPD hydrogen H1 and the guanines in positions 4 and 8. There is also an H-bond between ligand H2 and the O6 of guanine 10 (pose 2), but we cannot exclude that other intermolecular forces (for instance, hydrophobic interactions) contribute to the complex formation. Interestingly, the tPD aromatic moieties in two poses out of three lay almost parallel to the quartets of G4 (Figure 3), thus suggesting a partial π–π stacking of the polyphenol to the terminal G4 quartet (end-stacking). In our hypothesis, the partial end-stacking of tPD to the G-quartet could have a role in the experimental CD thermal behavior observed, as this interaction could reinforce the G4 stabilizing it. Our docking suggests that polydatin could bind the G4 structure by end-stacking as reported in the literature for other stilbene derivatives [63]. Interestingly, Esaki et al. [64] found that naphthalene derivatives are able to stack with the quadruplex G-quartet and afforded thermal stabilizations similar to those observed by us with polydatin, which are also comparable to those recorded for polydatin and resveratrol with the G4s tel26 and hTERT [30]. When tPD is bound to Pu22, it leads to the formation of a complex in which 18 hydrophobic/π–π stacking intermolecular interactions along with six intermolecular H-bonds sustain a trimeric structure of polydatin-(Pu22)$_2$, although it prevents Pu22 G4 dimerization with a full eight-floor coplanar system (Figure 4a). In this regard, other ligands of G-quadruplex DNAs were able to induce dimerization in monomeric G4-forming sequences, such as truncated c-myc promoter DNAs [61], leading also to G4 thermal stabilization [65].

4. Materials and Methods

4.1. Materials

All the reagents and solvents were of the highest commercially available quality and were used as received from Sigma-Aldrich (Merck S.r.l., Milan, Italy). Pu22 DNA sequence d[TGAGGGTGGGTAGGGTGGGTAA], purchased by Eurofins (Turin, Italy) in lyophilized and desalted form, was dissolved in nuclease-free bidistilled water and quantified by UV measurements of the absorbance at 260 nm at 95 °C using as extinction molar coefficient of 228,700 M^{-1} cm^{-1} (ssDNA, nn model, https://atdbio.com/tools/oligo-calculator, accessed on 3 March 2022). The DNA stock solution had a 200 µM concentration. Stock solutions of tPD ligand (kind gift of Prof. G Ravagnan) were prepared at 8 mM concentration in DMSO.

4.2. CD Studies

Circular dichroism (CD) spectra were registered with procedures similar to previous literature reports [66] on a Jasco J-810 (Jasco Europe S.R.L., Cremella, Italy) spectropolarimeter, equipped with a Peltier ETC-505 T temperature controller, in a Hellma (Milan, Italy) quartz cell with a light path of 0.1 cm. The spectra were recorded within the 240–320 nm wavelength range and corrected by subtracting the contribution of the solvents. All experiments were performed in PBS buffer (137 mM NaCl, 2.7 mM KCl, 10 mM Na_2HPO_4, and 1.8 mM KH_2PO4, pH 7.4; Sigma Aldrich, Milan, Italy), using 2.5 µM DNA (Pu22), diluted from the stock solution in water, and 125 µM tPD (50 equiv. respect to the DNA), diluted from the stock solution in DMSO.

4.3. CD Denaturation Studies

All G4-containing solutions were annealed by heating them at 95 °C for 5 min and then letting them slowly cool down to room temperature (over 16 h). The presented melting curves (obtained by recording CD_{265nm} vs. T in the 40–90 °C temperature range) are representative of three independent experiments. Melting temperature (T_m) values were determined as the temperatures relative to the minima of the 1st derivative plots of the denaturation curves. All experiments were repeated at least three times and all spectra were recorded in triplicate.

4.4. Molecular Docking

We conducted our blind molecular docking with the program HDOCK [50,51] using default parameters for all dockings and the PDB entry 6AU4 that is suitable for studies involving dimerization (selecting one of the two G4 monomers) of the Pu22 G4 structure [56]. The HDOCK server uses the iterative knowledge-based scoring function ITScore-PP to rank the top-ten poses provided after each docking run. The HDOCK score furnished by the program is an energy score whose values are listed as dimensionless, and larger negative numbers of the HDOCK score indicate stronger binding interactions between the interacting ligand/macromolecules, which was reported to correlate well to experimental binding affinities.

We used the 3D structure of the Pu22 DNA with the PDB (Protein Data Bank) ID: 6AU4 [56]. The 3D structure, including H-atoms, for the natural compound trans-polydatin, was retrieved by us from the PubChem database (https://pubchem.ncbi.nlm.nih.gov/, accessed on 8 November 2021). More details on the HDOCK docking server and on the procedures for docking experiments can be found at http://hdock.phys.hust.edu.cn/ (accessed on 9 November 2021). We analyzed the top-ranked pose (Top-1) and the top-three ranked poses for the complexes predicted by HDOCK according to the energy scores provided by the program as explained in the Results section. Ligand/DNA complexes were visualized by Discovery Studio 2021 software (Accelrys, San Diego, CA, USA) [67] that was used also for analyzing H-bonding between tPD and G4 DNA.

4.5. CD Predictions

The prediction of the CD spectrum of the monomeric Pu22 G4 structure was performed using the DichroCalc [45] web server starting from the PDB file of the NMR structure deposited with PDB ID 1XAV. At first, the 1XAV.pdb file was manually edited by replacing the unrecognized "DA, DC, DG, DT" text for deoxyribonucleotides with "A, C, G, T". Then, the edited PDB file was uploaded as the input file in DichroCalc obtaining the predicted CD spectrum file, which was edited with SpectraGryph 1.2 [68]. The predicted CD spectrum from the "ds" format was finally visualized in Jasco Spectra Manager (JASCO Corporation, Sendai, Japan).

5. Conclusions

Here, we described a combined approach including in silico (molecular docking) and experimental (CD binding assay/CD thermal denaturation) analyses, through which we verified that tPD can interact with Pu22, a G4-forming sequence related to the promoter region of the c-myc oncogene, stabilizing this DNA structure. The tPD anticancer activity previously observed in vitro correlates with its stabilizing effects on this cancer-related target. The interaction of tPD with the parallel quadruplex has been proven by CD, showing changes in the CD spectrum of this DNA secondary structure under our experimental conditions, especially in the characteristic positive band centered at 263 nm. Moreover, slight thermal stabilization effects on the G4 by tPD have been revealed by CD melting studies. The binding with the DNA structure has been described in more detail in silico by molecular docking, which suggests that the interaction of tPD with Pu22 G4 may take place through partial end-stacking to the terminal quartet involving deoxynucleotides placed in the external regions of the G4 and the sugar moiety of the ligand. Finally, the exploitation of the DichroCalc web-based server, normally used for the prediction of CD spectra of proteins, for the computation of CD spectra of Pu22 revealed the feasibility of the method for the predictions of CD spectra of G4 DNA.

Author Contributions: All authors contributed to the conceptualization, experimental design, methodology, data analysis, writing, and editing and reviewing of the article. All authors have read and agreed to the published version of the manuscript.

Funding: This research was supported by the Department of Pharmacy-University of Naples Federico II grant "Sostegno allo Sviluppo della Ricerca Dipartimentale" (N.B., M.T.).

Institutional Review Board Statement: Not applicable.

Informed Consent Statement: Not applicable.

Data Availability Statement: Not applicable.

Acknowledgments: We thank Antonietta Gargiulo for her technical assistance and help in the literature search. We also thank Glures s.r.l., a spin-off company of the Italian National Research Council (CNR), and Giampietro Ravagnan for kindly providing tPD.

Conflicts of Interest: The authors declare no conflict of interest.

Sample Availability: Not applicable.

References

1. Patel, D.J.; Phan, A.T.; Kuryavyi, V. Human telomere, oncogenic promoter and 5′-UTR G-quadruplexes: Diverse higher order DNA and RNA targets for cancer therapeutics. *Nucleic Acids Res.* **2007**, *35*, 7429–7455. [CrossRef] [PubMed]
2. Chen, Z.-F.; Qin, Q.-P.; Qin, J.-L.; Liu, Y.-C.; Huang, K.-B.; Li, Y.-L.; Meng, T.; Zhang, G.-H.; Peng, Y.; Luo, X.-J. Stabilization of G-quadruplex DNA, inhibition of telomerase activity, and tumor cell apoptosis by organoplatinum (II) complexes with oxoisoaporphine. *J. Med. Chem.* **2015**, *58*, 2159–2179. [CrossRef] [PubMed]
3. Qin, Q.-P.; Qin, J.-L.; Meng, T.; Lin, W.-H.; Zhang, C.-H.; Wei, Z.-Z.; Chen, J.-N.; Liu, Y.-C.; Liang, H.; Chen, Z.-F. High in vivo antitumor activity of cobalt oxoisoaporphine complexes by targeting G-quadruplex DNA, telomerase and disrupting mitochondrial functions. *Eur. J. Med. Chem.* **2016**, *124*, 380–392. [CrossRef]
4. Amato, J.; Pagano, B.; Borbone, N.; Oliviero, G.; Gabelica, V.; Pauw, E.D.; D'Errico, S.; Piccialli, V.; Varra, M.; Giancola, C. Targeting G-quadruplex structure in the human c-Kit promoter with short PNA sequences. *Bioconjugate Chem.* **2011**, *22*, 654–663. [CrossRef]
5. Esposito, V.; Galeone, A.; Mayol, L.; Oliviero, G.; Virgilio, A.; Randazzo, L. A topological classification of G-quadruplex structures. *Nucleosides Nucleotides Nucleic Acids* **2007**, *26*, 1155–1159. [CrossRef] [PubMed]
6. Neidle, S. The structures of quadruplex nucleic acids and their drug complexes. *Curr. Opin. Struct. Biol.* **2009**, *19*, 239–250. [CrossRef]
7. Onel, B.; Lin, C.; Yang, D. DNA G-quadruplex and its potential as anticancer drug target. *Sci. China Chem.* **2014**, *57*, 1605–1614. [CrossRef]
8. Siddiqui-Jain, A.; Grand, C.L.; Bearss, D.J.; Hurley, L.H. Direct evidence for a G-quadruplex in a promoter region and its targeting with a small molecule to repress c-MYC transcription. *Proc. Natl. Acad. Sci. USA* **2002**, *99*, 11593–11598. [CrossRef]
9. Seenisamy, J.; Rezler, E.M.; Powell, T.J.; Tye, D.; Gokhale, V.; Joshi, C.S.; Siddiqui-Jain, A.; Hurley, L.H. The Dynamic Character of the G-Quadruplex Element in the c-MYC Promoter and Modification by TMPyP4. *J. Am. Chem. Soc.* **2004**, *126*, 8702–8709. [CrossRef]
10. Rankin, S.; Reszka, A.P.; Huppert, J.; Zloh, M.; Parkinson, G.N.; Todd, A.K.; Ladame, S.; Balasubramanian, S.; Neidle, S. Putative DNA Quadruplex Formation within the Human c-kit Oncogene. *J. Am. Chem. Soc.* **2005**, *127*, 10584–10589. [CrossRef]
11. Fernando, H.; Reszka, A.P.; Huppert, J.; Ladame, S.; Rankin, S.; Venkitaraman, A.R.; Neidle, S.; Balasubramanian, S. A Conserved Quadruplex Motif Located in a Transcription Activation Site of the Human c-kit Oncogene. *Biochemistry* **2006**, *45*, 7854–7860. [CrossRef] [PubMed]
12. Huppert, J.L.; Balasubramanian, S. G-quadruplexes in promoters throughout the human genome. *Nucleic Acids Res.* **2007**, *35*, 406–413. [CrossRef]
13. Phan, A.T.; Kuryavyi, V.; Burge, S.; Neidle, S.; Patel, D.J. Structure of an Unprecedented G-Quadruplex Scaffold in the Human c-kit Promoter. *J. Am. Chem. Soc.* **2007**, *129*, 4386–4392. [CrossRef] [PubMed]
14. Chen, Y.; Agrawal, P.; Brown, R.V.; Hatzakis, E.; Hurley, L.; Yang, D. The Major G-Quadruplex Formed in the Human Platelet-Derived Growth Factor Receptor β Promoter Adopts a Novel Broken-Strand Structure in K+ Solution. *J. Am. Chem. Soc.* **2012**, *134*, 13220–13223. [CrossRef] [PubMed]

15. Francisco, A.P.; Paulo, A. Oncogene expression modulation in cancer cell lines by DNA G-quadruplex-interactive small molecules. *Curr. Med. Chem.* **2017**, *24*, 4873–4904. [CrossRef]
16. Ciribilli, Y.; Singh, P.; Spanel, R.; Inga, A.; Borlak, J. Decoding c-Myc networks of cell cycle and apoptosis regulated genes in a transgenic mouse model of papillary lung adenocarcinomas. *Oncotarget* **2015**, *6*, 31569. [CrossRef]
17. Zhang, Z.; Wu, Q.; Wu, X.-H.; Sun, F.-Y.; Chen, L.-M.; Chen, J.-C.; Yang, S.-L.; Mei, W.-J. Ruthenium (II) complexes as apoptosis inducers by stabilizing c-myc G-quadruplex DNA. *Eur. J. Med. Chem.* **2014**, *80*, 316–324. [CrossRef]
18. Platella, C.; Mazzini, S.; Napolitano, E.; Mattio, L.M.; Beretta, G.L.; Zaffaroni, N.; Pinto, A.; Montesarchio, D.; Dallavalle, S. Plant-Derived Stilbenoids as DNA-Binding Agents: From Monomers to Dimers. *Chem.–A Eur. J.* **2021**, *27*, 8832–8845. [CrossRef]
19. Chaudhuri, R.; Bhattacharya, S.; Dash, J.; Bhattacharya, S. Recent Update on Targeting c-MYC G-Quadruplexes by Small Molecules for Anticancer Therapeutics. *J. Med. Chem.* **2020**, *64*, 42–70. [CrossRef]
20. Sheikh-Zeineddini, N.; Safaroghli-azar, A.; Salari, S.; Bashash, D. C-Myc inhibition sensitizes pre-B ALL cells to the anti-tumor effect of vincristine by altering apoptosis and autophagy: Proposing a probable mechanism of action for 10058-F4. *Eur. J. Pharmacol.* **2020**, *870*, 172821. [CrossRef]
21. Mazzini, S.; Gargallo, R.; Musso, L.; De Santis, F.; Aviñó, A.; Scaglioni, L.; Eritja, R.; Di Nicola, M.; Zunino, F.; Amatulli, A.; et al. Stabilization of c-KIT G-Quadruplex DNA Structures by the RNA Polymerase I Inhibitors BMH-21 and BA-41. *Int. J. Mol. Sci.* **2019**, *20*, 4927. [CrossRef] [PubMed]
22. Chanvorachote, P.; Sriratanasak, N.; Nonpanya, N. C-myc Contributes to Malignancy of Lung Cancer: A Potential Anticancer Drug Target. *Anticancer Res.* **2020**, *40*, 609–618. [CrossRef] [PubMed]
23. Dallavalle, S.; Mattio, L.M.; Artali, R.; Musso, L.; Aviñó, A.; Fàbrega, C.; Eritja, R.; Gargallo, R.; Mazzini, S. Exploring the Interaction of Curaxin CBL0137 with G-Quadruplex DNA Oligomers. *Int. J. Mol. Sci.* **2021**, *22*, 6476. [CrossRef] [PubMed]
24. Georgiades, S.N.; Abd Karim, N.H.; Suntharalingam, K.; Vilar, R. Interaction of metal complexes with G-quadruplex DNA. *Angew. Chem. Int. Ed.* **2010**, *49*, 4020–4034. [CrossRef]
25. Brooks, T.A.; Hurley, L.H. Targeting MYC expression through G-quadruplexes. *Genes Cancer* **2010**, *1*, 641–649. [CrossRef]
26. Yan, S.; Zheng, Z.; Cui, Y.; Mi, Y.; Liu, H.; Zhao, X.; Luo, D. C-myc g-quadruplex stabilization and cytotoxicity of an oxadiazole-bearing ruthennium (II) complex. *Rev. Roum. De Chim.* **2021**, *66*, 423–433.
27. Platella, C.; Guida, S.; Bonmassar, L.; Aquino, G.; Bonmassar, E.; Ravagnan, G.; Montesarchio, D.; Roviello, G.N.; Musumeci, D.; Fuggetta, M.P. Antitumour activity of resveratrol on human melanoma cells: A possible mechanism related to its interaction with malignant cell telomerase. *Biochim. Et Biophys. Acta (BBA)-Gen. Subj.* **2017**, *1861*, 2843–2851. [CrossRef]
28. Quagliariello, V.; Berretta, M.; Buccolo, S.; Iovine, M.; Paccone, A.; Cavalcanti, E.; Taibi, R.; Montopoli, M.; Botti, G.; Maurea, N. Polydatin reduces cardiotoxicity and enhances the anticancer effects of sunitinib by decreasing pro-oxidative stress, pro-inflammatory cytokines, and nlrp3 inflammasome expression. *Front. Oncol.* **2021**, *11*, 680758. [CrossRef]
29. Lanzilli, G.; Cottarelli, A.; Nicotera, G.; Guida, S.; Ravagnan, G.; Fuggetta, M.P. Anti-inflammatory effect of resveratrol and polydatin by in vitro IL-17 modulation. *Inflammation* **2012**, *35*, 240–248. [CrossRef]
30. Platella, C.; Raucci, U.; Rega, N.; D'Atri, S.; Levati, L.; Roviello, G.N.; Fuggetta, M.P.; Musumeci, D.; Montesarchio, D. Shedding light on the interaction of polydatin and resveratrol with G-quadruplex and duplex DNA: A biophysical, computational and biological approach. *Int. J. Biol. Macromol.* **2020**, *151*, 1163–1172. [CrossRef]
31. Bai, L.; Ma, Y.; Wang, X.; Feng, Q.; Zhang, Z.; Wang, S.; Zhang, H.; Lu, X.; Xu, Y.; Zhao, E. Polydatin Inhibits Cell Viability, Migration, and Invasion Through Suppressing the c-Myc Expression in Human Cervical Cancer. *Front. Cell Dev. Biol.* **2021**, *9*, 587218. [CrossRef] [PubMed]
32. Mathad, R.I.; Hatzakis, E.; Dai, J.; Yang, D. c-MYC promoter G-quadruplex formed at the 5′-end of NHE III 1 element: Insights into biological relevance and parallel-stranded G-quadruplex stability. *Nucleic Acids Res.* **2011**, *39*, 9023–9033. [CrossRef] [PubMed]
33. Roviello, V.; Musumeci, D.; Mokhir, A.; Roviello, G.N. Evidence of protein binding by a nucleopeptide based on a thymine-decorated L-diaminopropanoic acid through CD and in silico studies. *Curr. Med. Chem.* **2021**, *28*, 5004–5015. [CrossRef] [PubMed]
34. Musumeci, D.; Mokhir, A.; Roviello, G.N. Synthesis and nucleic acid binding evaluation of a thyminyl l-diaminobutanoic acid-based nucleopeptide. *Bioorganic Chem.* **2020**, *100*, 103862. [CrossRef]
35. Musumeci, D.; Ullah, S.; Ikram, A.; Roviello, G.N. Novel insights on nucleopeptide binding: A spectroscopic and In Silico investigation on the interaction of a thymine-bearing tetrapeptide with a homoadenine DNA. *J. Mol. Liq.* **2022**, *347*, 117975. [CrossRef]
36. Amato, J.; Stellato, M.I.; Pizzo, E.; Petraccone, L.; Oliviero, G.; Borbone, N.; Piccialli, G.; Orecchia, A.; Bellei, B.; Castiglia, D. PNA as a potential modulator of COL7A1 gene expression in dominant dystrophic epidermolysis bullosa: A physico-chemical study. *Mol. BioSyst.* **2013**, *9*, 3166–3174. [CrossRef]
37. Pirota, V.; Platella, C.; Musumeci, D.; Benassi, A.; Amato, J.; Pagano, B.; Colombo, G.; Freccero, M.; Doria, F.; Montesarchio, D. On the binding of naphthalene diimides to a human telomeric G-quadruplex multimer model. *Int. J. Biol. Macromol.* **2021**, *166*, 1320–1334. [CrossRef]
38. Platella, C.; Capasso, D.; Riccardi, C.; Musumeci, D.; DellaGreca, M.; Montesarchio, D. Natural compounds from Juncus plants interacting with telomeric and oncogene G-quadruplex structures as potential anticancer agents. *Org. Biomol. Chem.* **2021**, *19*, 9953–9965. [CrossRef]

39. Oliviero, G.; Borbone, N.; Amato, J.; D'Errico, S.; Galeone, A.; Piccialli, G.; Varra, M.; Mayol, L. Synthesis of quadruplex-forming tetra-end-linked oligonucleotides: Effects of the linker size on quadruplex topology and stability. *Biopolym. Orig. Res. Biomol.* **2009**, *91*, 466–477. [CrossRef]
40. Oliviero, G.; Amato, J.; Borbone, N.; Galeone, A.; Varra, M.; Piccialli, G.; Mayol, L. Synthesis and characterization of DNA quadruplexes containing T-tetrads formed by bunch-oligonucleotides. *Biopolym. Orig. Res. Biomol.* **2006**, *81*, 194–201. [CrossRef]
41. Fuggetta, M.P.; De Mico, A.; Cottarelli, A.; Morelli, F.; Zonfrillo, M.; Ulgheri, F.; Peluso, P.; Mannu, A.; Deligia, F.; Marchetti, M. Synthesis and enantiomeric separation of a novel spiroketal derivative: A potent human telomerase inhibitor with high in vitro anticancer activity. *J. Med. Chem.* **2016**, *59*, 9140–9149. [CrossRef] [PubMed]
42. Arounaguiri, S.; Easwaramoorthy, D.; Ashokkumar, A.; Dattagupta, A.; Maiya, B.G. Cobalt (III), nickel (II) and ruthenium (II) complexes of 1, 10-phenanthroline family of ligands: DNA binding and photocleavage studies. *J. Chem. Sci.* **2000**, *112*, 1–17. [CrossRef]
43. Ricci, C.G.; Netz, P.A. Docking studies on DNA-ligand interactions: Building and application of a protocol to identify the binding mode. *J. Chem. Inf. Model.* **2009**, *49*, 1925–1935. [CrossRef] [PubMed]
44. Mulliri, S.; Laaksonen, A.; Spanu, P.; Farris, R.; Farci, M.; Mingoia, F.; Roviello, G.N.; Mocci, F. Spectroscopic and in silico studies on the interaction of substituted pyrazolo [1, 2-a] benzo [1, 2, 3, 4] tetrazine-3-one derivatives with c-myc G4-DNA. *Int. J. Mol. Sci.* **2021**, *22*, 6028. [CrossRef] [PubMed]
45. Bulheller, B.M.; Hirst, J.D. DichroCalc—Circular and linear dichroism online. *Bioinformatics* **2009**, *25*, 539–540. [CrossRef]
46. Kong, D.-M.; Wu, J.; Wang, N.; Yang, W.; Shen, H.-X. Peroxidase activity–structure relationship of the intermolecular four-stranded G-quadruplex–hemin complexes and their application in Hg^{2+} ion detection. *Talanta* **2009**, *80*, 459–465. [CrossRef]
47. Egbuna, C.; Kumar, S.; Ifemeje, J.C.; Ezzat, S.M.; Kaliyaperumal, S. *Phytochemicals as Lead Compounds for New Drug Discovery*; Elsevier: Amsterdam, The Netherlands, 2019.
48. Mechchate, H.; Costa de Oliveira, R.; Es-Safi, I.; Vasconcelos Mourão, E.M.; Bouhrim, M.; Kyrylchuk, A.; Soares Pontes, G.; Bousta, D.; Grafov, A. Antileukemic Activity and Molecular Docking Study of a Polyphenolic Extract from Coriander Seeds. *Pharmaceuticals* **2021**, *14*, 770. [CrossRef]
49. Artali, R.; Beretta, G.; Morazzoni, P.; Bombardelli, E.; Meneghetti, F. Green tea catechins in chemoprevention of cancer: A molecular docking investigation into their interaction with glutathione S-transferase (GST P1-1). *J. Enzym. Inhib. Med. Chem.* **2009**, *24*, 287–295. [CrossRef]
50. Yan, Y.; Zhang, D.; Zhou, P.; Li, B.; Huang, S.-Y. HDOCK: A web server for protein–protein and protein–DNA/RNA docking based on a hybrid strategy. *Nucleic Acids Res.* **2017**, *45*, W365–W373. [CrossRef]
51. Yan, Y.; Tao, H.; He, J.; Huang, S.-Y. The HDOCK server for integrated protein–protein docking. *Nat. Protoc.* **2020**, *15*, 1829–1852. [CrossRef]
52. Majumder, A.; Mondal, S.K.; Mukhoty, S.; Bag, S.; Mondal, A.; Begum, Y.; Sharma, K.; Banik, A. Virtual Screening and Docking Analysis of Novel Ligands for Selective Enhancement of Tea (Camellia sinensis) Flavonoids. *Food Chem. X* **2022**, *13*, 100212. [CrossRef] [PubMed]
53. Roy, A.; Chatterjee, O.; Banerjee, N.; Roychowdhury, T.; Dhar, G.; Mukherjee, G.; Chatterjee, S. Curcumin arrests G-quadruplex in the nuclear hyper-sensitive III1 element of c-MYC oncogene leading to apoptosis in metastatic breast cancer cells. *J. Biomol. Struct. Dyn.* **2021**, 1–17. [CrossRef]
54. Ji, D.; Juhas, M.; Tsang, C.M.; Kwok, C.K.; Li, Y.; Zhang, Y. Discovery of G-quadruplex-forming sequences in SARS-CoV-2. *Brief. Bioinform.* **2021**, *22*, 1150–1160. [CrossRef] [PubMed]
55. Stoddard, S.V.; Wallace, F.E.; Stoddard, S.D.; Cheng, Q.; Acosta, D.; Barzani, S.; Bobay, M.; Briant, J.; Cisneros, C.; Feinstein, S. In silico design of peptide-based SARS-CoV-2 fusion inhibitors that target wt and mutant versions of SARS-CoV-2 HR1 Domains. *Biophysica* **2021**, *1*, 311–327. [CrossRef]
56. Stump, S.; Mou, T.-C.; Sprang, S.R.; Natale, N.R.; Beall, H.D. Crystal structure of the major quadruplex formed in the promoter region of the human c-MYC oncogene. *PLoS ONE* **2018**, *13*, e0205584. [CrossRef]
57. Ambrus, A.; Chen, D.; Dai, J.; Jones, R.A.; Yang, D. Solution structure of the biologically relevant G-quadruplex element in the human c-MYC promoter. Implications for G-quadruplex stabilization. *Biochemistry* **2005**, *44*, 2048–2058. [CrossRef]
58. Jana, J.; Weisz, K. A Thermodynamic Perspective on Potential G-Quadruplex Structures as Silencer Elements in the MYC Promoter. *Chem.–A Eur. J.* **2020**, *26*, 17242–17251. [CrossRef]
59. Moriyama, K.; Yoshizawa-Sugata, N.; Masai, H. Oligomer formation and G-quadruplex binding by purified murine Rif1 protein, a key organizer of higher-order chromatin architecture. *J. Biol. Chem.* **2018**, *293*, 3607–3624. [CrossRef]
60. Marzano, M.; Falanga, A.P.; Marasco, D.; Borbone, N.; D'Errico, S.; Piccialli, G.; Roviello, G.N.; Oliviero, G. Evaluation of an Analogue of the Marine ε-PLL Peptide as a Ligand of G-quadruplex DNA Structures. *Mar. Drugs* **2020**, *18*, 49. [CrossRef]
61. Funke, A.; Karg, B.; Dickerhoff, J.; Balke, D.; Müller, S.; Weisz, K. Ligand-Induced Dimerization of a Truncated Parallel MYC G-Quadruplex. *ChemBioChem* **2018**, *19*, 505–512. [CrossRef]
62. Brinkley, R.L.; Gupta, R.B. Hydrogen bonding with aromatic rings. *AIChE J.* **2001**, *47*, 948–953. [CrossRef]
63. O'Hagan, M.P.; Peñalver, P.; Gibson, R.S.L.; Morales, J.C.; Galan, M.C. Stiff-Stilbene Ligands Target G-Quadruplex DNA and Exhibit Selective Anticancer and Antiparasitic Activity. *Chem.–A Eur. J.* **2020**, *26*, 6224–6233. [CrossRef] [PubMed]
64. Esaki, Y.; Islam, M.M.; Fujii, S.; Sato, S.; Takenaka, S. Design of tetraplex specific ligands: Cyclic naphthalene diimide. *Chem. Commun.* **2014**, *50*, 5967–5969. [CrossRef] [PubMed]

65. Bianchi, F.; Comez, L.; Biehl, R.; D'Amico, F.; Gessini, A.; Longo, M.; Masciovecchio, C.; Petrillo, C.; Radulescu, A.; Rossi, B.; et al. Structure of human telomere G-quadruplex in the presence of a model drug along the thermal unfolding pathway. *Nucleic Acids Res.* **2018**, *46*, 11927–11938. [CrossRef] [PubMed]
66. Pradeep, T.P.; Tripathi, S.; Barthwal, R. Molecular recognition of parallel quadruplex [d-(TTGGGGT)] 4 by mitoxantrone: Binding with 1: 4 stoichiometry leads to telomerase inhibition. *RSC Adv.* **2016**, *6*, 71652–71661. [CrossRef]
67. Pawar, S.S.; Rohane, S.H. Review on discovery studio: An important tool for molecular docking. *Asian J. Res. Chem.* **2021**, *14*, 86–88. [CrossRef]
68. Menges, F. Spectragryph-Optical Spectroscopy Software, Version 1.2. 8. 2018. Available online: http://www.effemm2.de/spectragryph (accessed on 1 August 2018).

MDPI
St. Alban-Anlage 66
4052 Basel
Switzerland
www.mdpi.com

Molecules Editorial Office
E-mail: molecules@mdpi.com
www.mdpi.com/journal/molecules

Disclaimer/Publisher's Note: The statements, opinions and data contained in all publications are solely those of the individual author(s) and contributor(s) and not of MDPI and/or the editor(s). MDPI and/or the editor(s) disclaim responsibility for any injury to people or property resulting from any ideas, methods, instructions or products referred to in the content.